Design Science and Innovation

Series Editor

Amaresh Chakrabarti, Centre for Product Design and Manufacturing, Indian Institute of Science, Bangalore, India

The book series is intended to provide a platform for disseminating knowledge in all areas of design science and innovation, and is intended for all stakeholders in design and innovation, e.g. educators, researchers, practitioners, policy makers and students of design and innovation. With leading international experts as members of its editorial board, the series aims to disseminate knowledge that combines academic rigour and practical relevance in this area of crucial importance to the society.

More information about this series at http://www.springer.com/series/15399

P. K. Nag · L. P. Gite

Human-Centered Agriculture

Ergonomics and Human Factors Applied

Springer

P. K. Nag
Environment and Disaster Management
Ramakrishna Mission Vivekananda
Educational and Research Institute
(Deemed University)
Kolkata, West Bengal, India

L. P. Gite
AICRP, Ergonomics and Safety
in Agriculture
Central Institute of Agricultural
Engineering (ICAR)
Bhopal, Madhya Pradesh, India

ISSN 2509-5986 ISSN 2509-5994 (electronic)
Design Science and Innovation
ISBN 978-981-15-7268-5 ISBN 978-981-15-7269-2 (eBook)
https://doi.org/10.1007/978-981-15-7269-2

This Springer imprint is published by the registered company Springer Nature Singapore Pte Ltd.
The registered company address is: 152 Beach Road, #21-01/04 Gateway East, Singapore 189721, Singapore

The book is dedicated to the few who spend their lives in feeding all of humanity

Preface

About 12 millennia ago, the humankind recognized that food and feed could be produced from the cultivation of plants. This discovery of humankind is the highest achievement in the history of time for subsistence and survival of eight billion people today. Dr. Norman Borlaug, the Nobel Peace Prize winner, aptly expressed—*Civilization, as it is known today, could not have evolved, nor can it survive, without an adequate food supply.* About one billion people sweat in agriculture to produce nearly 3 billion tonnes of food grains and other food commodities. Whereas corporatization and value-added commercial farming in the Western world witness incorporation of gigantic farm machinery, automation, and robotic application, the scenario is gloomy in Asia, Africa, and other developing countries. The predominant farm workforce belongs to small and marginal farm holdings, with their primary dependence on the traditional farming methods, by using animal power or through sectoral mechanization. Undoubtedly, improvement in life and living of the workforce bears the fundamental focus towards sustainable food security and the livelihood of the majority of the rural population.

The traditional farming practices are drudgery-prone. The newer farm machines, if not correctly designed and operated, can lead to a higher number of accidents and injuries. Besides, health hazards of the farmworkers arise from the exposure of workplace stress agents such as vibration, noise, dust, climatic stress, and toxic agrochemicals. The present book titled as *Human-Centered Agriculture* comes out of the authors' clear understanding that the farmers and farmworkers deserve a decent living with reduced drudgery, better health and safety, and productivity at the workplace. The sublime of the present contribution evolves from the prophetic message of W. B. Jastrzebowski, the Polish professor who pioneered the discipline of ergonomics and human factors, that made the proclamation *Work is the mother of all good ... Work enriches us, making us more like unto the divine. Man's forces and faculties evolve in the form of physical labour, aesthetic, rational, and moral (expressing labour, entertainment, reasoning, and dedication).*

An effort has gone into conceptualizing the content of the book, bringing an assimilated convergence of the knowledge base of physiology, ergonomics and human factors, agricultural engineering, and behavioural and health sciences.

Literature source materials have been drawn from extensive electronic searching of databases and manual searching of journals, proceedings, and standards in the related domain. In-depth scrutiny of literature gives a perception of scatteredness of the source materials in terms of the strength and consistency of evidence on the specific subject matter. Field experiments, observational studies, and design interventions on farming tools and practices were considered persuasive in selecting the source materials. Due to scatteredness, however, the selection of studies was partially purposive. Potentially unfamiliar numerical and statistical derivations have generally been avoided. Many national and international standards are quoted in related chapters despite the usual brevity in the public domain.

With the view of the multi-disciplinary readership, the book is organized into six major areas of importance, namely agriculture growth and development (Part I), fundamentals of ergonomics and human factors (Part II), farm development (Part III), health and safety (Part IV), ergonomics application in design (Part V), and health services in agriculture (Part VI).

Part I, *Agriculture growth and development,* brings a brief overview of world agriculture and workforce including women in farming (Chap. 1) and elaborates the conventional farming work methods and practices in small landholding (Chap. 2). Part II, *Fundamentals of ergonomics and human factors,* includes the conceptual understanding of the subject matter about work and workplace design in farming. The part embodies the historical emergence of the discipline of ergonomics and human factors and its gradual penetration in agricultural practices (Chap. 3), the workplace drudgery in farming activities (Chap. 4), the scope of work scheduling and work–rest cycles to mitigating drudgery and fatigue (Chap. 5), and human dimensional and strength compatibility requirements in the design of farm tools and machinery (Chap. 6).

Part III, *Farm development,* brings an analysis of the disparities of farm mechanization in different geographical regions. Discussion emphasizes the quantitative approaches in estimating the mechanization index and evaluating farm mechanization (Chap. 7). Part IV, *Health and safety,* covers an analysis of world scenario of farm accidents and injuries and accident prevention interventions (Chap. 8), and the gamut of health hazards prevailing in farming (Chap. 9). The cumulative health hazards arise due to the exposure to noise and vibration, climatic stress, and dust emanates from crop and cropping. A special section elucidates the issues of pesticide and chemical toxicity, including guidance on safety and health (Chap. 10).

Part V, *Ergonomics application in design,* covers a summary of ergo-design criteria for farm tools and machinery (Chap. 11). Chapter 12 deals with the specific application of ergonomics in the design of selected farm tools and equipment. The examples include hand tools and manual- and power-operated farm machines that are used by both men and women. Chapter 13 extends the application of ergonomics principles in the design of the tractor and power tiller operator's workplace. In summary, Part V provides illustrations of the application of the anthropometric and muscle strength data in the ergo-design of the workplace, farm tools, and machinery of representative interest. The professionals can use the concept and population-specific data in order to make the design of the farming devices more human-centred, safe, comfortable, productive, and user-friendly. Part VI, *Health services in agriculture,* is a composite account of the systems approach in occupational health and safety (OHS) services and management (Chap. 14). The contribution elucidates national, international, and sectoral OHSMS schemes and models, including the basic occupational health services, for development and adoption of OHS system in informal agriculture. A national-level model framework of OHS management in agriculture is suggested.

The professionals of allied agricultural sciences, ergonomics and safety, and the manufacturers of farm tools and machinery are welcome to suitably adopt and amend the ergo-design criteria to reflect conditions in particular farm machinery design and development. Needless to reiterate, any modification of criteria may require a thorough understanding of the interacting factors. Some guidance points have moved up and down from basic principles to practical hints, within a chapter and across chapters. This approach was intended to allow the smooth flow of information for readers to comprehend the subject of study effectively and use in a suitable intervention.

Communication among all concerned is a critical element in recognizing the expectations and requirements of farmworkers and innovating farming practices. A significant body of emphasis on ergonomics and human factors application is evident in sizeable mechanized farming in the West. The insights into solution-oriented intervention, along with the objective assessment of farmworkers' perspectives, call for systematic data, which is scanty or at most at the embryonic stage in developing countries. The All India Coordinated Research Project (AICRP) on Ergonomics and Safety in Agriculture of the Indian Council of Agricultural Research (ICAR) is a landmark initiative that spreads over 12 cooperating centres to research on ergonomics in agriculture. Similar federative endeavour in other geographic regions will form a strong foundation for effective ergonomics application.

Growth in agriculture will depend in large part on getting effective technologies into the hands of farmers. An obvious priority is to bring greater awareness among farmers and farmworkers, society, and governments about safety and health issues in agriculture. The vast majority of farmworkers in small landholdings are not well privileged to understand the research implications of ergonomics in agriculture, due to their lack of resources as well as knowledge constraints. Often a country imports farm machinery and technologies that are manufactured for user compatibility of another country. Due to country-wise differences in anthropometric and physiological parameters of people, many times, the technologies manufactured and found successful in one country become a reason for increased injuries and health hazards in another country. The concept of anthropotechnology put forward by Prof. Alain Wisner of CNAM (Paris) emphasized attention in manufacturing machines and exporting technologies to other countries.

The book evolves as an overview of agricultural ergonomics, and several examples have been drawn from the data of Indian farming scenario explaining the man–machine compatibility of the user population. Hope that this book serves as a handy reference to those engaged in research, health and safety, teaching and extension, designing and manufacturing, and policymaking on farming practices. The young students, professionals, and local support givers to the farming community may find an inspiration to reinvigorate the zeal for innovating better conditions of work at the farming scenario. Different stakeholders will find the book useful in adopting anthropotechnology concepts in the ergo-design of farm machinery and equipment. Aspiring that if even a tiny portion of the present contribution is found useful in creating sustainable agriculture, the authors will feel gratified and well repaid of their labour.

Kolkata, India
Bhopal, India

P. K. Nag, Ph.D., D.Sc.
L. P. Gite, M.Tech., Ph.D.

Acknowledgments

The authors are deeply indebted to the late Prof. A. C. Pandya, a renounced academician and former Director of the CIAE who asked decades back to write a book on agricultural ergonomics for multi-disciplinary users. He continually kept reminding the authors till his sad demise in 2017.

The authors reiterate that they are profusely endowed by the sagacity of the authorities of the Central Institute of Agricultural Engineering (CIAE, Bhopal), and the legacy of the ICAR (New Delhi) in continually engendering generating the knowledge base of contemporary importance. Both authors have grown up together with the researchers of the AICRP of ESA project. Dr. L. P. Gite served as the project coordinator for over 15 years and Prof. P. K. Nag was a reviewer since the inception of the project in 1996. The book cites examples and findings from the studies of ESA project.

Dr. P. K. Nag expresses gratitude to the National Institute of Occupational Health (Ahmedabad) of the ICMR (New Delhi) and Dr. L. P. Gite is grateful to the ICAR-CIAE for the support received in undertaking research programmes on ergonomics in agriculture. Dr. P. K. Nag expresses sincere gratitude to Swami Atmapriyananda, Vice-Chancellor of the RKMVERI (Belur), for extending kind support to bring the book to fruition. The authors extend heartfelt thanks to the colleagues of AICRP (ESA) and others, namely, Drs. C. R. Mehta, K. N. Agrawal, S. P. Singh, P. S. Tiwari, R. R. Potdar, A. Khadatkar, and Mr. J. Majumder (CIAE, Bhopal); Drs. K. Kathirvel and A. Krishan (TNAU, Coimbatore); Dr. S. K. Mohanty (OUAT, Bhubaneshwar); Prof. V. K. Tewari (IIT, Kharagpur); Dr. N. K. Chhuneja (PAU, Ludhiana); Dr. V. V. Aware (BSKVV, Dapoli); Dr. A. Kumar (IARI, New Delhi); Drs. K. N. Dewangan and P. K. Pranav (NERIST, Nirjuli); Drs. A. K. Mehta and S. S. Meena (MPUAT, Udaipur); Er. Sukhbir Singh; Drs. D. K. Vatsa and Neena Vyas (CSKHPKV, Palampur); and Drs. T. Mondal and J. Sen of RKMVERI (Kolkata) for their valuable contribution on the subject. Thanks are due to the entire team of Springer (New Delhi) for their untiring patience to make this publication possible.

The authors express their gratitude to Dr. (Mrs.) Anjali Nag and Mrs. Ujjwala Gite, without whose support and encouragement this work could not have seen the light of the day.

P. K. Nag, Ph.D., D.Sc.
L. P. Gite, M.Tech., Ph.D.

Contents

About the Authors

Prof. P. K. Nag, Ph.D., D.Sc. is currently associated with the Ramakrishna Mission Vivekananda Educational and Research Institute (Deemed University), Belur Math, Kolkata. Prof. Nag is the former Director of the National Institute of Occupational Health (Indian Council of Medical Research) Ahmedabad, and Regional Occupational Health Centers (Bangalore and Kolkata). He specializes in the areas of Ergonomics and Human Factors, Environmental Health and Management, Climate Change, Occupational Health and Safety, and has teaching and research experience spanning four decades. He has written several books and reports and published 175 research articles in reputed journals. Prof. Nag has professional affiliations with numerous national and international agencies in different capacities.

Dr. L. P. Gite, M.Tech., Ph.D. superannuated from the Central Institute of Agricultural Engineering (Indian Council of Agricultural Research), Bhopal, India as a Project Coordinator and Emeritus Scientist. He has been profoundly involved to promote about ergonomics principles in the design and application of agricultural tools and equipment, and strengthen such research in agricultural engineering, and allied institutes, and Standards bodies. He has authored multiple books and about 55 research papers which were published in reputed national and international journals.

About the Authors

Part I
Agriculture Growth and Development

Chapter 1
World Agriculture—Human Beings and Farming

Evolution of the Farming Sector

About twelve millennia ago, humankind moved into the Neolithic era and accidentally recognized that food and feed, and daily necessities such as fibre could be produced from the cultivation of plants. This discovery of humankind is the highest achievement in the history of time that has led to the food and fibre supply to feed and clothe over seven billion people today. About the time the population began aggregating into villages, cultivation of selected food crops along the river beds in fertile valleys and farming practices became the way of life (Kislev et al. 2004). An enormous variety of crop outputs, such as rice, wheat, maize, soybeans, oilseeds, cotton, sugar, and plantation crops products like tea and coffee, is produced from various crop production processes. The Neolithic revolution of moving away from hunting for survival to farming was evident in different geographic regions, namely, West and southwest of the Caspian Sea, Central America and Thailand along the Burmese border. In about 9750 BC, agriculture was practiced along the Burmese border to produce peas, beans, cucumbers, and water chestnuts. Perhaps, this was about 2,000 years before the real form of agriculture discovered along the Caspian Sea and Central America. About 7,500 years ago, millet and sorghum and other local crops were domesticated, in parts of China and the Sahel region of Africa, New Guinea, and Ethiopia. Gradually wheat, rye, and oats were found in other areas with the migration of people. Maize and beans were the staples in Central America.

The human endeavour to mitigating hunger through food production mostly continues as a primitive form of industry. Teeming millions of agricultural workforce are engaged in farming, with primary dependence on the traditional manual methods, by using animal power or through sectoral mechanization. Cultivation of family-based small holdings is the principal pattern in the developing nations. In contrast, farms are much more abundant in the developed world where large scale agricultural production contributed to value-added commercial operations, including food produce processing, marketing, and distribution (Loftas 1995). From the mid-twentieth century, a series of intermediary technological advances in machinery

P. K. Nag and L. P. Gite, *Human-Centered Agriculture*, Design Science and Innovation,
https://doi.org/10.1007/978-981-15-7269-2_1

development, methods of farm practices, and improvisation in irrigation systems to provide water to crops has dramatically changed the food scene of the world. The fertilizer industry evolved for the need to replenish the soil. The pesticide industry emerged with the need to control weeds, insects, and rodents. Mechanization in agriculture speeded up various farm activities. Introduction of chemicals—pesticides and fertilizers contributed to controlling pests and improving the quality of arable land. Biotechnological advances in new varieties of crops emerged as the green revolution to Asian nations and the regions of North America.

Socio-economic Issues

The world population has grown from about 2.5 billion in 1950 to 7.5 billion people in 2017 (FAO 2018), and the projected estimate is that the total population would reach 9.6 billion by 2050. The continuing surge of the human population and dangerous climate impacts in agriculture (ADB 2009) brings challenging pressure on meeting the demand for food energy and nutrients. Indicatively, agriculture gradually takes the shape of an enterprise for subsistence food supply to the community and income generation through agri-business activity.

The dietary supply is an indicator or an estimate of the food energy for human consumption at the national level. The dietary amount does not yield information on the affordability, access, or use of specific population groups. However, it allows determining whether the food supply contains enough dietary energy to meet the population needs of a country and the scope to improve the availability of dietary energy to the population. The world scenario indicates that the average nutritional energy supply had increased from 2716 kcal/person/day in 1999–2001 to 2904 kcal/person/day in 2015–17 (FAO 2018). The available caloric intake from grain ranges from 23% in northern America to about 60% in developing Asia and North Africa.

In 2017, global food grain production was about 2.7 billion tonnes of which cereals, including wheat and rice, constitute the primary commodity for staples. Of nearly 50,000 edible plant species, a mere 15 of them provide 90% of the world's food energy intake. Three grains, namely, rice, maize, and wheat, are the staples to more than 4 billion people. About 1 billion people in developing nations use starchy roots and tubers as staples, such as cassava, sweet potatoes, potatoes, yams. Another group of primary food crop for the requirement of starch and protein is the pulses (legumes) that comprise dry beans, peas, chickpeas, and lentils. The coarse grains are also consumed and used for animal feed. The horticultural crops—fruits and vegetable fulfil the needs of the nutrients (minerals, vitamins, and antioxidants) of the human body. Some legumes, such as soybeans and groundnuts, are oil crops. Crops like coconuts, sesame, cottonseed, oil palm, and olive make vegetables oil. Maize and rice bran also produce vegetable oil. Fibre crops (cotton, jute, hemp, flax) provide clothes. The plantation crops, such as natural rubber, palm oil, cane sugar, beverages (coffee, cocoa, and tea), tobacco, and bananas, are varied commodities for human consumption.

Farming Sector and Workforce

The growth and development of agriculture emphasize the health of the production base and the producers. Improvement in life and living in agriculture bears the fundamental focus towards sustainable food security and the livelihood of the majority of the population (Ayyappan and Chandra 2010). This population includes people engaged in farming, hunting, fishing, and forestry, for self-subsistence and also those who are in commercial enterprises. The global estimate indicates about 935 million people are involved in agriculture, and this sector represents nearly 28% of the world totaled 3305 million workforces (FAO 2018; ILO 2018). The most significant component of this workforce is in the developing countries and transitional economies. Total agricultural employment (Table 1.1) in Asia is more than 2/3rd of the combined total of the world's farming population.

The world's total non-agricultural population grew from 2.2 to 4.4 billion people between 1980 and 2011. The staggering growth is nearly five times greater than that of the growth of the agricultural population (Fig. 1.1). The growth pattern primarily attributes to the massive increase in the world's total population, which nearly doubled from 3.1 to 7 billion people between 1961 and 2011. There has been a vivid trend of migration of the workforce from agriculture to other non-agricultural occupational pursuits, such as construction and other menial jobs. Increasing the intensity of mechanization in the agricultural sector is also a recognizable phenomenon in

Table 1.1 World population and agriculture data

Particulars	World	Asia	Africa	America (North & South)	Europe	Oceania
Population (million)	7550.3	4504.4	1256.3	1006.8	742.1	40.7
Rural population (million)	3373.5	2257.8	718.0	192.8	193.2	11.8
Total workforce (million)	3305.3	1905.7	451.3	466.0	380.5	14.8
Workforce in agriculture (million)	935.6	618.4	229.8	42.8	27.5	0.5
Employment in agriculture (%)	26.7	25.7	53.2	9.4	5.9	3.3
Women workforce in agriculture (%)	27.5	28.2	55.9	4.8	4.6	1.7
Area harvested, crops (million hectares)	1168.9	619.5	255.8	298.1	186.7	24.7

Source FAO 2018, World Food and Agriculture—Statistical Pocketbook 2018

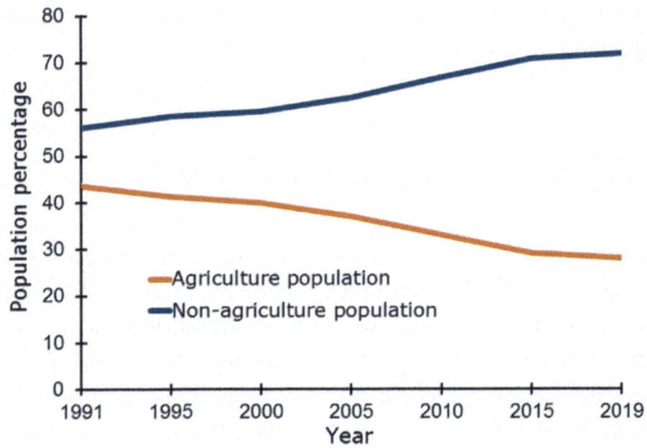

Fig. 1.1 Global population percentage in agricultural and non-agricultural sector

compelling the agrarian workforce to move away from rural settings in search of other areas of employment.

Many Asian countries have a dominant endeavour towards modernizing agricultural operations in certain areas and crops, and much of the food surpluses in these countries might partly attribute to the increasing introduction of mechanization and high-yielding species of crops such as rice and wheat. In general, however, the landholdings in Africa and Asia are small at the range of 2–5 hectares split into smaller plot sizes. Thus labour-intensive family-based farming becomes compulsive, where every member of the poverty-stricken families is engaged in activities throughout seasons. In peak working seasons, and particularly during the harvesting season, partial employment of wage labourers in specific farm activities is an established pattern in tropical and subtropical regions of the world. For obvious reasons, during the off-season, the farmworkers seek employment in non-farm enterprises. Despite regulations banning child labour, this has been a part of agriculture throughout its history. Poor living and working conditions of wage labourers in plantation and other commercial agriculture bear significant concerns of livelihood and aggravated health problems (Martin 2016).

On the other hand, northern America employs only a tiny fraction of the population in agriculture, where large farms are running up to hundreds of hectares. Landholdings in Europe have been increasing in size, as there is a vast migration from agriculture. Eastern Europe carries a history of socialized farming. In the former USSR, for example, the farm size was more than 10,000 hectares. In recent decades, however, there has been a changing pattern of farming in Eastern Europe as the countries are moving towards market economies.

Women in Agriculture

Globally, the engagement of women as a workforce in farming currently accounts for about 28% of the total agricultural workforce. Whereas in the regions of the developed world, a negligible number of women (about 4%) are engaged in farming (Nierenberg and Burney 2012). FAO documented gender issues of farm women in developing countries. Women represent about 20% of the farm workforce in Latin America. In sub-Saharan Africa, women dominate as farm workforce (56%). Whereas the women's asset ownership is limited, they contribute to producing up to 80% of the farm output. In the near East and North Africa, only 15% of land holdings belong to women, though women farmers have increased by about 15% since 1980. Regional war conflicts, substantial gendered migration patterns, and increased incidences of HIV/AIDS resulted in an increased contribution of women in farming and the overall issue of food security.

Regional statistics of the farm workforce in East and Southeast Asia are grossly dominated by the populous China that represents about 48% of women as workforce in farming. Indonesia employs a significant fraction of women workforce in agriculture and men folks in rubber plantations. In the countries in South Asia, Bangladesh shows extensive engagement (60%) of women farmers. There has been a significant increase in the participation of women in farming in Pakistan. In populous India that predominantly represented by the farming population (nearly 60%), women represent about 37% of the farm labour force. This percentage is higher than the world average.

Agricultural Development

There have been recognizable distinctions in the characteristics of industry and agriculture. One may argue that agriculture is also an industry with a marginal primitive image. Agrarian methods and practices vary across national boundaries. Advances in agricultural management, production, and distribution system are indications of the transformation of traditional agriculture to industrial bases. Accordingly, the sector of agriculture may be summarized into three types (Chamber et al. 1991), as given in Table 1.2.

Industrial agriculture encompasses industrialized countries of the West (temperate climate) and specialized sectors of the tropical countries. The green revolution agriculture represents the well-endowed areas in the tropics, primarily irrigated plains and deltas of Asia and parts of Latin America and North Africa. Resource-poor agriculture covers hinterlands, drylands, forests, mountains, hills, near deserts, and swamps, e.g. Deccan Plateau in India, uplands of Southeast Asian, and Latin American and most areas of Sub-Saharan Africa, as identified by the Brundtland Commission. Douglas N. Ross cited from Lapedes and Lapedes (1977) that '*... the task of the world's food supply systems in the last quarter of the 20th century is immense. While the specter*

Table 1.2 Three types of agriculture

Particulars	Industrial	Green Revolution	Resource-Poor Agriculture
Main locations	Industrialized countries and specialized enclaves of Asia	Irrigated, stable rainfall, high potential areas in Asia	Rainfed areas, hinterlands, areas of Sub-Saharan Africa
Main climatic zone	Temperate	Tropical	Tropical
The dominant type of farmers	Highly capitalized family farms	Large and small farmers	Small and poor farm households
Use of purchased inputs	Very high	High	Low
Farming system	Simple	Simple	Complex
Environmental diversity	Uniform	Uniform	Diverse
Production stability	Moderate risk	Moderate risk	high risk
Current production (% of sustainable output)	Far too high	Near the limit	Low
Priority for production	Regulate production	Maintain production	Raise production

of Malthus's predictions remains, there is no inexorable process for its materialization. Starvation and malnutrition are not necessary conditions of human existence. What happens depends upon government policies - from population programs to food production incentives - coordinated with business actions, and upon a sustained willingness to address the problem'. Undoubtedly, the situation will continue to demand actions in the coming decades. The national policy for intensification of land use, multiple cropping, farm mechanization, and the fertilizer-responsive high-yielding varieties of rice and wheat helped to bolster yields of crops (Searchinger et al. 2014). There are well-identified farm management approaches through climate-resilient cropping systems, IPM, biotechnological inputs, expansion of irrigation potentials, and support services for farmers.

The critical factors that determine growth performance in agriculture include (a) the country's per capita GNP, (b) the supply of per capita arable/cropped land, (c) Gini coefficient about the distribution of landholdings among agricultural producers, (d) the ratio of irrigated area to total arable land, indicating the availability of input resources such as the timely supply of water and other inputs, (e) the share of agricultural exports to agricultural GDP, and (f) adoption of improved methods and technology of production (Table 1.3). Further, agricultural growth in South and Southeast Asian countries, in particular, is a significant determinant of the overall economic growth of the nations.

The effort of addressing the food concerns of the early the mid-1960s for feeding the teeming millions of starved and semi-starved is significantly crowned with success to increase the production of primary cereals and to attain self-sufficiency of the food-deficit countries (ADB 1977). Throughout rural Asia, the ratio of farm population to farmland continued to rise. Increasing population pressure on land has led

Table 1.3 Some salient characteristics of population and agriculture of some countries

Country	Total population (million)	Rural population (million)	Share of agriculture in total expenditure (%)	Area harvested (Gross cropped area) (million ha)	Irrigated area (% of the total)	Employment in agriculture (%)	GDP per capita (USD, PPP)	Share of agriculture in GDP (%)	Total cereal production (million tonnes)
China	1441	601	9.9	180.6	NA	18.1	13644	8.3	618
India	1339	872	5.26	194.3	36.8	43.4	6096	15.4	297
United States	325	60	0.56	103.6	NA	1.7	53399	0.9	476
Indonesia	264	117	2.71	44.2	NA	31.8	10766	13.9	110
Pakistan	197	117	1.90	21.2	50.5	42.3	4857	24.7	43
Bangladesh	165	94	5.74	15.9	59.7	41.1	3319	14.2	53
Philippines	105	59	4.30	14.5	NA	27.0	7233	9.6	25
Thailand	69	32	6.46	18.9	NA	30.87	15706	8.2	32
Myanmar	54	36	5.37	17.1	NA	51.3	5305	24.8	28
South Korea	51	9	2.37	1.5	NA	4.9	35020	2.2	6
Malaysia	32	8	2.04	7.3	5.1	11.4	25669	8.4	3
Nepal	30	24	9.53	4.9	NA	72.3	2298	27.0	9
Taiwan	24	NA	12.95	0.8	NA	5.0	22400	1.8	1.3
Sri Lanka	21	18	5.65	2.3	NA	27.5	11446	7.8	3

Note NA (not available)
Source FAO 2018; NDC 2016; CIA 2018

to a varied pattern of farm activity and production management. These countries are now at different stages of development and resource endowment. The main features of agriculture in the region are the high density of rural population, the prevalence of small-scale farms, the scarcity of land and water resources for agricultural uses, and the dependence on monsoon rainfall patterns on which most cultivation depends. Countries like South Korea and Taiwan are relatively poorly endowed with natural resources but have made a remarkable transition from an agrarian-based economy to a modern manufacturing and service economy.

The enormous progress in the food supply in the South and Southeast Asia are vivid examples of the importance of agricultural development. Most countries in the region are overwhelmingly agricultural economies, having abundant agrarian resources and large rural workforce. The non-agricultural sector, such as formal industrial enterprises, absorbs only a small portion of the fast-growing workforce. The expansion of agriculture in terms of quality of production in traditional food-deficit countries, such as Indonesia, India, and Sri Lanka, markedly reduced their dependence on food imports. Indonesia, formerly the largest rice importer in the world, had a significant surplus food production since the 1990s. Populous India, for example, though have a large industrial sector, over 2/3rd of the population remains in rural communities and depends mainly on agriculture and the informal industry for income and employment. India had to import her food in the 1960s, but attained self-sufficiency in food grain production by the middle of the 1980s. The informal sector is the 2nd largest employer, after agriculture, that survives on low productivity, with subsistence-level wages.

Farm Power Availability

Cereals that include wheat, rice, maize, barley, rye, oats, and millet make up the more significant part of food crop production for human consumption. The top five crops produced in the world in 2016 (FAO 2018) are sugar cane, maize, wheat, rice, and potatoes (Table 1.4). In comparison to 2006, there has been a substantial increase in crop production in 2016. In a decade, maize production increased by nearly 50% and sugar cane as 33%. However, the increase in rice production was moderate (16%) in 2016, compared to the level in 2006. The external factors, such

Table 1.4 The top five crops produced in 2016 (thousand tonnes)

	2006	2016
Sugar cane	1,417,376	1,890,662
Maize	707,932	1,060,107
Wheat	614,538	749,460
Rice (paddy)	640,706	740,961
Potatoes	297,111	376,827

as rising incomes and urbanization, have a notable influence on dietary consumption in terms of higher protein, fats, and sugar. There has been a shift away from crops, like wheat and rice, towards increased growth in coarse grains, oilseeds, livestock, and biofuel production.

Evidence is affirmative of the positive correlation between food grain productivity and the farm power availability in a country or region. Singh et al. (2015) suggested a linear function, as

$$\text{Foodgrains productivity (tonne/ha)} = 0.55 + 0.82(\text{farm power availability, kW/ha})$$

That is, the farm power input has a significant relationship to achieving higher crop production, subject to taking into account the composition of the sources of farm power in various farm operations. The food grain production in India increased from 0.71 tonnes/ha in 1960–61 to 2.11 tonne/ha in 2013–14, while farm power availability increased from 0.30 kW/ha to 2.02 kW/ha during the same period.

The North American and the European nations are generally countries with record-high penetration of mechanization, that is, about 1,300 tractors per 100 km^2 of arable land. Low farm mechanization persists in the developing world, for instance, the penetration rate of tractors in Sub-Saharan Africa is as low as 2.24 per 100 km^2 of arable land (Blade 2018). Vast resources of agriculture play an essential economic role in the regions of Latin America. However, the farming expanse in the areas is taking place much at the expense of tropical forests. In both the Middle East and Africa, per capita, food production has been in decline. In the Middle East, the principal limiting factor is the non-availability of water for agricultural activity.

Water Scarcity

Generally speaking, an area is said to be experiencing water stress when annual water supply falls below 1,700 m^3 per person. A region is said to face water scarcity when the amount falls below 1,000 m^3 per person. The absolute water scarcity is when supply drops below 500 m^3 per person. Agriculture is one of the most water-intensive sectors (Gleick 2009). Agricultural water withdrawal accounts for 44% of total water withdrawal among the OECD countries; the withdrawal by agricultural-based economies like India amounts to 74%.

Inefficient water use in agriculture leads to the overexploitation of groundwater resources as well as the depletion of the natural flow of major rivers, e.g., the Ganges in India and the Yellow River in China. Efficient use of water in agriculture has been possible from the application of various options, such as (a) growing an array of crops suited to local conditions, especially in drought-prone regions and (b) practicing agroforestry to build secure root systems and reduce soil erosion. Besides, maintaining healthy soils, through the application of organic fertilizer or growing cover crops to retain soil moisture, and adopting irrigation systems, drip lines to

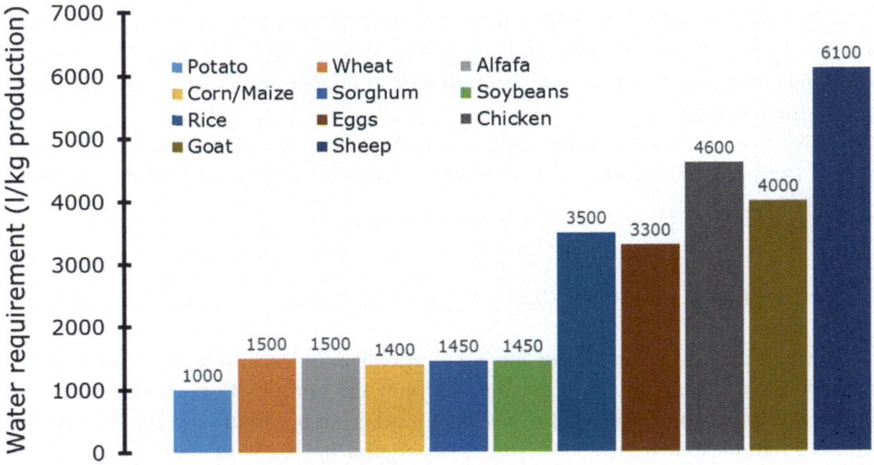

Fig. 1.2 Requirement of water in the production of different crops and animal products

deliver water directly to plants' roots are the water scarcity mitigating approaches. For example, contrary to the method of conventional rice production, the System of Rice Intensification (SRI) increases crop yields with about 20–50% less use of water. With the gradual shifting of preference of the type of food consumption from predominantly starch-based diets to meat and dairy, there is an apparent perceptible impact on water consumption over the last decades, and the scenario is likely to continue in the future (FAO 2018). Producing different crops and animal products requires a substantial amount of water (Fig. 1.2). Producing 1 kg of rice, for example, requires about 3,500 l of water, while 1 kg of beef needs some 15,000 to 70,000 l of water.

Scope and Challenges in Agriculture

Despite crowning success in agricultural methods and practices throughout the world, about one billion people are still living with chronic hunger (Gourmelon 2014). The world agriculture will continue to explore multiple avenues of farming in the backdrop of eco-geological constraints, socio-economic compulsions, resource considerations, and regional demographic characteristics. On the one hand, the emphatic corporatization of farming in the western world evolves from heavy farm mechanization, automation, and robotic application. On the other hand, the present dismal scenario of small and marginal farm holdings in Asia and other developing countries will continue, with dependence on labour intensive, medium, and small-scale farm development initiatives. In medium-sized farms, power operated as well as animal-drawn and manually operated equipment might be used. Since the cultivable land is highly fragmented and primarily distributed among small and marginal farmers, animal power, as well as human power, will be in extensive use. On many of these small farms, suitable power operated machines may also be used on custom hiring basis for various operations.

While the extensive application of agricultural technologies and machinery will bring higher production of grains, fruits, vegetables, and other crops, the potential health, and safety risk of farmworkers cannot be ignored. Needless to mention that during the process of adoption of machines, the potential risk of accidents and loss of lives gets aggravated. It is a fact that danger is a part of the life of agricultural workers. The cost of the accidents and injuries is not well accounted, due to the non-availability of systematic data of farming injuries. Considering the vast workforce (about 935 million) involved in agriculture all over the world, the cost of deaths and injuries due to agricultural accidents can be very high, and utmost attention is required for minimizing these accidents. Besides injuries, health problems also arise due to exposure of workers to various stress agents such as vibrations, noise, dust, chemicals, and thermal stresses.

Growth in agriculture will depend in large part on getting effective technologies into the hands of farmers to raise agricultural productivity for its sustainability. More efforts will be needed to efficiently use land, water, and energy resources (IFPRI 2019). The role of the farmworkers will be gradually shifting from as a source of muscle power in traditional farming to the machine operator, with the large-scale incorporation of agricultural technologies and machinery. There is an obvious priority to bring in greater awareness among farmers and farmworkers, society, and governments about safety and health issues in agriculture. More attention will emerge for a comprehensive approach to fostering the healthy and productive role of farmers in rural, peri-urban, and urban areas. This approach is built on the recognition that both urbanization and rural transformation are necessary to achieve sustainability in agriculture, the vitality of the rural economy, to drive food security and make a contribution to the macro scale in the national, regional, and global value chains.

References

ADB. (1977). *Rural Asia: Challenge and opportunity: Report on the 2nd Asian Agricultural Survey*. Asian Development Bank. New York: Praeger.

ADB. (2009). Building climate resilience in the agriculture sector in Asia and the Pacific. Asian Development Bank and International Food Policy Research Institute. ISBN 978-971-561-827-4, Philippines.

Ayyappan, S. & Chandra, R. (2010) Indian agriculture- an economic perspective. *Think India Quarterly* 13(3):61–78.

Blade, S. (2018). Goldstein Research. 2018 Global Agricultural Tractors Market Outlook, 2016–2024. Goldstein Research.

Chambers, R., Pacey, A., & Thrupp, L. A. (Eds.). (1991). *Farmer first—farmer innovation and agricultural research*. London: Intermediate Technology Publications.

CIA. (2018). *The world factbook*. Washington, DC: Central Intelligence Agency.

FAO. (2018). *World food and agriculture- statistical pocketbook 2018* (p. 2018). Rome: Food and Agriculture Organization of United Nations.
Gleick, P. H. (2009). *The World's water*. Pacific Institute, 486.
Gourmelon, G. (2014). *Chronic Hunger Falling*. World Watch Institute: But One in Nine People Still Affected. Vital Signs.
IFPRI. (2019). *Global food policy report* (p. 2019). Washington, DC: International Food Policy Research Institute.
ILO. (2018). Employment in agriculture (% of total employment) (modeled ILO estimate), ILOSTAT database, International Labour Office, Geneva. Data retrieved September 2018.
Kislev, M. E., Weiss, E., & Hartmann, A. (2004). Impetus for sowing and the beginning of agriculture: ground collecting of wild cereals. *Proceedings of the National Academy of Sciences, 101*(9), 2692–2695.
Lapedes, D. N., & Lapedes, D. N. (1977). *McGraw-Hill encyclopedia of food, agriculture & nutrition* (No. C TX 349. M32). McGraw-Hill.
Loftas, T. (1995). *Dimentiosn of Need*. Roma: An Atlas of Food and Agriculture.
Martin, P. L. (2016). *Migrant workers in commercial agriculture*. Geneva: International Labour Office.
National Development Council (NDC). (2016). *Taiwan statistical data book*. Taiwan: ROC.
Nierenberg, D., & Burney, S. A. (2012). *Investing in women farmers*. Vital signs: Global trends that shape our future.
Searchinger, T., Hanson, C., Ranganathan, J., Lipinski, B., Waite, R., Winterbottom, R., & Dumas, P. (2014). Creating a sustainable food future. A menu of solutions to sustainably feed more than 9 billion people by 2050. World resources report 2013–14: interim findings. *World Resources Institute (2014)*.
Singh, R. S., Singh, S., & Singh, S. P. (2015). Farm power and machinery availability on Indian farms. *Agricultural Engineering Today, 39*(1), 45–56.

Chapter 2
Manpower Utilization: Working Methods and Practices in Small Land Holdings

Introduction

Crop production began more than 10,000 years ago when domestication of plants became essential to supplement the natural food supply. The age-old methods of crop production prevail in small and marginal landholdings. The farming activities include (a) preserving seeds of the crop plants; (b) destroying weeds grown on the land; (c) preparing the seedbed by churning the soil; (d) planting at an appropriate season; (e) protecting the crops from pests and insects; and (f) gathering, processing, storing, and distributing farm products. Besides crop production and processing, other activities include, for example, animal raising, dairy management, and agroforestry. Extensive human involvement in diverse functions is the essential requisites concerning different classes of agriculture in many geographic regions, as described in Chap. 1. In the farms that are highly mechanized and automated, the crop production activities differ due to the virtual minimization of human labour and animal power. Based on distinct climatic and geographical features and methods of cultivation, the farm crops include field crops, horticultural crops, and plantation or perennial crops.

Field Crops

This group includes cereals and millets (e.g. maize, sorghum, rice, wheat), pulses and legumes (e.g. pigeon pea, chickpea, soybean), oilseeds (e.g. groundnut, rapeseed, mustard, sunflower), fibre crops (e.g. cotton, jute), sugar crops (e.g. sugarcane, sugar beet), and fodder crops. While rice is grown in wetland, upland, and also in the semi-upland system of cultivation, all other plants are grown under an upland system of cultivation.

P. K. Nag and L. P. Gite, *Human-Centered Agriculture*, Design Science and Innovation,
https://doi.org/10.1007/978-981-15-7269-2_2

Upland and Semi-upland Cultivation

In this system, the land is repeatedly ploughed and harrowed to get the desired tilt. The seed is sown when optimum soil moisture condition is available (i.e. at the onset of rains or after irrigation) either by broadcasting or in lines with the help of plough or seed drill. Weeding is done with the help of hand tools (Fig. 2.1a, b), animal-drawn equipment, or power-operated equipment. In some areas, applying weedicides helps to control weeds in the cropped field. Plant protection chemicals are applied with manually operated/power-operated sprayers, dusters, and other appliances to protect the crops from pests and diseases. After the maturity of the crops, harvesting is done with a sickle or other hand tools or animal-/power-operated equipment. Upland cultivation is followed for most of the plants like wheat, sorghum, pearl millet, soybean, pigeon pea, groundnut, cotton, rapeseed, and mustard. For rice also, this system is followed in some regions where irrigation/rainwater is not abundantly available.

The basic system of semi-upland farming is similar to that of upland crops, and it is followed for rice crops in the tracts, which depend on rains and do not have adequate irrigation facilities. In semi-upland cultivation, the rainwater is impounded in the field when the crops are 5 to 6 weeks old, and the ground is ploughed crosswise with a light plough. Ploughing of an area with the standing crop serves multiple purposes, namely, weeding, thinning, and interculture of the crop.

Wetland Cultivation

The system of wetland cultivation is primarily followed for the rice crop, in which the land is thoroughly ploughed and puddled with 5 to 6 cm of standing water in the field. Puddling aims to get a soft soil so that the transplanted seedlings can establish themselves quickly, to control weeds, and to reduce the percolation rate so that water can be impounded in the field for a long time. The ploughing and puddling can be done by animal-drawn equipment or power-operated equipment, such as power tillers and tractors. After puddling, the field is levelled up by planking. Then the sprouted seeds are directly sown in the puddled field, or seedlings grown in the nursery are transplanted (Fig. 2.1c). After transplanting operation, a 2 to 3 cm level of water is maintained for about a fortnight till the seedlings are established. After that, the water level is maintained at about 5 cm till a week before harvesting of the crop. In a broadcast crop, weeding is done three or four times by hand or manual-operated weeders at intervals of 20 to 25 days. In a transplanted crop, weeding and interculture operations are done within 3 to 4 weeks of transplanting. After the maturity of the crop, water is drained out of the field, and harvesting is done.

Fig. 2.1 Field crops—**a**, **b** upland and **c** wetland cultivation

Horticultural Crops

The group under horticulture crops includes fruit crops, vegetable crops, and flower crops. The fruit crops are banana, mango, lemon, orange, grape, guava, papaya, pineapple, sapota, pomegranate, cashew, custard, apple, apricot, peach, pear, plum,

Fig. 2.2 Horticulture and plantation crops—**a** tea garden, **b** apple orchard, **c** banana plantation, **d** coconut plantation, **e** orange orchard

and strawberry. These perennial crops can be grown in small as well as large areas (Fig. 2.2b, c, e). Vegetable crops, such as potato, cauliflower, cabbage, peas, tomato, chilli, beans, brinjal, lady's finger, and gourds, are generally short duration crops and produced in succession throughout the year. The vegetables may be grown in kitchen gardens or on a field scale, and accordingly, the type of cultivation and equipment used also varies. Some essential flower crops, such as rose, chrysanthemum, and marigold, are grown in specific regions, primarily for commercial purposes. Tools and equipment in raising these crops are usually termed as horticultural tools and equipment.

Plantation Crops

Some crops like tea, coffee, rubber, coconut, areca nut, palmyra, black pepper, and large cardamom are perennial types and grown in plantations (Fig. 2.2a, d). The cultivation system and the operations involved in the production and processing of plantation crops differ from field crops. Methods, practices, and equipment used in plantations are crop-specific.

Hill Agriculture

Hill agriculture mainly includes valley cultivation, terrace cultivation, and growing crops and orchards on slopes (Fig. 2.3). A large percentage of women workforce is engaged in hill agriculture as compared to other sectors. Because of the constraints of terrain and pathways, taking machinery to the field is a problematic proposition. The workers manually carry tools/equipment to the ground. Manual material handling activities, i.e. to bring inputs to areas and to gather and transport produce are drudgery

Fig. 2.3 Hill agriculture

prone. In general, the productivity of the workers is less, and safety in work is of paramount importance (CIAE 2014).

Fig. 2.4 Seedbed preparation **a** animal-drawn plough, **b** tractor-operated implement

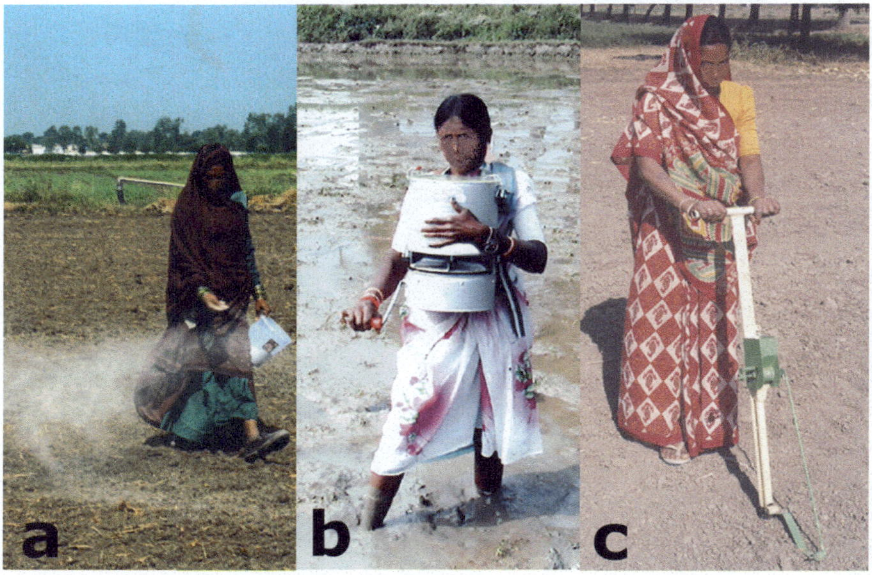

Fig. 2.5 **a** Manual fertilizer broadcasting, **b** fertilizer applicator, and **c** manual dibbling

Agricultural Operations

Crop cultivation involves the preparation of the soil, i.e. ploughing, tilling, breaking of clods, harrowing, weeding, manure spreading, seed sowing, planting, care of seedlings, pest control, top dressing and irrigation (Nag 1998). Brief information about various operations involved in raising crops is given herewith (Table 2.1). The information about the work method/tool as well as for productivity indicators, i.e. field capacity/output per hour is collated from Hopfen and Biesalski (1953); Pandey

Fig. 2.6 Tractor operated **a** seed-cum-fertilizer drill and **b** pneumatic planter

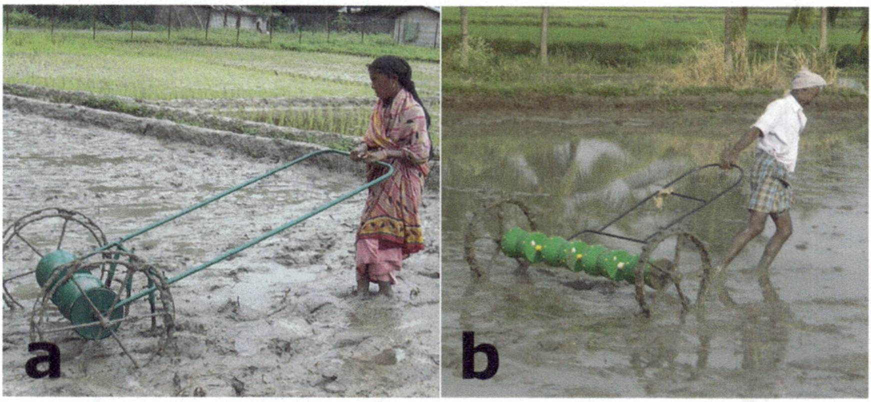

Fig. 2.7 Manual rice seeder **a** two seed hopper and **b** four seed hopper

et al. (1997); Varshney et al. (2004); Ojha and Michael (2005); Singh (2007); Singh et al. (2007), ICAR (2012); Gite (2017); and Mehta et al. (2018).

The farm activities embodied herewith are primarily observed in small and marginal farm holdings. One may vividly witness such situations in the Southeast Asian regions, where agriculture practices have a primary dependence on human and animal power sources. Though small farm holdings might dominate in a geographical area, it does not deter the farm mechanization. The labour force migration from rural to urban settlements, from farming to other sectors of employment, leads to a compulsive gradual introduction of machinery, and mechanical and electrical power sources in various farm operations. The bringing of changes is visible in other farm

Fig. 2.8 Weeding **a** hand hoe, **b** wheel hoe weeder, **c** Mandwa weeder

Fig. 2.9 Knapsack spraying

activities, including processing, greenhouse and nursery operations, and food product storage.

Fig. 2.10 Harvesting operation, using the sickle

The grain/fruit processing operations, such as peeling, polishing, milling, and oil extraction, are carried out to prepare value-added products from the agricultural produce. For many of the processing operations, manually-operated machines are available, e.g. manually -operated potato peeler and slicer units, and vegetable washers. Nowadays, most of the processing operations are carried out with power-operated machines, e.g. grain mill, dal (pulse) mill, and power-operated vegetable washers. The output of a power-operated grain mill/dal mill ranges from 40 to 100 kg/h. Grading of fruits is primarily a manual operation. Fruit graders are also available. Oil extraction is traditionally carried out using mechanical oil expeller operated by animal power, and the process is fast replaced by using a solvent extraction method performed in large plants.

Greenhouse and nursery operations vary depending on the purpose, i.e. to grow rare and exotic plants, production plants, or seedlings. The typical nursery crops are trees and shrubs, whereas the greenhouse crops include flowers, vegetables, and herbs. Generally, the hardy plants are grown outside, and the less hardy plants are propagated and raised inside in the greenhouse at controlled climatic conditions. The activities in the greenhouse include putting the soil into small pots; planting the

Fig. 2.11 Harvesting operation—**a** self-propelled reaper windrower, **b** tractor-operated reaper windrower, and **c** combine harvester

Fig. 2.12 Threshing of paddy crop, **a** manual beating, and **b** animal treading

Fig. 2.13 Padel threshing—**a** without safety guard and **b** with safety guard

Fig. 2.14 Power threshers

seeds to the containers, watering, fertilizing, and applying pesticides to the plants; trimming or thinning the plants as needed; and transporting the plants or product from the greenhouse. Soil filling and planting operations have become mechanized in large production greenhouses. Nowadays, a timer-controlled sprinkler or piping system allows the easy way of watering the plants. The output/field capacity in nursery operations varies based on the process and the tool/equipment used.

The storage of farm produce in the proper environment and structures bear importance for safekeeping of grains and seeds for later use. Vegetables and fruits are stored in cold storages to preserve the nutrient contents and to protect those from getting spoiled. Storage bins, silos and godowns, and warehouses are used for storing grains. The storage bins have about 1 to 10 quintal capacity and silos of 10 to 100 quintal capacity. The operations in grain storage involve loading/unloading, grain/bag

Fig. 2.15 **a** Manual winnowing and **b** hand-operated winnower

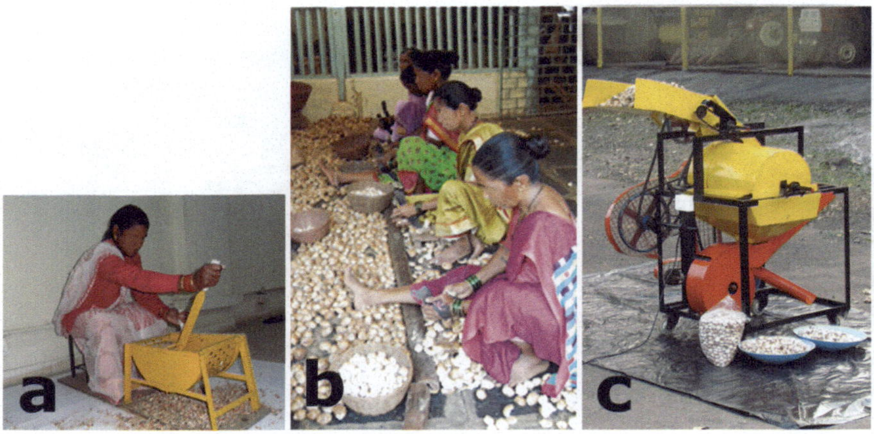

Fig. 2.16 **a** Groundnut decorticator in sitting mode, **b** manual arecanut dehusking, and **c** power-operated arecanut dehusker

handling, maintaining a proper environment, and fumigation. In cold rooms (fruit and vegetable storage), maintaining a suitable climate is very important.

Particular crops like mushrooms require a different set of procedures for their cultivation. The output/field capacity varies based on the operation and the tool/equipment used. The typical white button mushroom is grown on compost consisting of a fermented mixture of manure, straw, and gypsum in a proper environment. Mushrooms appear in flushes at an interval of weeks, and harvested by hand-picking or mechanically. After 3 to 6 blooms, the growing room is steam sterilized, cleaned, and disinfected for the next growing cycle.

Fig. 2.17 Modes of MMH and transportation

A range of manually operated machines has been developed based on the needs of the farmers and other stakeholders. Due to economic constraints, the marginal and smallholding poor farmers might not be capable of buying large machines. However, these farmers can make use of custom hiring centres/service providers to get farm machines for various operations (Mehta et al. 2014). Manually guided self-propelled machines or power-operated machines are becoming popular on the farm, and that would imply the role of the workers to emerge as the controller of the devices. Today, about 1/5th of male workers serve as the controller of tools and methods, and about the remaining 4/5th as a source of power (Mehta et al. 2018). In the case of women workers, however, they serve merely as the source of power in various in-field and out-field activities. This situation is in the changing path, and more and more men and women will work as controllers of machines in the farms in coming decades. As the role of women workers will be getting significant in the future, the use of women-friendly tools and equipment will help in small farm mechanization to a great extent. Suitable design of power-operated/self-propelled machines, as well as skill upgradation of the workers in operating these machines, will help in increasing the efficiency of farm operations as well as productivity and profitability on farms.

Table 2.1 Farm operations and tools/equipment used in small land holdings

Brief description	Work method/tools	Productivity indicators (Field capacity/output)
Operation: Seedbed preparation (Tillage) (Fig. 2.4)		
Tillage is done to prepare a seedbed, to eliminate competition from weed growth, and to improve the physical condition of the soil. This process helps the destruction of native vegetation, weeds, or the sod of the previous crop. It helps in the removal, burial, or incorporation of crop residues in the soil or manure mixing. The tillage on drylands helps in conserving soil moisture and prevents surface runoff and wind erosion. A desirable seedbed is one that is mellow to keep the soil particles in close contact with the seed, and the seedbed is free from trash and vegetation that would interfere with seeding. Old implements for tillage were hand tools made of wood or iron, or stone to dig the soil. The application of the power of domestic animals made it possible to develop improvised implements.	Manual methods in seedbed preparation, primarily for cereal production, include the use of hand tools (hoe, spade), wooden plough pulled by draught animals (ox, buffalo, camel, mule), levelling plank, and blade harrow. Over the generations, there has been improvisation in the design of plough. Some wooden ploughs used in seedbed preparation weigh more than 40 kg. The steel mouldboard plough (weighing less than 20 kg) is smoother in shape and design (Fig. 2.4a) The mouldboard plough breaks, loose, or shears off the furrow slice by forcing a triple wedge through the soil. It inverts and partly pulverizes the soil. Disc ploughs are used in soils that are too dry and hard for easy penetration of mouldboard ploughs. The tractor-operated machines include mouldboard plough, disc plough, cultivator, harrow, subsoiler, rotavator (Fig. 2.4b) The harrows equipped with spikes, blades, and disc points are the secondary tillage implements used to break clods, destroy weeds, and agitate soil surface to improve moisture absorption.	Ox-drawn plough/blade harrow: 0.025 to 0.05 ha/h; Tractor-drawn implements: 0.2 to 0.6 ha/h in different operations. Note: ha/h (hectare/hour)
Operation: Sowing/planting and fertilizer application (Figs. 2.5, 2.6 and 2.7)		
Sowing includes placing of seeds at the desired depth and distance in the seedbed. The conditions necessary for the germination of seeds are ample supplies of moisture and oxygen, a suitable temperature, and specific light conditions.	The sowing and planting include manual broadcasting of seeds, use of seed drills, planters, and transplanters. *Broadcasting:* The seeds are broadcasted and then covered with a harrowing operation. The broadcasting operation is generally done manually by hands	Manual broadcasting by hand: 0.1 to 0.2 ha/h; Seeding using a plough: 0.02 to 0.08 ha/h
	Seeding behind a plough: This is a method of putting seeds by hand into the furrows that opened with a wooden plough or a tyned implement.	

(continued)

Table 2.1 (continued)

Brief description	Work method/tools	Productivity indicators (Field capacity/output)
	Drilling: In this method, seeds are placed in soil by a seed drill or seed-cum-fertilizer drill. Mechanical or pneumatic metering mechanisms are used for the digging of seeds. Here seeds can be placed at a uniform depth, row-to-row distance, and the desired rate. The drills may be manual, animal-drawn, or tractor-drawn units. It may vary in size from single row units to multi-row machines. Combination grain and fertilizer drills are suitable for sowing and fertilizer application simultaneously. Many times, manure/fertilizer application is made by manual broadcasting by hand or by using manually operated broadcasters/applicators (Fig. 2.5a, b). Animal-operated broadcasters are in use in some areas. Tractor-operated broadcasters/drills are used worldwide. In these devices, for fertilizer application, mechanical metering devices are used. In the case of rice, pregerminated paddy seeds are put in wetland using rice seeder (Fig. 2.6a).	Manual seed drill/seed-cum-fertilizer drill: 0.03 to 0.05 ha/h; Animal-drawn seed drill/seed-cum-fertilizer drill: 0.04 to 0.06 ha/h; Tractor-drawn seed/seed-cum-fertilizer drill: 0.2 to 0.4 ha/h
	Dibbling: The method is used where the supply of seeds is limited, or the seeds are expensive. Placing of seeds is done manually or with the help of an implement known as dibbler (Fig. 2.5c).	Manual dibbler: 0.01 to 0.03 ha/h
	Planting: It includes putting seeds at a desired distance within a row. A planter can maintain row-to-row distance as well as a plant-to-plant range in a row (Fig. 2.6b). The planting method is used for some crops, like maize, cotton, sugarcane, and potato. In the case of sugarcane, it includes placing of cane sets of about 30 cm length in-furrow and covering them with soil. The planters may be semi-automatic or automatic, i.e. either animal-drawn or tractor-operated. A combination implement like lister planter can do tillage and planting simultaneously. Crops like potatoes are planted on ridges in irrigated conditions. Corn and cotton are planted in the bottom of furrows in semi-arid areas. Surface or level seedbed planting prevails in the areas where moisture conditions are favourable. In high rainfall region, row crops are often planted on elevated beds with a lister or bed former.	The field capacity of planting equipment depends on the power source used and the type of crop

(continued)

Table 2.1 (continued)

Brief description	Work method/tools	Productivity indicators (Field capacity/output)
	Transplanting: Worldwide, the rice crop is primarily sown by transplanting method. Usually, pregerminated seeds are broadcasted densely on a puddled field. The seedlings are grown up to about 15 to 20 cm height in about 3 to 4 weeks. These are uprooted and transplanted to a puddled ground either by hand transplanting or with manually operated or power-operated transplanters. Generally, the spacing of planting is 15 × 15 cm, 20 × 20 cm, or 25 × 25 cm. The spacing allows easy access to remove weeds, sufficient sunlight, and aeration for growing plants. It also helps to reduce humidity between the plants and discourage the breeding of insects. The transplanting method applies to tobacco and vegetable crops, like brinjal, chilli, and tomato.	Traditional rice transplanting: 0.004 ha/h/person; Manual rice transplanter: 0.017 ha/h; Self-propelled 8-row rice transplanter: 0.09 to 1.2 ha/h
	Budding/Grafting: Some of the ornamental/horticultural plants are grown by adopting the process of budding or grafting. In budding, a healthy bud of the desired variety is inserted into the stem of another plant. In grafting, a small plant called a scion plant of the desired type is put with an already existing plant having roots (called as rootstalk). Special tools such as a budding knife and grafting knife are used for these operations.	
Operation: Weeding and inter-cultivation (Fig. 2.8)		
A weed is an undesirable plant or only as a *plant out of place*. Weeds cause losses in crop yields, impair the crop quality, harbour plant pests, and increase irrigation cost. Weeds emerge along with crop seedlings, and if not controlled in the early stages of crop growth, these may cause a reduction in yield varying from 10 to 60% depending on the intensity and kind of weeds. The weeds are removed from the cultivating land to get the best utilization of the soil and fertilizers. It is challenging to keep the crop weed-free throughout the growing season. Usually, the crops are kept weed-free for the first 30 to 45 days after sowing. Of the total working hours involved in crop production, about 15 to 20% is spent in removing weeds,	Weeds are controlled by manual weeding, either by pulling out the weeds by hand or with hand tools like hand hoe or wheel hoe in drylands and with rotary weeder in wetlands (WSSAN 2006). In the drylands, the workers squat on the ground with one or two legs flexed at the knee, and the weeds are removed using a hand hoe (Fig. 2.8a). In wetlands, the workers bent forward (stooping posture) to remove the weeds by hand. Some of the simple weeders are a Dutch hoe, grubber weeder, long-handled spade, projection finger-type weeder, sweep-type weeder, twin wheel hoe, dryland weeder, and cycle hoe (Fig. 2.8b, c). Animal-drawn hoes are also used in some regions. Tractors operated as well as self-propelled weeders are available now for weeding and interculture operations in drylands. For rice crops in wetlands, self-propelled or manually guided power-operated weeders are used for weeding operation.	Hand hoe: 0.005 ha/h per person; Wheel hoe: 0.01 to 0.02 ha/h per person; Cono weeder for wetland rice: 0.02 to 0.03 ha/h per person; Animal-drawn hoes: 0.08 to 0.12 ha/h per person; Tractor-drawn weeder/self-propelled weeders:0.1 to 0.3 ha/h

(continued)

Table 2.1 (continued)

Brief description	Work method/tools	Productivity indicators (Field capacity/output)
Operation: Plant protection (knapsack spraying) (Fig. 2.9)		
Chemical application is the commonly used method of plant protection from various pests and diseases. It involves either spraying of a liquid mixture of chemical (pesticides) and water or a dusting of chemical powder on crop plants.	For pest management, the efficient use of equipment becomes necessary to secure a uniform and optimal application of a pesticide on a target substrate. Different types of hand-/manual- or power-operated sprayers and dusters are available (Fig. 2.9). For tall tree crops and orchards, a portable power sprayer is available. The workers should be well acquainted with spraying safety procedures in carrying out spraying or dusting operations. Non-compliance with the method of safe operation may bring in a multitude of health hazards to the sprayer operators, and people close to the spraying area.	Manual knapsack sprayer: 0.08 to 0.12 ha/h; Power-operated knapsack sprayer: 0.2 to 0.3 ha/h
Operation: Irrigation		
Irrigation is a prerequisite for intensive cropping, particularly in arid and semiarid regions. It includes the utilization of water from wells, canals, and tanks for raising successful crops. A significant component in irrigation is the lifting of water from the water resources and reservoirs to irrigate the fields.	For generations, indigenous devices have been in use for lifting water. Most of these devices were either manually operated, e.g. swing basket, counterpoise water lift, or animal powered, e.g. leather bucket, Persian wheel. Lifting water for irrigation by different manual methods is physically strenuous. Significant technological improvisation has taken place in the irrigation devices. The use of pump sets, either electrically powered or engine powered, is common. In general, improvement in irrigation requires not only better pump sets but also improved distribution systems such as sprinkler, drip, trickle, and subsurface irrigation	A 5 hp pump set can irrigate about 0.4 ha/h with 6 cm of water

(continued)

Table 2.1 (continued)

Brief description	Work method/tools	Productivity indicators (Field capacity/output)
Operation: Harvesting (Figs. 2.10 and 2.11)		
The gathering of crops upon maturity, or the practice of harvesting, and the quantity and quality of the harvest gathered from the field, orchard, or vineyard is a measure of farm productivity. Harvesting of the field crops mainly includes cutting of plants near ground level or cutting of the desired portion of the plant. Harvesting of vegetable crops depends on the type of plant. Root vegetables, such as carrot, radish, ginger, are uprooted, washed, and processed or sold in the market. Vegetables, such as cabbage, cauliflower, spinach, and peas, are generally harvested by hand with or without tools. At places, machines are used in some of these operations. Harvesting of fruit crops is generally done by manual plucking. In some countries, machines are used to assist the workers in fruit plucking operation. In the case of harvesting of fibre crops like cotton, the cotton stalks are uprooted manually or using cotton stalk puller	For many staple crops, like rice and wheat, a time-tested hand tool called sickle is used (Fig. 2.10). A sickle is a curved, serrated edge blade attached to a wooden handle. A massive need for this simple tool led to development of several varieties of sickles in different geographic regions and countries. Other hand tools, like a scythe, machete, and cradle, are used as harvesting tools. Due to lack of hand, arm, and leg protection, harvesting accidents, lacerations, and incised wounds are often reported from the paddy, wheat, and sugar cane farmworkers. Harvesting machines include mower, reaper, chopper, baler, etc. Power-operated reapers and reaper windrowers are used for harvesting of field crops in some regions (Fig. 2.11a) Animal-drawn reaper is available in some places for the harvesting of wheat crop. Harvesting by combine harvester (self-propelled or tractor operated) is frequent in areas where intensive cultivation is practised, and the labour shortage is acute. Combine harvester cuts the standing plant and also does threshing and winnowing operations simultaneously (Fig. 2.11b, c). The combines, however, produce substantial clouds of dust that are responsible for air pollution in the surroundings. In maize harvesting, the cobs are plucked from the plants and are kept for drying. Sickle, or a machine cuts the stalks. In the case of sorghum, the earheads are cut first, and then the plants are cut or vice versa. For cotton crop, the cotton is collected in 3 to 5 pickings by hand as the balls mature. After the last picking, the cotton stalks are removed manually or cotton plant puller or by using animal-/tractor-drawn plough or harrow. Harvesting of potato and sugar beet is done by a blade harrow or digger, which may be animal or tractor-operated. For groundnut, the vines are either pulled manually or removed using digger, and then the pods are separated. Tractor-operated/self-propelled machines are also used at some places for cotton picking. In the case of sugarcane, harvesting is done manually by cutting the cane plants at the bottom with a knife, and then the leaves of the plants are removed using the cutter or sugarcane stripper. Tractor-operated/self-propelled sugarcane combine harvesters are also available.	Sickle: 0.01 to 0.015 ha/h per person; Self-propelled reaper windrower: 0.15 to 0.25 ha/h; Combine harvester: 0.8 to 1.2 ha/h for grain crops

(continued)

Table 2.1 (continued)

Brief description	Work method/tools	Productivity indicators (Field capacity/output)
Operation: Threshing (Figs. 2.12, 2.13 and 2.14)		
Operation includes separation of grains from the earheads/cobs or termination of the desired portion from the plant. For maize, threshing includes removing the husk and then separating the grains from the cobs either by beating with a stick or using maize shellers. Maize shellers may be manually operated or power driven. Combination equipment like dehusker sheller has also been developed and is commercially available. For groundnut, the pods are separated from vines by beating the plant or by using some equipment or thresher.	The primitive way of threshing paddy crop was to rub the earheads between feet or beat them on a wooden plank or drum, to separate grains from the pinnacles (Fig. 2.12a). In some areas, the crop is spread on the floor and trampled by animals (animal treading) (Fig. 2.12b) Alternatively, a wooden roller (called as Olpad thresher) is pulled on the spread crop to carry out the threshing operation. Also, in other crops, the traditional way of threshing is to beat the earheads, cobs, or the crop stalks with sticks. Trampling by animals or a tractor is also done to separate the grains from the crop earheads. A pedal-operated thresher in which a drum is rotated by pedalling or treadling action is also commonly used for threshing operation (Fig. 2.13a, b). The output of power threshers (Fig. 2.14) is significantly more than manual methods. The power threshers carry out threshing and winnowing operations simultaneously and deliver clean grains at the output end.	Pedal-operated rice thresher: 35 to 50 kg of paddy/h; Power thresher: 4 to 20 quintal/h depending on the power source used
Operation: Winnowing/cleaning and grading (Fig. 2.15)		
Winnowing follows the threshing operation, whereby to get clean grains from the mix of chaff/straw and grains. The process involves the separation of grains from the chaff using natural air current or by creating air current using a hand, pedal, or motor-driven fan.	Commonly used manual methods are the following—(a) the entire content is thrown up in the air. By differential momentum, the grain and chaff get separated and (b) the material is dropped from a matted bamboo frame tray from a height of more than 2 m in natural or human-made air current (Fig. 2.15a). The grains being heavy fall straight on the ground, whereas the chaff/husk/straw being lighter goes along with the air and falls away from the grains. Manual as well as power-operated winnowers are also available and used widely.	Manual Winnower:1.5 to 2 q/h; Power winnowers: 2 to 10 q/h depending on the power source; Note: quintal (q)
	Most grains require cleaning before storage or use. Cleaning is done either manually using screens, hand- or pedal-operated cleaning equipment (Fig. 2.15b). The power-operated cleaners/cleaner graders are also available. Power-operated de-stoners are also used widely to separate small stones/sand particles/clay particles from the grains.	Cleaner grader: 2 to 20 q/h depending on the power source

(continued)

Table 2.1 (continued)

Brief description	Work method/tools	Productivity indicators (Field capacity/output)
Operation: Decortication, hulling, Dehusking (Fig. 2.16)		
Decortication operation involves breaking of shells and removing kernels/seeds from the pods.	The traditional practice of decortication for crops like groundnut and castor was to break the pods with hand/fingers. Now, a large number of manual devices (Fig. 2.16a, b), as well as power-operated equipment, are available for the purpose.	Manual decorticator: 0.3 to 0.5 q/h; Power-operated decorticator: 0.5 to 5 q/h
Shelling/hulling In some grains, like paddy, soybean, the outer seed coat or husk is loosely attached to the inner portion. The paddy husk gets separated by pounding or with mechanical means. Removal of such type of husk from seeds is called shelling or hulling.	The manual device and also, power-operated equipment are available for this operation, e.g. traditional rice huller, power-operated huller, soybean dehuller, rice mill.	Traditional rice huller: 4 to 6 kg/h; Power-operated hullers (20 to 100 kg/h).
Dehusking For grains, such as pigeon pea, the seed coat or husk is tightly attached with the inner part with glue or some other compounds. Removal of shell/husk requires a process called dehusking. In the case of crops like areca nut, the fruits are dried, and then dehusking is carried out.	For the dehusking of pigeon pea, the grains are soaked overnight in water and then sundried. Afterwards, power-operated dehuskers are used for dehusking as well as milling operation. For areca nut, manually operated as well as power-operated dehuskers are available (Fig. 2.16c, d).	Hand-operated areca nut dehusker: 4 to 8 kg/h; pedal- and power-operated dehusker: 10 to 25 kg/h
Operation: Manual materials handling and transportation (Fig. 2.17)		
In farm activities, manual materials handling (MMH) is frequent. The workers carry raw materials and implements to farm and produce from farm to home/store. MMH tasks often carry a potential risk of musculoskeletal stress and strain and back injuries. Frequent complaints of back disorders are primarily attributed to strenuous physical activity in awkward postures. Since agriculture in most developing countries falls under the informal sector of the economy, the health issues associated with MMH tasks require an understanding of suitable intervention to mitigate the hazards of work.	MMH tasks cover actions such as holding, lifting, lowering, turning, and carrying of loads, and applying push and pull forces in the operation of various equipment (Nag and Hsiang 2000). Many times, these MMH tasks are the significant causes of overexertion and musculoskeletal injuries in the places of work. Earlier, the grain bags were of 100 kg weight, and the workers had to lift and carry them from one place to another. Nowadays, the use of 50 kg of grain bags has become common. Transport of agricultural materials from one place to another is done physically by carrying loads for short and medium distances. At locations, wheelbarrows, manual carts, or animal carts are used for medium distances. In the case of long distances, generally, tractor-trailers are used for material transport (Fig. 2.17).	Head load—15 to 25 kg; Grain bags—50 kg capacity; Wheelbarrow capacity—50 to 100 kg; Animal carts—5 to 10 quintal capacity; Tractor-trailer—20 to 50 quintal capacity

References

CIAE. (2014). *Research highlights 2012–14. All india coordinated research projects on ergonomics and safety in agriculture*. Technical Bulletin No. CIAE/2014/17. Bhopal: Central Institute of Agricultural Engineering.

Gite, L. P. (2017). *Final report. Emeritus scientist project on development of ergonomical design guidelines for agricultural tools, equipment and workplaces*. Bhopal, India: ICAR-Central Institute of Agricultural Engineering.

Hopfen, H. J., & Biesalski, E. (1953). *Small farm implements*. Rome: Food and Agriculture Organization of the United Nations.

ICAR. (2012). *Handbook of agricultural engineering*. New Delhi, India: Indian Council of Agricultural Engineering.

Mehta, C. R., Gite, L. P., & Khadatkar, A. (2018). Women empowerment through agricultural mechanization in India. *Current Science, 114*(9), 1934–1940.

Mehta, C. R., Chandel, N. S., Senthilkumar, T., & Singh, K. K. (2014). Trends of agricultural mechanization in India. *Economic and Social Commission for Asia and the Pacific (ESCAP) Policy Brief*, *2*.

Nag, P. K. (1998). Manual operations in farming, In: ILO Encyclopaedia of Occupational Health and Safety (4th Ed.), Chapter 64, *Agriculture and natural resource based industries*, 64.23–64.29, (ILO, Geneva).

Nag, P. K., & Hsiang, S. (2000). Ergonomics in manual materials handling tasks. *ICMR Bulletin, 30*(8), 23–31.

Ojha, T. P., & Michael, A. M. (2005). *Principles of agricultural engineering* (Vol. I). New Delhi, India: Jain Brothers.

Pandey, M. M., Majumdar, K. L., & Singh, G. (1997). *Farm machinery research digest*. Bhopal: Central Institute of Agricultural Engineering.

Singh, S. P. (2007). Ergonomical evaluation of manually operated rice farming equipment with farm women. *Final Report Project*, (532). Bhopal, India: CIAE.

Singh, S. P., Gite, L. P., Agarwal, N., & Majumder, J. (2007). Women friendly improved farm tools and equipment. *Technical Bulletin No. CIAE/2007/128. Bhopal: Central Institute of Agricultural Engineering (CIAE).*

Varshney, A. C., Tiwari, P. S., Narang, S., & Mehta, C. R. (2004). *Data book on agricultural machinery design*. Book No. CIAE/2004/1. Bhopal, India: Central Institute of Agricultural Engineering.

WSSAN, Watershed Support Services and Activities Network. (2006) Weeders—A reference compendium, Secunderabad, p. 28.

Part II
Fundamentals of Ergonomics and Human Factors

Chapter 3
Perceiving Agricultural Ergonomics and Human Factors

Introduction

Ergonomics is an age-old discipline with a new image in recent times. In the treatise entitled *Rys ergonomji czyli nauki o pracy* (An Outline of Ergonomics, or the Science of Work-based upon the truths drawn from the Science of Nature), published in Przyroda i Przemysl, Poznari, (Nature and Industry), Poland in 1857, W. B. Jastrzebowski, a natural scientist, created the foundations of ergonomics (cover page of the journal, Fig. 3.1). The Science of Work was ventured to call Ergonomics, deriving from Greek ergon (epyov)—work and nomos (νομος-)—principle or law. The word '*Work*' expresses its full and integral sense that not merely as physical labour or toil, but tangible, aesthetic, rational, and moral work, emphasizing on labour, entertainment, reasoning, and dedication. In other words, Work is performed by Man's forces and faculties that have been endowed by our Maker for the common good. This prophetic message conveys a shiny sense of personal dignity on work as worship.

World War II may be taken as the real beginning of the present-day Ergonomics. Apart from a century of hibernation, perhaps the term ergonomics was independently reinvented by the mid-1950s, as the discipline that assembles knowledge on individual's properties and capabilities to apply in the design of products, workplaces, equipment, and achieve the benchmark in work design and (re)design. The scope of the discipline has become evident from the storytelling by Edholm and Murrell (1973) about the history of the Ergonomics Research Society. Today, ergonomics, as an interdisciplinary area of study, has gained emphasis in the academic curricula and research of other disciplines, including agricultural engineering that applies engineering S&T to farm production and processing. The professionals may take cognizance of the terms, such as ergonomics, human factors, human factors engineering, human ergology, and their interchangeable use, viewing them as synonymous. Towards expanding the scope of application and sustainability of the discipline of ergonomics in multi-disciplinary perspectives, there has been a continuing debate and endeavour to delineate its limits and variations.

P. K. Nag and L. P. Gite, *Human-Centered Agriculture*, Design Science and Innovation,
https://doi.org/10.1007/978-981-15-7269-2_3

Fig. 3.1 Jastrzebowski's writings on the Science of Work, published in the Polish journal 'Przyroda i Przemysl, Poznari'

In Aristotelian terms, it is necessary to determine the real nature of an entity by indicating the genus that it includes and the specific differences that set it distinctively apart. In other words, a technical or scientific definition, to be useful, must be directed towards for whom it is intended so that the broader meaning of the explanation is interpretable to the academics and practitioners. It should describe, expound, interpret, and use metaphor to clarify precise meaning. In this process, however, it is not unlikely that several synonyms or subtle words very close in a sense or definition that allows some shade of difference maybe infringing upon its meaning and explanation. New words (or combinations of words) are continually being introduced in this interdisciplinary context, with different perspectives and comprehension. Therefore, it is also not unlikely that the new expression might raise the debate of its precise meaning and reference. This contribution is a modest endeavour to compile the set of words and phrases used by the eminent authorities in expressing the essential nature, meaning, and significance of things on the subject of ergonomics (and human factors). Beyond its academic interest, this analysis bears importance for all allied professionals to arrive at a unified understanding and contribute to its sustainability as a growing science and technology.

Definitions

A search of volumes of literature, books and materials, databases, and source-based web searches, using keywords, yielded a variety of viewpoints on the subject of interest. These are scrutinized, culled, compiled, and embodied herewith. Several terminologies and definitions emerged independently and also in combinations, e.g. ergonomics, applied/industrial ergonomics, human factors, human factors engineering, human engineering, engineering psychology, applied experimental psychology, human factors psychology, biomechanics, and the like. For factual verification, the verbatim expressions of different related publications were taken, sorted, and analysed based on the authors' comprehension of synonymy or distinctions. Accordingly, to bring in better clarity, the compilations are categorized by a primary description of the discipline.

The compilation and analysis suggest that despite the interchangeable use of the terminologies, there are distinctive differences regarding the domain of study, objectives encompassed, and the categories of classification of the definitions. By targeting the consensus building of a unifying expression of discipline, obvious need is to define the classes and objectives, and the domains to describe the body of knowledge. Three main categories, namely, ergonomics, human factors, and human factors engineering, appeared to be amply illustrated by many eminent researchers, and therefore these are tabulated in that order (Table 3.1).

The readers may note that Table 3.1 begins with the definition agreed upon by the expert group of the International Ergonomics Association (IEA 2000). Till that time, the subject matter has been variedly expressed, with considerable shades of difference in meaning, emphasis, and interpretation. Several experts emphasized the scope and application of ergonomics to the relation between man and his occupation, equipment, and environment and, particularly, the application of anatomical, physiological, and psychological knowledge to the problems arising the man–machine incompatibility (Parker 1989). There are also contributions from other subjects such as medicine, sociology, and cybernetics. Galer (1987) viewed ergonomics as an area of study and application that devoted to the problem of fit between user and machine, tool or product. Primarily, ergonomists measure human characteristics and human function, and establish the way that the human body and mind work in the solution of practical problems in the design and manufacture of products and systems.

In a similar context, the goal of human factors is directed to applying knowledge in designing systems that work, accommodating the limits of human performance and exploiting the advantages of the human operator in the process. Hawkins (1987) described human factors in the context of (a) people in their working and living environments, (b) people's relationship with machines and equipment, with procedures and situations about them, and (c) also about the people-to-people relationship in the working environment. Its twin objectives can be seen as the effectiveness of the system regarding the safety and efficiency, and well-being of the individual. In other words, human factors are a psychology/engineering partnership for increasing system efficiency. In contrast, ergonomics incorporates a more substantial anatomical/physiological/medical component and pays attention to health and well-being and efficiency. Reference has been drawn to the role of humans in complex systems, the design of equipment for human use, and the development of a working environment for comfort and safety (Salvendy 1987). The basis for the field of human factors is found virtually in many branches of life sciences, including biology, anthropology, physiology, neurology, and psychology (Christensen 1987).

Human factors engineering is the scientific knowledge about human behaviour in specifying the design and use of a human–machine system, to improve system

Table 3.1 Diversity in defining ergonomics and human factors

Definition	References
Ergonomics	
The scientific discipline concerned with the understanding of interactions among humans and other elements of a system, and the profession that applies theory, principles, data, and methods to design in order to optimize well-being and overall system performance	IEA 2000
It describes in one word many and varied applications of sciences like anatomy, physiology, and experimental psychology to the problems of fitting the job to the worker. The discipline aims at improving industrial efficiency by designing equipment to ensure that its operation is within the limits of the mental and physical capacities of most ordinary people, taking into account the effects on the performance of the complexity of the job, and environmental conditions	Murrell 1965
The study refers to the design of the system covering psychology, physiology, and anatomy that are relevant to the design of human tasks, workplaces, machines, and environments. It calls for an organized approach to the business of decision-making in any design context with what we regarded as a proper emphasis on human factors	Poulton 1966
The study of the relationships between men and machines, referring to the psychological, biological, and cultural traits to adapt tools and jobs to meet the needs of men and of choosing suitable persons for particular jobs or machines	Singleton et al. 1967
A study of man's behaviour towards (a) fitting the demands of work to the efficiency of man, and to reduce stress; (b) designing machinery, and installations that can be operated with high efficiency, accurately, and safely; (c) working out proportions and conditions of the workplace to ensure correct body posture; and (4) adapting ambient environmental conditions to match man's physical requirements	Tichauer 1978
The study devotes to alleviating the rigours of the workplace and to improving the persons' performance on the job. It applies those sciences relating human performance to the improvement of the work system, tools and equipment, workplace and workspace, and the immediate environment	Grandjean 1980
The study is related to the nature of the man himself, to his abilities, capacities, and limitations in his working environment. The environment encompasses the ambient working environment, tools and materials, methods of work, and the organization of one's work, either as an individual or in a working group	Park 2014; Grandjean1982

(continued)

Table 3.1 (continued)

Definition	References
Human Factors	
The study refers to the relationship between the capabilities and limitations of men, the characteristics of machines to designing, and evaluating man–machine systems. The human factors as a professional discipline (variously called human engineering, biotechnology, engineering psychology, or ergonomics) apply in the development and evaluation of man–machine systems to improve the performance of men in operating and maintaining their machines. Thus, system operations meet specified performance requirements	Bennett et al. 1963
The area of study focuses on improving the productivity of the operator by taking into account human characteristics in designing systems. That it refers to how personnel perform to use the equipment, the effect of that performance on other system elements, and the impact of the overall system upon its personnel elements. The study is concerned with designing manufactured objects so that people can use them more effectively and creating environments that are better suited for human living and work. The study extends to virtually every consideration of the human in the system, e.g. reasons for being in the system, functions and tasks, the design of jobs for various personnel, training, and evaluation. That is, to improve (a) human performance by increased speed, accuracy, and safety, and less energy expenditure and fatigue; (b) use of human resources through minimizing the need for specialized skills and aptitudes; and (c) comfort and acceptance by the user/operator. It reduces training and loss of time and equipment as accidents due to human errors are minimized	Meister 1971
A broad field of study concerned with the design, maintenance, operation, and improvement of operating systems in which human beings are components, such as industrial equipment, automobiles, healthcare systems, recreational facilities, consumer products, and the general living environment	Huchingson 1981
A branch of S&T that includes what is known and theorized about human behavioural and biological characteristics that can be applied to the specification, design, evaluation, operation, and maintenance of products and systems to enhance safe, effective, and satisfying use by individuals, groups, and organizations	Goldenson 1984
Research and engineering studies of human factors in technology seek to realize greater recognition and understanding of man's characteristics, needs, abilities, and limitations when the procedures and products of technology are being designed	Alexander 1986

(continued)

Table 3.1 (continued)

Definition	References
Human Factors Engineering	
The study concerns the systematic approach to address problems of human–machine interaction and to arrive at practical solutions. The scientific basis is on data gathering and experimentation to yield precise information about human capabilities so that machines can be built to fit humans	Behan and Wendhausen 1973
A body of knowledge concerns the design of both products and processes, and equipment used in manufacturing to maximize their ability to be used comfortably, safely, and effectively by human beings. The study refers to the concept of designing machines and machine systems, associated work methods, and environments for the safety, comfort, and productiveness of human users and operators. The knowledge is drawn from multi-disciplinary science, including social or physiological sciences, such as anatomy, anthropometry, applied physiology, environmental medicine, psychology, sociology, and toxicology, as well as parts of engineering, industrial design, and operations research	Spencer 1985
The area of study covers the capabilities and limitations of human performance about the design of machines, jobs, and other modifications of the physical environment. It seeks to ensure that humans' tools and environment are best matched to one's physical size, strength, and speed and the capabilities of the senses, memory, cognitive skill, and psychomotor preferences	Chapanis 1986
The area of study deals with (a) any man/machine combination as a total system to ensure that the operational requirement of the equipment does not exceed human abilities; and (b) the human performance tolerance, and thereby to ensure optimal speed, accuracy, and quality of performance and eliminate hazards to operating personnel and maximize the comfort of the operator	Sanders 1988

efficiency by minimizing human error (Adams 1989). Chapanis (1971) aptly distinguished that ergonomics seems more physiologically oriented than that of the understanding of the North American counterpart on human factors engineering. Needless to mention that in contrast to ergonomics and human factors, human factors engineering places emphasis on the design as the medium to effect change on an end system for effective human use. The focus is to integrate the framework of systems engineering with the human sciences in optimizing the relationship between people and their activities to achieving effectiveness in the man–machine–environment system. The human factors design guidebook (Woodson 1981) is an illustrative account of how to resolve human–product interface problems and solutions wherever or whatever they are. That is, to make (a) the user's contribution to product output as efficiently as possible, (b) the combined user-product involvement as safe as possible, (c) to minimize the stress that a product imposes on the user during its operations and maintenance, and (d) to maximize acceptability of the product. In other words, the designer should make the design fit to the user, as opposed to trying to make the user match the design.

A large volume of literature in the '60s and '80s discussed the speciality discipline of *Human Engineering*. Specifically, it attempts to analyse the factors that help a man perform his job with speed, accuracy, and efficiency, which is allied to the safety and comfort of the operator. Mostly, human engineering addresses to eliminate the danger of making an operator the bottleneck of this man–machine system. It encompasses the design of human tasks, man–machine systems, and specific items of man-operated equipment for the useful accomplishment of the job in complex man–machine systems. Often, this branch of study has been referred to as engineering psychology, as a hybrid of engineering and psychological knowledge (Chapanis 1959). It seeks to understand how human performance is related to task variables and to formulate theory and principles of human performance that can be applied to the design of the physical conditions and environments, machines, and equipment about human capabilities, learning capacities, efficiency, and comfort (Fitts 1963). In essence, human engineering and engineering psychology call for broad interdisciplinary knowledge, for the optimum of man–machine functional interrelationship, regarding efficiency, reliability, and cost-effectiveness.

Literature shows a clear chronological trend towards broadening of considerations and opening up perspectives in respective applications. The focus of ergonomics/human factors dramatically moved away from the initial emphasis on militarily oriented man–machine systems environment to the expanded panorama of manufacturing and consumer-oriented environment. The synonymy of ergonomics and human factors has been vividly emphasized by the researchers in the 90s, as quoted from the respective publications (Table 3.2). The word ergonomics implies the study of a man at work, while word human factors presuppose the study of the man about equipment and environment (Salvendy 1985). The application of the discipline has been broadened to the relationships between man and his occupation, machine, and the environment in the broadest sense, including work, play, leisure, home, and other micro- and macro perspectives (Nag 1996). Insight brings out twofold objectives of the discipline, i.e. (a) to enhance the effectiveness and efficiency with which work and other human activities are carried out and (b) to maintain or improve specific desirable human values (e.g. health, safety, satisfaction).

The categorical statements have been aired (Edwards 1988) about synonymy of ergonomics and human factors, through integration and synthesis of biological and behavioural sciences with engineering, in optimizing the relationship between people and their activities. The federative science and technology of man at work form the foundation for development and adjustment of technology for rational utilization of human potentials under the favourable environmental and social conditions. Both micro- and macro-ergonomics developments (Hendrick 2002) broaden the scope of understanding at human–machine–environment–organization relationships (Nag 1996). A schematic is given herewith for clarity (Fig. 3.2). The unique conceptualization involves the socio-technical system components, such as the technological and the personnel subsystems and the external environments, including the structure and climate of the organization. The positive interactions of the subsystems at all

Table 3.2 Emphasis on Synonymy of Ergonomics and Human Factors

Synonymy	References
The study that describes analyses, measures, predicts, and controls the real world of systems functioning operationally. In other words, the research goal for ergonomics/human factors is to study the relationship between the personnel subsystem and the other system elements, including the terminal system output	Meister 1985
The study explores human characteristics and capabilities—physical, psychological, and cognitive—to the design of products, processes, and environments to improve well-being and optimize productivity	Pelsma et al. 1987
Consideration of human characteristics, expectations, and behaviours in the design of the things people use in their work and everyday environments in which they work and live. That is, human functions in the (a) design and creation of human-made objects, products, facilities, and environments that people use; (b) development of procedures to perform work and other human activities; (3) provision of services to people; and (4) evaluation of things in terms of suitability for people to use	McCormick and Sanders 1982; Nag 1996
It is recognized as an interdisciplinary engineering discipline and applied science that concerned with designing safer, more productive, easier-to-use equipment and environments, and with selecting and training people to deal with existing environments	Rubinstein and Hersh 1987
The scientific study of the relationship between humans and their work, that is, what we know about the way people respond to design systems and tools that help make people more productive and happier	Thomas 1984
A branch of S&T theorized about human behavioural and biological characteristics, with a repository of data and principles that are applied to the specification, design, evaluation, operation, and maintenance of products and systems for safe, effective, satisfying use by individuals, groups, and organizations	Christensen 1988

possible interface levels are the necessary conditions for organizational effectiveness. Thus, in simplest terms, ergonomics or human factors can be defined as a study of a person and his/her working environment. The term environment includes tools and materials used, the physical and ambient conditions, and the organizational structure under which the person works. The aim here is to increase the output and productivity, to enhance comfort and efficiency, and to minimize the occupational health

Fig. 3.2 System components in ergonomics/human factors design and development

and safety issues in the work situation. The final goal of ergonomics/human factors is to heighten the quality of work life of human beings all over the world.

Frontiers of Ergonomics Application

Explaining about the application of ergonomics is like reinventing the wheel. Often, however, the practitioners in Ergonomics are face to face with stark realities, that is—when, where, and how to use ergonomics? This may demand to redefine ergonomics with clarity, i.e. it deals with the analysis of problems of people in their real-life situations. Ergonomics is essential and useful in those work situations when people are present, such as when people operate production systems and when people repair and service equipment, and the like. Ergonomists try to design the relations and conditions of real-life situations, aiming to *harmonize people's demands and capacities, claims and actualities, and longings and constraints* (Rohmert 1985).

Typically, the very foundations of the present-day practices of ergonomics have been rooted mainly in the correction and prevention of problems in military and industry, e.g. correcting the existing challenges for the elimination of aeroplane crashes in the army during the World War II. At a later phase, the concept of prevention of the issues received gradual attention. Later on, the application of ergonomics spread in other areas encompassing all walks of life, including man–machine interface analysis and design of tools and equipment in the industry, commercial airplanes, workshops, and offices. The considerations include the study of human needs, characteristics, abilities, and skills as applied to design, production, management, maintenance, and in some cases, recycling or destruction of high quality and human products or services with a given physical/chemical environment and social surroundings.

At its generality, ergonomics practices have an impact in three distinctive areas (Fig. 3.3), namely, (a) health and safety issues of the workers, (b) productivity, cost of production and organizational efficiency-related fields, and (c) human comfort, physico-mental and social well-being. When the health and safety issues of the workers are in jeopardy, the primary concern is to correct those related problems, the cost, and comfort issues being the secondary concerns. Organizational efficiency and cost of production-associated areas are those where the application of ergonomics ensures systems effectiveness, thereby minimizing the operating cost and increasing productivity. This may have been achieved by weighing the criteria of health and safety concerns, yet reducing compromise on the loss of productivity and increased operating costs. On the other hand, the intangible issues like physico-mental and social well-being, and human comfort problems are a humane concern for men and women in a work environment, which must be weighed along with the health and safety aspects.

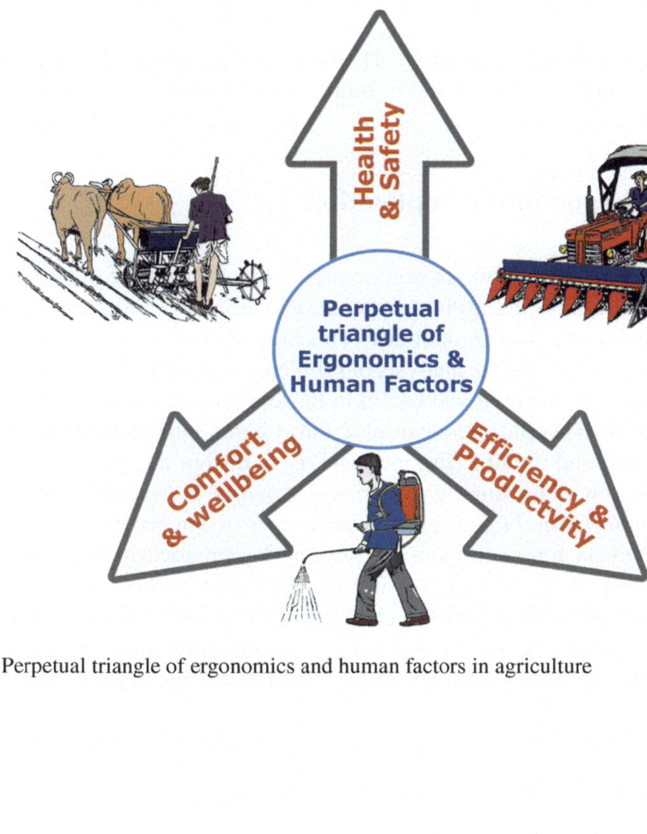

Fig. 3.3 Perpetual triangle of ergonomics and human factors in agriculture

The safety and health, and productivity and human comfort issues are central since most sectors of social employment are squarely confronted with the problems of work organization and human health. Therefore, it requires investigation of human beings, real living and working conditions, and a precise understanding of the functional necessities of the work environment. This broader philosophical view is fundamental for the growth and development of the discipline. In the context of agricultural activity, besides many other influencing variables, the work environment itself is a bottleneck, whether it is open-field activity or in animal confinements. The implication of the physical and chemical environment comes to the attention of ergonomics practitioners. In the tropical climate, the severe heat load is a substantial health hazard. Therefore, actions may be taken to improve the situation, e.g. (a) allow workers to continue work in a comfortable work area, (b) suitable body cooling may minimize heat stress, (c) change the work timings to avoid high heat load, (d) slow down the pace of the work, and (e) increase the rest periods between the spells of work. For such kinds of situations, the ergonomics approach is to develop and establish safety and health standards for environmental exposure, with an ultimate objective to ascertain the associated health risks and to recommend the limits of physiological as well as psychophysical strains. In situations, health problems are not anticipated, but concerns are related to the slow pace of work, decreased production, and other issues of work organization. The concept of ergonomics brings solutions which may not change anything in the current operation or task contents of the workers but may attempt to improve cost and productivity justified factors.

Concepts and Practices of Agricultural Ergonomics

Since ergonomics is centred on the *Man* in man–machine environment complex, the original idea is the *human performance,* i.e. productive implications of human efforts, and secondly, the conducive interrelationship of the man–machine environment system as a whole. The component *machine* includes equipment, tools, and accessories of the local work environment. The third concept is that of the work system with a broader boundary of work situation, including field and household activities, occupational and recreational activities, and work organization design and management. Last but not least, the concept of ergonomics practices is on *the improvement* of skill of oneself by optimizing human motions and also the conditions and system of work. That is, the improvement in the work system must be measurable in quantitative or qualitative terms.

Here we shall refer to *agricultural ergonomics* to elucidate its use in farming. Agricultural ergonomics emerges as a potential discipline for use in farming methods and practices relating to human performance and improvement of the work system (Fig. 3.4). One may perceive the implication of ergonomics based on its generalized principles. However, there are lacunae for practical application in agriculture by following any defined rule of thumb. By the authors' comprehension of the subject

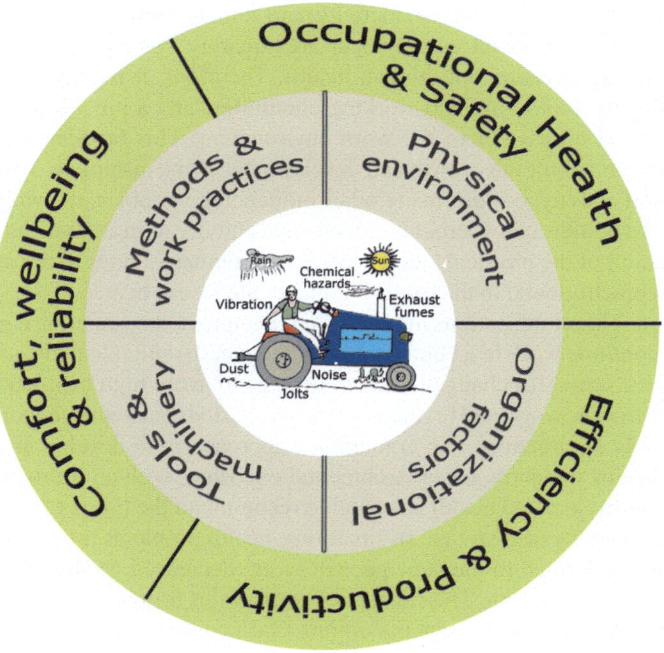

Fig. 3.4 Man–machine–environment system in agriculture

matter, this contribution discusses certain philosophies and frameworks for the application of ergonomics in the infrastructure of traditional and modern agriculture. The importance of ergonomics and human factors has been realized by all concerned in human resources management in industry and agriculture. However, there are particular fundamental distinctions between industry and agriculture. One may argue that agriculture is also an industry with a marginal primitive image. Gradual advances in agricultural management, production, and distribution system are indications of the transformation of traditional agriculture to industrial bases. Wide-ranging variations in agriculture across national boundaries suggest that ergonomics application would indeed differ with the farming practices.

The authors intend to emphasize the ergonomics implications that may have a long-ranging strategic direction for sustainable agricultural development (Fig. 3.5). It stands on the belief that ergonomics contributes to the social goals by the protection of workers' health, and to the economic goals in terms of productivity and quality of work. The role of ergonomics has often been elaborated in providing answers to the problems of work, work method, equipment, and form of cultivation. The issues may be well known, or they may only be recently identified. In other extreme cases, some specific problems are yet undiscovered entirely. When ergonomics practices are useful to provide answers to questions, it carries immense value for renewed work

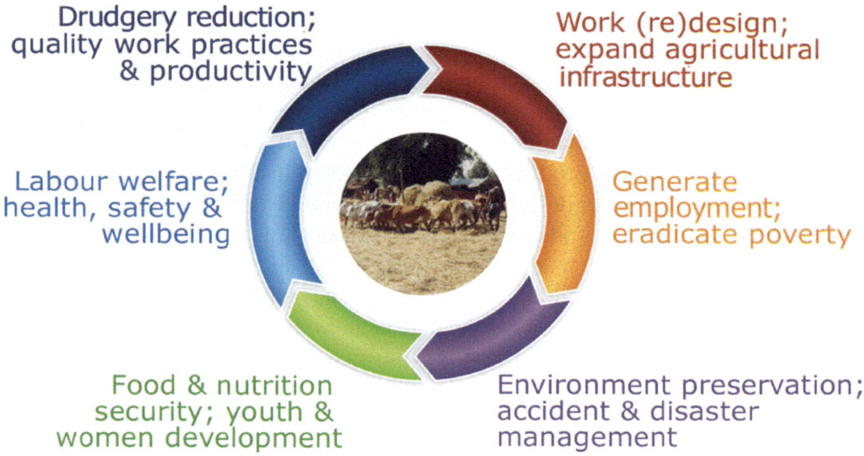

Fig. 3.5 Ergonomics implication to strategic sustainable agricultural development

organization in farming and to the people involved. The practitioners in ergonomics must realize the tall order and its possible implications.

There are two generally accepted methods for assessing the implications of ergonomics in agriculture. One is tangible, i.e. about cost–benefit ratio and considered as the measure of significant importance. The decision concerning the resolution of a problem often becomes more accessible to make, if the choices are put regarding cost–benefit of a given improvement. The other method is intangible, which is heavily dependent on those choices of importance, such as human health, comfort, and safety. The options that cannot be expressed in any form of quantitative scale are likely to be ignored as of less significance. However, the role of ergonomics is to emphasize arguments on both tangible and intangible issues for the substantial benefit to the workers.

Ergonomics has a vital role in generating a great deal of information from traditional farming systems and applying the modern concepts of ergonomics to improve work conditions and workplaces. Advanced technology is gradually finding its place in agriculture, with still large-scale dependence on family-scale farming. With the introduction of advanced technology, ergonomics becomes essential for its successful application. There are countless illustrations, however, on workers' benefits concerning measures taken for increased work comfort and safety. Thanks to the liberal farm management groups for continued thrust on educational curricula in different branches of agricultural sciences towards the promotion and application of ergonomics.

The critical point here is that the significant benefit to agriculture is to have a safe, healthy, and productive worker. Since the social costs due to ill health and injuries are real and substantial, there is significant perception and recognition that ergonomics can serve as a dominant scientific discipline for socio-technical development in farming. Other circumstances prevailing in farming activity are injuries and accidents. Proper attention to ergonomic aspects in the design and operation of machines, e.g. mechanical threshers, may reduce accidents.

There are inherent organizational problems to execute strategies in the informal sector of agriculture that is vastly spread over in different geographical regions of the world. It is emphasized that if any cost factor of a particular improvement can be compensated by increased productivity, there is every justification for introducing the development. The proper design of manually operated tools and equipment is one related area to the cost of production and work efficiency. There may be situations that the solutions or changes made may not bring about any significant improvement in productivity and also not directly related to physical health concerns. The decision to implement the desired solution is subject to examination and judgment of the cost benefits and priority on better comfort and physico-mental health. Suitable design alternatives in the farm machinery may minimize the stresses on the operator. One such case is the tractor seat, which is a significant problem area—optimal design of the position concerning other articulated segments of the machine may reduce transmission of harmful vibrations to the human body and minimize the discomfort of the operators. Therefore, the ergonomics practitioners in agriculture realize the complexity to suggest improvements, and the importance of the suggestion will decide on the scope of implementation.

Philosophy of Ergonomics Application

The philosophy of ergonomics application is generic to the concept of technology application in any sphere of human activity. In its more specific terms, we briefly state below the tendencies and philosophies:

1. *Work is worship,* and this is fundamental of the ergonomics discipline. It carries a value system for one's work, and thus promotion of work is the highest human objective. The policy decisions on *agricultural development* primarily emphasize the soil, water, and other natural resources for development, with fair neglect to the greater asset of *human energy.* Concepts of ergonomics open up avenues for the *rural generation* of multiple varieties of services and employment, and in turn, bring about useful human involvement. In the populous Southeast Asian and African countries, ergonomics has a special meaning, which has often been ignored in the West. For example, in India, if the natural and human resources are suitably optimized, developed, and harnessed, the country's food requirements can be met effectively despite the increased population. The ergonomics strategy on human resources planning for the large agricultural sector aims at providing a renewed quest for work excellence and valuing *the dignity of labour,* even amid large-scale deprivation. Gradually, dissemination of ergonomics know-how can bring about the enormous benefit to farmers and rural artisan communities for the betterment of work.

2. There is a need to devise an *appropriate* solution to a problem that can be implemented and workable. The term *workable* is chosen for the immediate implementation of a solution that is warranted to the context. Although the best *solution* or *technology* is expected, it may not be available or attainable at the immediate need due to problems of technology development, like technology transfer issues, geographical barriers, economic constraints, and unavailability of technology. Therefore, when a solution is achieved to control a problem in a working system, which may not be the best possible solution, it is of value to the system and the workers involved. The ultimate objective, here, is the improvement itself, and not merely the use of a tool or technique to resolve a problem. Therefore, many simple, common-sense solutions in the agricultural context raise the status and importance of ergonomics as a scientific discipline of practicality, i.e. the solutions that are workable and can be implemented for improvement.
3. The level of technology applied is often immaterial, i.e. in the real-world situation, the ergonomics solutions may or may not be technically sophisticated. Continuing to the earlier view, the emphasis is placed that ergonomics brings about solutions appropriate to a context, irrespective of whether the best technology is available at one's immediate disposal. Ergonomics practitioners are confronted with the problems of farming accidents. While it is true that statistical analysis of the accident and injury rates provides an account of the prevalence and occurrence of incidents and personal damages as a consequence of farm operations, it may be more important to target our efforts in making improvements to the job. The authors have been tempted to cite the example of the ergo-design efforts in mechanical threshers to avoid accidents. Poor feeding systems and workplace arrangements were responsible for the majority of the mutilating hand injuries at the threshing machines. A detailed analysis of body dimensions and placement of feeding chutes at the proper height was necessary to allow shorter people to reach the feeding area and also tall people the required clearance to avoiding hitting the threshing cylinder. Such a need solving approach is advocated in both the industrialized and industrializing nations.
4. *Time is money* is an all-time truth. This common saying has critical importance as regards to ergonomics practices. Timeliness of deriving a solution minimizes human health hazards and loss of production. When an ergonomics solution is proposed, many ergo-design principles and criteria are assimilated and adopted. The process is similar to any design methodology being followed in engineering disciplines. It is always a costly absurdity to delay the total solution, with the argument for seeking up the current technical knowledge. The ergonomics strategy favours the choice of immediate resolution of incompatibility to a man–machine work system.
5. The machinery/technologies from big industrial corporations introduced in rural farming, very often are not well supported for services and maintenance at the local level. A tractor or a power tiller, which comprises multiple machine components assembled in such a way that it is always a problem for easy access to the part needing service. The lack of accessibility has become a legend, yet changes are prolonged to come. This is one more well-known case where

improvement is recognized but is not implemented due to other seemingly more important reasons. The situation suggests building up technical capability among the local artisans for the maintainability of farm machinery. The specialists must emphasize the useful recommendations for the solution to the problems and take the right ergonomics design decision.

6. The best ergonomics concept may not always be chosen for some reason. The farm work practices, methods, tools, and equipment, including the organization of work, which has been in use for decades, are not ergonomically designed. Take the instances of hand tools that are primarily intended for right-handed persons and are often being used by the left-handed users. Due to the non-availability of the devices for the left-handed persons, one may be completely unaware of any difference that might be possible for right- and left-handed operation. Everybody accepts the problem of the unsuitability of the whole gamut of farm tools. Under such circumstances, when a specially designed device is made available for right- or left-handed operators, there is an unlikelihood, if not a remote possibility, that better and specific tools will replace all tools. Apart from the influences of factors like *relearning,* the economically poor farming community has a constant fear of likely endangering themselves to a new tool that has not been well time tested.
7. The organization design in agriculture has inherent difficulty, as far as unorganized working sectors are concerned. Wherever agriculture is brought under organized operation through cooperative movements or through corporate endeavour, it is always based on multiple criteria, other than ergonomics alone. Therefore, the agricultural scientists, engineers, and farm managers are continuously confronted with various design alternatives and organizational factors for allocating and managing scarce resources of land, human, animal, and mechanical power. One must realistically trade-off the constraints of space, time, and money for the effective organization of farming. Two significant issues are at the forefront of discussion in all sectors of agriculture development, i.e. the demographic trend of the working population and rapid mechanization in agriculture. The practitioners in agricultural ergonomics are required to view the whole perspective, and with this understanding alone, sound farm strategy, including ergonomics criteria, may be proposed.

Typical Areas of Ergonomics Application in Agriculture

Addressing to the agricultural problems may demand different approaches and methodologies. There is a frequent use of the ergonomics checklists to identify the issues and the related application. With fundamental concerns, i.e. the study of people, the issues can be determined by taking into account the type of body system that is affected. Since different problems affect different body systems, the grouping of the issues concerning body systems is useful. The typical areas of ergonomics study and application in agriculture are briefly explained herewith.

1. *Anthropometry*—The study of body dimensions, including physical and body composition characteristics, is required to be considered to deal with a material conflict between the person, and the equipment one uses as well as some aspects of the work area. To understand the principles of anthropometry, one must be conversant with the anatomical sites and bony prominence of the human body. The problems of workplace and workspace are the most common anthropometric problems, and these problems fall under man–machine physical compatibility. That is, someone is too large or too small to fit the machine. By obviating these problems, often better equipment can be designed and workplace organized. Information on human body composition regarding chemical constituents, such as body fat, solids, and intracellular and extracellular workspace, are required to aid in studying rational functions under extreme stresses, like acceleration, altitude, temperature, vibration, or radiation.
2. *Work Physiology*—The type of work decides the stress on the human body. An analysis is made frequently about the energy delivery and the strain on the cardiovascular and respiratory system due to muscular work. The stress may arise from the physical efforts and the demand of the muscles, and that gets reflected in the form of cardiorespiratory strains. Similarly, environmental factors, like heat stress in the tropical environment, may result in the need for extra blood flow to dissipate the bodily heat load. The increased circulatory load is also associated with higher energy demand. When there is a stagnation of blood flow like in static muscular contraction, the localized muscular fatigue is the result of the accumulation of acidic metabolites. Mention may be made that the localized discomfort is a major limiting factor in continuous operation of manually operated agricultural implements. The endurance to work depends on the cardiorespiratory system efficiency and the concurrent development of the skeletomuscular structures. This information is used in establishing work organization principles like work/recovery cycles, shift work, or standard for an allowable load of day's work. Designing tools and jobs to conform to a permissible level of energy demand is an approach widely accepted for various applications.
3. *Biomechanics*—The skeletomuscular structures determine the range, strength, and speed of human movements, including response behaviour to physical forces such as acceleration and vibration. This information grouped under *Biomechanics* is useful in avoiding injuries on the job, in tool design, in the workplace and task layout, and the protection of personnel against mechanical forces. The range, strength, and speed of body movements are analysed by various biomechanical and psychophysical techniques and methodologies. When a muscle strength problem is identified, the information on the strength characteristics of different muscles might help in assessing the severity of the problem. Accordingly, alternative solutions can be obtained.
4. *Physical and Ambient Environment*—The environmental problems are those which are external to the workers. These include heat and cold stress, lighting, noise, vibration, and issues of organic as well as inorganic dust in agriculture. For problems between the man and the environment, changes in the external environment are the obvious solutions. However, due to its limitation of practicality,

some degree of individual adjustment, e.g. work adaptation, may also be possible on the other end.

5. *Cognitive Issues*—The manipulative and motion economy problems of human action are characterized by the ability to perform the desired movement of the body parts or the difficulty in completing tasks at the required speed and accuracy. These problems usually show up in complex tasks, requiring precise control of equipment. It may be a fact that the agricultural machinery and methods of work are far from complex work nature. However, due to many overload problems and relatively low skill levels, the workers are often subjected to manipulative problems and risks of making frequent work errors. Accident behaviour of farmers with the agricultural machinery, particularly tractors and threshers, is mainly associated with manipulative errors at a fatigued state of the person and partly due to design-induced errors. The cognitive problems typically show up in some form of operator's errors concerning the limits of short- and long-term memory. The perceptual problems associated with vision and hearing are also common cognitive difficulties. The situations that violate the human mental qualities to function in sequential order, human errors are inevitable to follow. One should not associate these situations with the motivational factors since even the highly motivated persons are not devoid of the risks in design-induced operating errors.
6. *Drudgery Reduction*—*Agriculture is a profession by compulsion, and not by choice*, is a general perception all over the world. This is because most of the agricultural activities involve drudgery. In some activities, there is severe drudgery to workers. Drudgery is the overall result of many factors mentioned above, namely, improper tool and equipment design and excessive force requirement in operation, awkward postures and movements during operations, adverse physical and ambient environment, and boredom due to repetitive type of work. Therefore, ergonomics interventions to reduce drudgery are very necessary to make the life of agricultural workers better.
7. *Accidents and Safety*—Agriculture is considered to be one of the most hazardous professions in developed as well as developing countries. Accidents happen due to machine factors, crop factors, environmental factors, or human errors leading to injuries. The cost incurred due to these accidents to the workers as well as society and country is enormous. Therefore, ergonomics has a huge role to play in minimizing accidents in agriculture by making the machines safer and reducing the injuries to workers.

Often lamented by the ergonomics professionals that their work does not receive the attention and consideration that it deserves. Interested readers may refer to collected articles in Theoretical Issues in Ergonomics Science (Hollnagel 2001) and Ergonomics (Stanton and Stammers 2008), predicting the future of ergonomics regarding expected developments and effects on the content of the discipline or specific regions. There is no denying that ergonomics is inherently imbibed with a strong value proposition (well-being) and interactivity with other stakeholder groups. In essence, the systems performance concerns with the interaction of humans with their physical, organizational, and social environment (Carayon 2006). That covers the performance of the work system (including worker, working methods/practices, and working environment) or the product/service system (consumer—product user or business goods or service receiver, including the situation where the product is used or service received). With the rapid technological transition, the traditional *Science of Work,* as indicated by the etymology of the word ergonomics, has undergone a sea change. Ergonomics/human factors, today, are concerned with all kinds of activities that go beyond (paid) work and activities carried out by a wide range of users (e.g. customers, patients) in domestic, leisure and sport, transport, and other environments.

Limitations of Application

To apply ergonomics in agriculture, different questions may require a varying depth of knowledge, including a thorough understanding of the multifacet field of ergonomics, and its inherent strength and limitations. There may be a question that merely requires an answer. One might ask about the equipment dimensions or the width of the entry port at a particular workplace, like a tractor cab. If the question is self-contained, available quantitative answers are easy to incorporate. It is possible to provide the solutions based on the standard anthropometric data source for which no additional expert opinion is required.

Some questions require judgment or prior experience to make a suitable recommendation, like how much force a worker should apply to operate a machine or how much weight should a man may regularly carry during the whole workday? Such questions require complex decisions with detailed information, like the physical status of the worker, pregnancy state (in case of women workers). Some issues need extensive experience and in-depth analysis to form the answer; an example of this type of question would be—*if the open-field agricultural work in summer causes a heat stress problem, what remedy can be proposed considering the economic viability of the solution?* A similar question has been explained earlier, which readers may find useful. These questions require a good understanding of the situation to be analysed. The questions are always location-specific, and there is a need for additional information about the task, the people, human adaptability, and the environment. A beginner in ergonomics may not be able to answer this kind of question at its appropriate context, since instant answers may not be available. The solution depends on the quality of information and the experience of the practitioners to deal with such a problem.

A comprehensive view has emerged that by fitting the environment to the human, two related system outcomes (performance and well-being) can be achieved (Wilson et al. 2009; Neumann and Dul 2010). The performance dimension covers, for example, productivity, efficiency, effectiveness, quality, innovativeness, flexibility, safety (systems) and security, reliability, sustainability, whereas the well-being dimension covers, for example, health and safety, satisfaction, pleasure, learning, and personal development. Both of these dimensions are intertwined and strongly connected (Pot and Koningsveld 2009). Despite the recognition of the fundamental objective of ergonomics to contribute to human performance and well-being by establishing and implementing basic system design, in general, the discipline has not yet found widespread acceptance and application. Views emerged (Dul et al. 2012) that (a) the relevant stakeholders are not aware of this knowledge; (b) the experience is not readily available in design projects; (c) the knowledge is not sufficiently singled out; and (d) due to its multi-disciplinary nature, the understanding is multi-faceted and ambiguous.

The arguments are convincing; however, the professionals in rural farming and other agro-based small enterprises may not find those compelling. Decision-making of the front-line system users (small and landless farmers, for example) are farthest away in their mindset and preparedness and pay no attention to ergonomics/human factors issues. Inevitably, ergonomics/human factors, taken as synonymous, have excellent potential to contribute to agricultural engineering to the design of tools, equipment, and other systems (Nag 2014). The current state of knowledge of ergonomics/human factors applications in agriculture vividly indicates the challenges it faces in terms of system readiness of its market and in the supply of quality applications. The India initiative under the aegis of the All India Coordinated Research Project on Ergonomics and safety in Agriculture (Gite and Singh 1996; CIAE 2014) stands apart by its distinctive role on the apt mandate of *application of ergonomics principles and anthropometric data for increasing productivity, reducing drudgery, and minimizing accidents and occupational health problems of workers in agriculture and allied activities*. This would beckon others to recognize its value proposition and the relevance of the knowledge base to problem-solving in agriculture.

The authors of the present contribution, by virtue of years of involvement in the area of ergonomics in agriculture, have taken the privilege to express views on the subject. The man–machine interface knowledge in the perspectives of the agricultural tenets demands inputs at multiple levels. These encompass (a) design of farming tools and machinery; (b) methods and conditions of work in farm practices, such as tillage, harvesting of crops; (c) optimization of farm resources, such as human resources, animal power, and mechanical/electrical power; (d) climatological factors that influence humans and agriculture; and (e) monitoring health, diseases, accidents, and injuries. Besides, other compelling aspects are (a) soil and water management for crop irrigation and livestock production; (b) physical and chemical properties of materials used in crop protection; (c) management of animal waste and agricultural residues; (d) rural technology for the processing of farm produce; (e) S&T of ergonomics in rural areas for sustainable livelihood; and (f) disaster mitigation and development endeavours, and the like.

The sustainability of the man–machine interface technology in the context of application would require demonstrating its value more successfully to the primary stakeholders (e.g. vastly populated poor farmers) and also prove undoubted feasibility to stakeholders of systems control (e.g. systems designers, financial institutions, regulatory bodies). This calls for substantial improvement in grass-roots communication and the building of partnerships with stakeholders. The (macro) ergonomics approach is positioned on a conceptual framework of health and well-being, along with providing a bridge to a stronger emphasis on livelihood through performance-oriented outcomes of the rural sector of work.

References

Adams. J. A. (1989). *Human factors engineering*. New York: McMillan Publishing. Air Force Systems Command. (1977). Air Force systems command design handbook 1–3, *Human factors engineering*, (3rd ed.).

Alexander, D. C. (1986). *The practice and management of industrial ergonomics*. Prentice-Hall.

Behan, R. A., & Wendhausen, H. W. (1973). Some NASA contributions to human factors engineering: A survey. SP-5117, NASA, Washington DC.

Bennett, E., Degan, J., & Spiegel, J. (1963). Human factors in a technological society. In E. Bennett, J. Degan, & J. Spiegel (Eds.), *Human factors in technology* (pp. 3–11). New York: McGraw-Hill.

Carayon, P. (2006). Human factors of complex sociotechnical systems. *Applied Ergonomics, 37*(4), 525–535.

Chapanis, A. (1959). *Research techniques in human engineering*. Johns Hopkins University Press.

Chapanis, A. R. (1971). The search for relevance in applied research. In W. T. Singleton, J. G. Fox, & D. C. Whitfield (Eds.), *Measurement of man at work* (pp. 1–14). London: Taylor & Francis.

Chapanis, A. R. (1986). Human-Factors Engineering. *The New Encyclopaedia Britannica (Vol 21)* (15th ed., pp. 227–229). Chicago: Encyclopaedia Britannica.

Chapanis, A. (1979). Quo vadis, ergonomia. *Ergonomics, 22*(6), 595–605.

Christensen, J. M. (1987). The human factors profession. In G. Salvendy (Ed.), *Handbook of human factors* (pp. 3–15). New York: John Wiley & Sons.

Christensen, J. M. (1988). Human factors definitions. *Human Factors Society Bulletin, 31*(3), 8–9.

CIAE. (2014). *Research highlights 2012–14. All India coordinated research project on ergonomics and safety in agriculture*. No. CIAE/2014/176. Bhopal, India: Central Institute of Agricultural Engineering.

Dul, J., Bruder, R., Buckle, P., Carayon, P., Falzon, P., Marras, W. S., et al. (2012). A strategy for human factors/ergonomics: developing the discipline and profession. *Ergonomics, 55*(4), 377–395.

Edholm, O. G., & Murrell, K. F. H. (1973). *The ergonomics society: A history, 1949–1970*. London: Ergonomics Research Society.

Edwards, E. (1988). Introductory overview. *Human factors in aviation* (pp. 3–25). Academic Press.

Fitts, P. (1963). Human factors engineering: Concepts and Theory. The University of Michigan Engineering Summer Conferences, University of Michigan. *Ann Arbor, MI*.

Galer, I. A. (Ed.). (1987). *Applied Ergonomics Handbook*. Butterworth-Heinemann.

Gite, L. P., & Singh, G. (1996). All India Coordinated research project on human engineering and safety in agriculture. *Agricultural Engineering Today, 20*(1–4), 106–109.

Goldenson, R. M. (Ed.). (1984). *Longman dictionary of psychology and psychiatry*. White Plains, NY: Longman Incorporated.

Grandjean, E. (1980). *Fitting the task to the man: An ergonomic approach*. Taylor & Francis.

Grandjean, E. (1982). Ergonomics is a study of man's behavior in relation to his work, Proc. Inter. Erg. Association.

Hawkins, F. H. (1987). Vision and visual illusions. *Human factors in flight*, (pp. 100–123) 1st Ed. England: Gower Technical Press Ltd.

Hendrick, H. W. (2002). *An overview of macroergonomics* (pp. 1–23). Macroergonomics: Theory, methods, and applications.

Hollnagel, E. (2001). The future of ergonomics (guest editorial). *Theoretical Issues in Ergonomics Science, 41*(2), 219–221.

Huchingson, R. D. (1981). *New horizons for human factors in design*. McGraw-Hill College.
IEA. (2000). The discipline of ergonomics. *International Ergonomics Association (IEA)*. http://www.iea.cc/01.
McCormick, E. J., & Sanders, M. S. (1982). *Human factors in engineering and design*. McGraw-Hill Companies.
Meister, D. (1971). *Human factors: Theory and practice (Book on human factors engineering covering systems design requirements and interface equipment for man machine interaction implementation)*. New York: Wiley-Interscience,. 418 p.
Meister, D. (1985). *Behavioral analysis and measurement methods*. Wiley-Interscience.
Murrel, K. H. F. (1965). *Ergonomics: Man in his working environment* (p. 496). London: Chapman & Hall.
Nag P. K. (1996). Work design: An ergonomics perspective, 1-23, In: *Ergonomics and Work Design (Emerging Issues in Organizational Sciences)*. Ed. P.K. Nag, New Delhi: New Age International, p396
Nag, P. K. (2014). Ergonomics sustainability: A relook into the definitions for utility in agricultural engineering, In: *Emerging Technology Trends in Agricultural Engineering*, ETTAE 2014, 01-12, New Delhi: Excel Pub.
Neumann, W. P., & Dul, J. (2010). Human factors: Spanning the gap between OM and HRM. *International Journal of Operations & Production Management, 30*(9), 923–950.
Park, K. S. (2014). *Human reliability: analysis, prediction, and prevention of human errors*. Elsevier.
Parker, S. P. (1989). *McGraw-Hill dictionary of scientific and technical terms*. McGraw-Hill Book Co.
Pelsma, K. H., Rylko, H. M., & McGee, K. (1987). *Ergonomics sourcebook: A guide to human factors information*. Ergosyst Assoc/the Report Store.
Pot, F. D., & Koningsveld, E. A. (2009). Quality of working life and organizational performance-two sides of the same coin? *Scandinavian Journal of Work, Environment & Health*, 421–428.
Poulton, E. C. (1966). Engineering psychology. *Annual Review of Psychology, 17*(1), 177–200.
Rohmert, W. (1985). Ergonomics and manufacturing industry. *Ergonomics, 28*(8), 1115–1134.
Rubinstein, R., & Hersh, H. (1987). *The human factor: Designing computer systems for people* (pp. 502–509). Morgan Kaufmann Publishers Inc.
Salvendy, G. (1985). Has ergonomics the same meaning in Europe and North America? In I. D. Brown, R. Goldsmith, & M. A. Sinclair (Eds.), *Ergonomics international 85* (pp. 97–98). London: Taylor & Francis.
Salvendy, G. (Ed.). (1987). *Handbook of human factors*. New York: John Wiley & Sons.
Sanders, M (1988). Untitled. In J. M. Christensen, D. A. Topmiller, & R. T. Gill (Eds.), *Human Factors Definitions Revisited, Human Factors Society Bulletin, 31*(10), 7–8.
Singleton, W. T., Easterby, R. S., & Whitfield, D. C. (Eds.). (1967). *The human operator in complex systems*. London: Taylor & Francis.
Spencer, R. H. (1985). *Computer usability testing & evaluation*. Prentice-Hall, Inc.
Stanton, N. A., & Stammers, R. B. (2008). Bartlett and the future of ergonomics. *Ergonomics, 51*(1), 1–13.
Thomas, J. C. (1984). Organizing for human factors. In Y. Vassilou (Ed.), *Human factors and interactive computer systems* (pp. 29–46). Norwood, NJ: Ablex Publishing.
Tichauer, E. R. (1978). *The biomechanical basis of ergonomics: Anatomy applied to the design of work situations*. John Wiley & Sons.
Wilson, J. R., Ryan, B., Schock, A., Ferreira, P., Smith, S., & Pitsopoulos, J. (2009). Understanding safety and production risks in rail engineering planning and protection. *Ergonomics, 52*(7), 774–790.
Woodson, W. E. (1981). *Human factors design handbook*. New York: McGraw-Hill.

Chapter 4
Energy Cost of Human Labour in Farming

Introduction

Traditionally, the demand for physical activities in a professional sphere is determined by the complexity of the jobs performed. Despite the growing trend of farm mechanization in the developed as well as in the developing world, manual labour as a source of power continues to dominate the vast sectors of agriculture. In the agriculture-dependent countries, like in India, the ratio of farmworkers as a power source to the workers as farm machine operators may be about 80:20. With overdependence on machinery, the nature of the involvement of the farm workforce in the developed world has undergone substantial change. Nowadays, more and more tractor and farm machines have also been introduced in all sectors of agriculture in the developing world. When it comes to women workers in agriculture in developing countries, most of them represent only as a source of power, since fewer women operate the farm machines.

There are two significant types of work stressors in manual labour, namely, those related to (a) the cardiovascular and respiratory system and (b) the musculoskeletal system. This chapter focuses on describing the stresses on the cardiovascular and respiratory systems. Due to human-oriented self-employment, the vast farm activities are not labour saving, vis-a-vis the daily energy demand. Traditional farming activities are high energy demanding as an integral part of human behaviour. It involves socioeconomic and cultural components and is dependent on the type of work performed, personality, body build, and physical fitness. Stresses of overwork culminate in multiple health risks among the farmworkers. Given the bread-and-butter compulsions, primarily in the smallholdings in the developing world, farmworkers have been compelled to engage in strenuous physical activity, ignoring the dangers associated with it.

P. K. Nag and L. P. Gite, *Human-Centered Agriculture*, Design Science and Innovation,
https://doi.org/10.1007/978-981-15-7269-2_4

Energy Production, Liberation, and Utilization

The physiological mechanism of energy production and utilization is briefly described herewith. The fundamental quality of human beings is one's ability to perform mechanical work by moving different body segments and by exerting forces. This ability depends on the skeletal muscles, where the stored chemical energy is transformed into mechanical work during muscular contraction. The physical action may be static or dynamic work (Fig. 4.1). The *static work* is associated with isometric contractions, i.e. the length of the muscles is unaltered, and there are no movements of the joints. Therefore, in the physical sense, no external work is performed in static work. The essential characteristics in static work are that the maximum tension produced by a muscle group, and for the length of time, the tension is maintained (endurance) depending on the functional properties of the muscles and one's motivation to sustain the stress. The *dynamic work* is associated with isotonic contractions, i.e. the muscles shorten and the joints move, thus producing external work. In

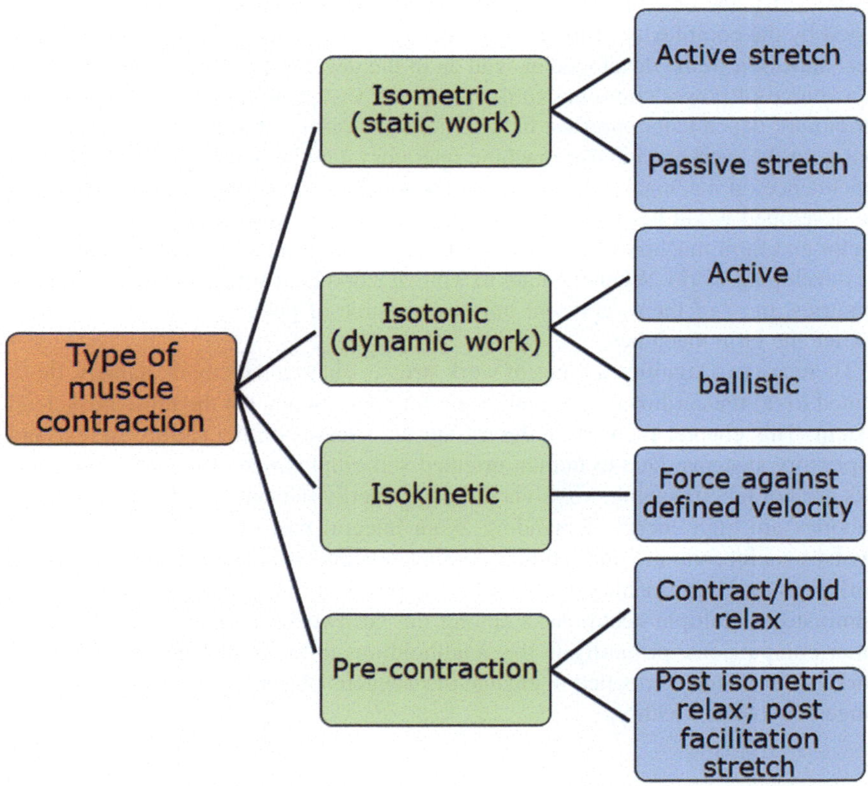

Fig. 4.1 Static and dynamic work

dynamic work, rhythmicity facilitates the energy delivery mechanism. In essence, the endurance and power output are more facilitatory in dynamic work, comparing to the static work, which is strainful and less efficient in energy utilization.

The activities of daily living are the combinations of two types of muscular work or by the continuous shift from one type to the other. The pattern of muscle activity changes continuously as the individual muscles are called into play or are relaxed under the overall control of the central nervous system. In principle, it is easy to understand the presence of two types of work; however, differentiating them or ascertaining the quantitative extent of involvement of static or dynamic components in actual practice is a challenge. In the maintenance of posture and execution of the physical activity, the extent of the presence of either static or dynamic components of muscular contractions is a matter of interest for evaluation of the possible state of fatigue of muscles.

Energy Liberation

Energy demands are replenished through intake of food, i.e. carbohydrate, fat, and protein. The energy is released by the process of oxidation and is transferred and stored in the cells as high-energy phosphate compounds, such as adenosine triphosphate (ATP). This represents the high-energy bond towards energy transfer, as needed to cover the functional requirements, like muscle contraction. A schematic outline of the chemical cycle of energy liberation of a glucose molecule and the metabolic pathway is illustrated herewith (Fig. 4.2). The food in the form of glucose diffused inside the cells undergo the process of oxidation (glycolysis) and finally completes (Krebs cycle) to form carbon dioxide. Simultaneously, the oxidation of food molecules generates the reduced, energy-rich, mobile electron carriers NADH and FADH2 that drive the formation of ATP through the reactions of the electron transport chain. Besides, the substrate level phosphorylations that occur during glycolysis and the Krebs cycle form a small amount of ATP. By splitting ATP into adenosine diphosphate (ADP) and phosphate (P), a quantum of energy is released for muscular work (i.e. one mole of ATP, by splitting into ADP and ~P, yields about 29.3 kJ (7 kcal). In reverse, to resynthesize ATP, energy is required from exergonic processes, mainly through the oxidation of nutrients and anaerobic glycogenolysis.

The energy required for phosphorylation of ADP to ATP is liberated by the oxidation of foodstuffs, which are thus broken down into carbon dioxide and water. That is, the energy released by oxidation for every gramme (g) of (a) glucose is equivalent to 15.5 kJ, (b) sucrose—16.7 kJ, (c) starch—17.1 kJ, (d) animal fat—39.3 kJ, (e) proteins—23.4 kJ, and (f) ethyl alcohol—29.7 kJ. The primary energy sources (carbohydrates, fats, and proteins) are interchangeable in terms of energy transformation, within certain limits. Carbohydrates and fats are appreciably used as fuels for muscular work. The energy available to the body is equal to the gross energy derived from the fuel minus the losses via urine and faeces. The losses are accounted for by applying the Atwater factors as 16.7 kJ (4 kcal) per g of protein and 37.6 kJ

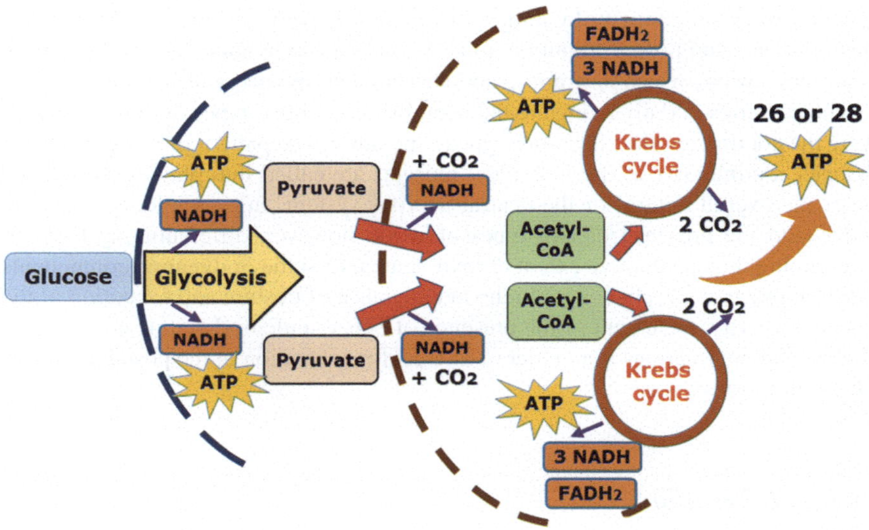

Fig. 4.2 Energy liberation from a glucose molecule, and metabolic pathway in the human body

(9.4 kcal) per g of fat. The Atwater factors 16.7 kJ (4 kcal) for per g of carbohydrate and 29.7 kJ (7.1 kcal) for per gm of ethyl alcohol, respectively. The energy derived from carbohydrates and ethyl alcohol is nearly entirely available for the body.

Accounting Energy Intake

A dietary survey is a fair assessment of the daily energy expenditure of a person or a homogeneous group of population. Dietary accounts are based on the registration of food consumption, from the recording of current intake and recall of past intake by interview. The record of current food intake is undertaken by (a) *weighed inventory method*, i.e. the number of foodstuffs served and the leftover on the served plate; (b) *precise weighing,* i.e. weighing of the edible parts of ingredients used in the food; and (c) *household measures,* i.e. the measure of cups and spoons of food consumed. The recalling of past intake by interview is noting of dietary patterns, including the description of kinds and amounts of food consumed.

Since people cannot always regulate daily food intake by one's energy expenditure, there are limitations on the usability of dietary records as indicators of energy expenditure on a day-to-day basis. For the recording of current intake, recording period of 1 week, and in recall methods, recall of 24 h is preferred. A reliable dietary pattern may be obtained by both the recording and the recall methods, and that supplemented with interviews of dietary history.

Anaerobic Glycogenolysis

The biological system allows yielding energy to resynthesize ATP in the skeletal muscles, and in other organs and tissues in the event of momentary insufficient oxygen supply. The anaerobic enzymatic breakdown of glycogen via pyruvate to lactate comes into operation to yield partial energy, depending on one's ability to tolerate the acute acidosis that might develop in the muscles and blood. The other smaller sources of energy that come into operation in rising demand are through the high-energy phosphate bonds other than ATP, mainly creatine phosphate. The quantity of energy available from anaerobic sources and utilized in a physical work can be estimated from the measurement of oxygen debt, after the cessation of work. The kinetics of oxygen demand during recovery has been explained as an exponential decay process, distinguishing between *alactoid* (fast) and *lactoid* (slow) components of the recovery phases. That is, the content of lactic acid in blood immediately on the cessation of work indicates the amount of energy that may be derived from anaerobic glycogenolysis.

The O_2 uptake or the energy demand of any muscular work increases during the initial work phase, and that gets levels off as the oxygen demand reaches the level adequate to meet the requirement of the body. At the cessation of the work, the oxygen uptake gradually decreases, and the cumulative deficit of the work phase and the debt of the recovery phase are paid off, usually termed as the O_2 deficit and O_2 debt. The total O_2 demand for work and the fractions of oxygen deficit and debt may be determined from the planimetric area of the graphical plots of VO_2 against the time duration of work and recovery, taking an approximate duration of 15 min recovery. However, the quantitative account of the contribution of the *alactoid* and *lactoid* component is not enough from the above-stated approach.

In a presumption of super-imposition of two or more decay processes to be analysed, complex orthogonal polynomial functions such as Prony's method of approximation may be used to best fit to the *f(X)*, the fraction of oxygen demand to *alactoid* and *lactoid* components (Nag 1984), as shown below:

$$f(x) = C_0 + C_1 e^{a}1^{X} + C_2 e^{a}2^{X} + \cdots C_n e^{a}n^{X} \quad (4.1)$$

$$f(x) = C_0 + C_1 \mu_1^X + C_2 \mu_2^X + \cdots C_n \mu_n^X \quad (4.2)$$

In Eq. (4.1), a_1, a_2, a_n are expected to have negative real parts, and C_0, C_1, C_2, C_n are the coefficients of the respective components. Equation (4.2) is equivalent to the above, where μ^k is equal to e^{ak}. Taking into consideration of only fast and slow parts of the recovery process, $n = 2$ may be assigned. The detailed procedure to compute μ's and C's is available in Hildebrand (1956), and its application to resolve the exponential forms for the resultant total O_2 debt, due to fast and slow processes, is given in Nag (1984).

The total O_2 utilization during work and recovery is the energy cost of work. The oxygen deficit during maximal work is the indicator of energy reserve (e.g. about 6 l for average Indian farmers). The reserve will remain undisturbed as long as the work is aerobic, and the rate of oxygen consumption is approximately within the range of 1.0–1.25 l/min. For example, any excess oxygen demands of 0.5 to 0.7 l/min in a given farming task (i.e. the required rate of O_2 consumption) would be 1.50 to 1.89 l/min, and therefore the energy reserve will last roughly for 10 to 12 min. Thus, for the disproportionate increase in respiration would cause fatigue unless work breaks replenish the reserve. By subtracting the resting O_2 consumption from the assumed working O_2 consumption of 1.0 to 1.25 l/min, the remaining oxygen uptake of 0.8 to 0.9 l/min can replace the energy reserve in approximately 6 to 7 min of a rest break. The derivation of the work–rest cycle at an O_2 consumption of 1.50 to 1.89 l/min would be 10:7, that is, 10 min of work followed by about 7 min rest (Nag and Pradhan 1992).

Some Important Definitions

Heart rate (HR) versus Pulse rate (PR)—Heart rate is the number of ventricular beats per minute as counted from records of the electrocardiogram, blood pressure curves, or heart rate monitor. Heart rate is the rate at which the heart muscles contract and relax when the blood exits and enters the heart. The pulse rate is the rate at which the arteries contract and relax when the heart ejects the blood. Pulse rate can be measured at specific points in the body, such as neck and wrist, by feeling the pulse (palpation). In healthy individuals, the pulse rate is similar to the heart rate.

Resting heart rate (HR_{rest}) is measured when the person is at rest, i.e. sitting comfortably in a chair for about 5 to 15 min and that the person does not have a residual cardiac load. Generally, an average healthy adult has a resting heart rate of 60–80 beats per minute. Usually, the resting heart rate of women is higher than males. *Working heart rate* (HR_{work}) is measured when the person is doing work or activity. However, for measuring the working heart rate, readings are taken when the person attains steady state, meaning that the heart rate is stabilized in about 5 to 7 min work when the work severity is moderate. The mean of the heart rate values during the work about gives the working heart rate for that activity. *Work pulse or* ΔHR is the difference between the number of working heart rate and resting heart rate of the person in beats per minute.

Total work pulse (cardiac cost) is the entire extra heartbeats (over rest) required to carry out an activity by a person and is calculated by multiplying the work pulse (beats/min) by the duration of work in minutes. *Total recovery pulse (recovery cost)* is calculated by summing the extra heart rate beats of each minute (over rest) after cessation of the work till the heart rate comes to or near the resting level. *Oxygen consumption rate (l/min)* is the volume of oxygen (at 0 °C, 760 mm Hg, at standard temperature and pressure dry, STPD) extracted by a person from the inspired air per min.

The Utility of Energy Expenditure Data

The farm activities vary with the agro-climatic regions, the type of crops, and seasons of cultivation. There is an emphasis on the basic needs of systematic collection and collation of human energy expenditure data to quantify and standardize farm activities. Assessment of the level and pattern of energy expenditure has various applications from the epidemiological, physiological, and pathophysiological viewpoints, such as

- To provide a basis for nutritional counselling for metabolic balance in healthy life and pregnancy, and also the extent of one's energy demands that reflects on the functional life and productivity of a person;
- To ascertain the age- and gender-related effects on the routine physical activity, including recreational activities that are needed to attain fitness with optimum health; accordingly, to evaluate the permissible level of activity for persons with coronary insufficiency;
- To determine the relative influence of new farm technology on the pattern of human energy expenditure in traditional, semi-mechanized, and mechanized farming; and
- To study the job demands and occupational workload about one's capacity to perform work, thereby to judge the suitability of farm activities in terms of the level of severity, and reorganize/redesign work and work methods to reduce undue physical strains.

A bible of human energy expenditure data presented by Durnin and Passmore (1967) was an eye-opener about the levels of energy demand during work in the diverse occupational sphere. Undoubtedly most references have been drawn from studies of the developed world, with fair neglect to scenarios of the developing world and only a cursory reference to the farming sector. As such, reviews on the farming activity of Indian farmworkers are patchy, apart from some reports of localized emphasis (Ramanathan and Nag 1982). A bulletin (Gite and Singh 1997) gives some account of energy expenditure in different farming activities.

Measurement of Energy Expenditure

Over the decades, the researchers have adopted different methods for the assessment of human energy expenditure. The analysis of energy expenditure is known as *calorimetry* by which the production of human energy is measured in terms of heat, as the most conveniently determined form of energy. Calorimetry has two types—*direct and indirect*. Since the human body expends all the energy in carrying out the biochemical reactions, the amount of energy utilization can be measured in terms of foodstuffs consumed and the products formed. In the case of direct calorimetry, both exothermic and endothermic energy forms are converted into heat units and then measured. Direct calorimetry (Fig. 4.3) is a tedious, time-consuming, and cumber-

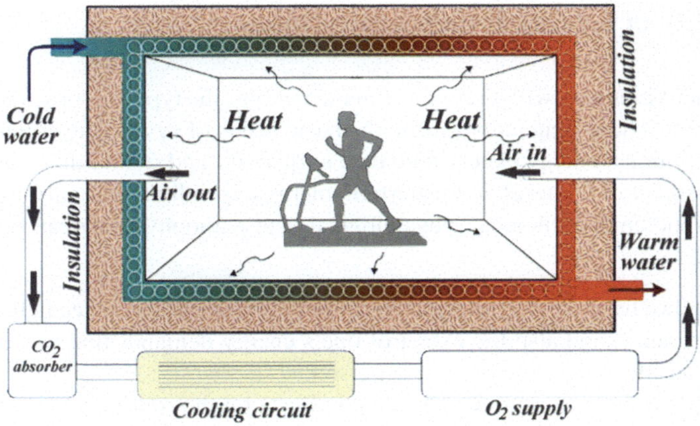

Fig. 4.3 Direct calorimetry

some procedure. Moreover, the apparatus required is expensive to construct and operate (Harris and Benedict 1918).

Indirect calorimetry is most preferred, in which the energy expenditure is determined from the measurement of oxygen uptake through the lungs and carbon dioxide produced by the body. The amount of carbohydrates, fat, and proteins that have been consumed and metabolized in the human body is measured from the amount of oxygen used (VO_2), carbon dioxide produced (V_{CO2}), and nitrogen excreted in the urine.

Indirect calorimetry is performed in two ways, either as *closed-* or *open-*circuit methods. In the *closed-circuit* process, a person is cut off from the open air and close circuited to breathe pure O_2. The CO_2 content of the expired air is removed continuously by passing the air through CO_2 absorbent (e.g. soda lime). A decrease in the gas volume is related to the rate of O_2 consumption and CO_2 production, from which the rate of energy expenditure is calculated. In the *open-circuit* method, a person breathes open air, while the expired air is collected in a bag (e.g. Douglas bag) which is emptied through a gasometer for volume measurement. A sample of the expired air is analysed with the aid of a gas analyser, such as Haldane Gas Analyser for oxygen, carbon dioxide, and nitrogen content.

The olden manual methods of gas analysis by any of the gas analysers (Haldane, Scholander, Llyod gas analysers) are time-consuming and require an individual skill. Different auto-gas analysers have also been used in laboratories conveniently. Each of the methods mentioned above and equipment (Fig. 4.4) has different practical applications. The specific advantages and limitations of the methods govern its use. Open-circuit indirect calorimetry has been preferred for field use, due to portability of equipment, and under defined conditions, the process is reasonably accurate. In stationary laboratory situations, e.g. in hospitals, a metabolic cart may be used to assess the energy requirements during some form of physical work. Portable equipment has some inherent limitation of breathing resistance when the rate and

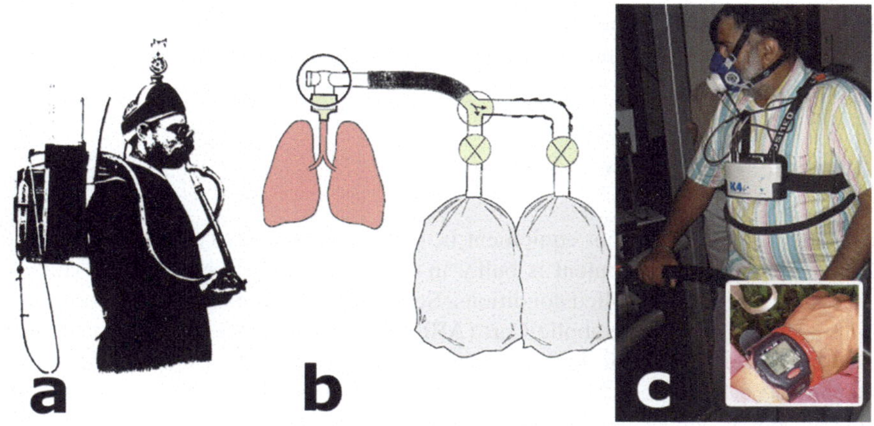

Fig. 4.4 Old and modern equipment for indirect calorimetry. **a** Kunz calorimeter, **b** Douglas bag method of oxygen consumption, and **c** K4b2 oxygen analyser with inset of polar heart rate meter

volume of breathing are high or very low. Therefore, the use of the equipment calls for proper selection of the breathing range for desirable accuracy.

The fraction of gases present in the inspired and expired air determines the O_2 consumption and production of CO_2. Accordingly, the energy expenditure is calculated in terms of the calorie equivalent of one unit of oxygen utilization that varies with food consumption (carbohydrate, fat, and protein). The detailed procedure for the measurement of VO_2 is available in Consolazio (1963). In the process of estimating energy expenditure, one may come across different approaches; for example, in indirect calorimetry, the respiratory exchange ratio can be measured, using a metabolic cart, and determined O_2 consumption and CO_2 production. The respiratory exchange ratio is essentially the ratio between the amount of CO_2 produced in metabolism and O_2 used, comparing to gaseous concentration in room air. The ratio greater than or equal to 1.15 is often used as an alternative endpoint criterion of VO_2 max (aerobic capacity) testing. Whereas the RQ (respiratory quotient) is the ratio of CO_2 production and O_2 consumption in the tissues. This quotient is indicative of the relative contribution of carbohydrates and lipids to total energy expenditure.

Heart Rate Monitors

A heart rate monitor is a device one wears to measure and display the heart rate (beats/min) continuously. Electrode sensors in a chest strap detect each heartbeat and transmit the data to a receiver display such as a watch or other accessory. Such monitors are considered to be as accurate as of the electrocardiogram. Polar wireless personal heart rate monitors are usually favoured for its easy use among the athletes.

Many heart rate monitors save and display the working heart rate on a graph compared to time, speed, elevation, and other aspects measured during the work.

Metabolic Measurement Systems

The metabolic measurement equipment is, namely, laboratory type and field type. The laboratory-type equipment is bulky in size and is used for metabolic studies in the laboratory or controlled conditions. Some of the examples of laboratory-type equipment are Max-II Metabolic Cart (AEI Technologies), Metalyzer (Life Max), Vista-MX2 VO_2 Metabolic Measurement System (VacuMed), Parvo Medics True One 2400(Parvo Medics), MC-TA-200 V Metabolic Cart (iWORX), and Aerosport TEEM 100 Portable Metabolic Measurement System (Aerosport, Inc.).

The equipment for field use is also called ambulatory type and is useful for collecting metabolic measurement data of a person during the habitual activity, without causing any hindrance. Examples of this type of equipment are Aerosport KB1C (Aerosport, Inc.), Oxylog (PK Morgan) (Harrison et al. 1982), Cortex Metamax 3B (Life Max), COSMED K4B2 (COSMED), VO_2 Master Pro-Metabolic Measurement System 17350 (VO_2 Master Health Sensors Inc), and VmaxST portable metabolic measurement system (VIASYS). These equipment give oxygen consumption as well as the heart rate of a person continuously during work or at rest.

Basal Metabolic Rate

In simpler terms, one estimates the basal metabolic rate (BMR) or standard metabolic rate (SMR), and that gets multiplied by an activity factor to calculate total daily energy expenditure. The resting energy expenditure is usually measured under standard basal conditions, i.e. after protein starvation for 3 days and at least 12 h of complete fasting, at rest, lying down at room temperature, i.e. after night rest and before breakfast. The individual characteristics, such as gender, age, and body size, influence the BMR/SMR. The levels are highly correlated with the fat-free body mass or lean body mass (LBM). Obese persons have a lower BMR than that of thin persons. Often Weir's (1949) formula is used in indirect calorimetry relating to metabolic rate (energy expenditure) and is based on the measurement of the volume of expired air (V_E) in litre/min) at BTPS, and the fraction of O_2 in expired air (F_{EO2}), i.e.

$$\text{Energy expenditure (kJ/min)} = 20.57\ V_E[(20.93 - F_{EO2}) * 0.01]$$

In the case of attaining the total daily energy expenditure in work and rest, the time and motion study techniques are usually employed to separate elements of the work performed. The energy expenditure for each aspect of work and the time spent in each component give the total energy expenditure during a period of activity. The balance between the energy content of the whole foodstuffs consumed and excreta to that of the total energy expenditure determines a person's energy balance. In case of strenuous physical activity, the length of time needed to obtain an energy balance between intake and expenditure is influenced by several factors, e.g. metabolic stature of a person, food habits, and the extent of accumulation of body fat. The persons who maintain the body's energy balance are able to keep stability in body weight. Even a marginal difference between energy intake and expenditure may result in a noticeable change in body weight within months. In energy expenditure studies among farming populations, determining the energy content of excreta is not always feasible, although the energy loss through this route may not be negligible. There may be a considerable inter-individual variation in the extraction of energy content from the consumed foodstuffs. Also, indications of body breakdown may be noted from the excretion of nitrogenous compounds in faeces, urine, and sweat.

Energy Expenditure Values

The human energy expenditure is expressed in SI units. The commonly used unit *kilocalorie* (kcal) is equivalent to 4.185 kilojoules (kJ), which correspond to the consumption of about 200 ml of O_2. The energy expenditure during 24 h is usually expressed as megajoules (MJ). The rate of work may be expressed in as kilogramme-metre/min (kgm/min or kpm/min) or in watt (W). That is, one kpm/min or one kgm/min is about 0.164 W. The mechanical efficiency of the work of a person is determined from the measurement of the net amount of work done and the amount of energy spent on the given work.

The weighted average of the coefficient of energy equivalent of a litre of O_2 may be obtained from the consumed mixed nutrients. Generally, there is a small difference in the factor of energy equivalent to differences in food habits of a sample population. When the RQ is about 0.90 as usually observed during farm work, variation in the factor of energy equivalent is small. Therefore, energy expenditure may be suitably expressed in units of O_2 consumed, corresponding to 5 kcal for one litre of O_2. With the reference of farmworkers in India, the power equivalents of male and female worker were calculated from the available literature (Nag et al. 1980a; Nag and Chatterjee 1981; Gite and Ganeshsan 2014), and the values are 60 W for a man and 50 W for a woman.

Besides, the energy expenditure data may be expressed in terms of net energy expenditure by subtracting the BMR or SMR (i.e. the energy requirements for maintaining the body at apparent rest) from the measured energy cost of an activity. The above expression may appear unconvincing to some readers, as the energy cost of work, to a great extent, depends on the initial value. When the BMR or SMR is relatively high, the condition influences the gross energy cost of work. For example, the BMR at low climatic warmth may be elevated to that of comfortable ambient environment. Nevertheless, the gross energy cost at the above conditions is unlikely to be different. The expression of energy expenditure as multiples of BMR (Mets)

Table 4.1 Categorization of physical activities in terms of Mets (Seliger et al. 1974)

Category	Symbol	Mets %	Description
1	LS	110	Lying sleeping
2	L	120	Lying (reading, talking)
3	S	150	Sitting or standing without unusual movement
4	SP	300	Sitting or standing with very light movement or activities
5	1	500	Activities with low intensity
6	2	800	Activities with middle intensity
7	3	1000	Activities with high intensity
	x		Other activities where the class is unknown and exact description needed

takes into account the individual differences in body size, age, and gender. The unit Mets may also be used to grade work intensity, in terms of multiples of the unit (Table 4.1).

ICRP (1975) report of the Task Group on Reference Man proposed the formula to estimate energy expenditure (E, kcal/24 h), as follows:

$$\text{E (for men)} = 66.5 + 13.8\ \text{Body weight (kg)} + 5.01\ \text{Body height (cm)} - 6.8\ \text{age (years)};$$
$$\text{E (for women)} = 66.5 + 9.6\ \text{Body weight (kg)} + 1.80\ \text{Body height (cm)} - 4.7\ \text{age (years)}$$

The energy expenditure is often expressed per unit of body surface area (BSA) in situations referring to the environmental influences. The rate of heat transfer per unit time between the human body and environment is proportional to the evaporative surface area required in dissipating the internal heat in the form of sweating and non-sweating perspiration. Therefore, the BSA expression has been applied in thermoregulation and environmental heat load assessment. There is a limitation in the BSA-associated expression since the surface area is calculated from formulae (e.g. Du Bois and Du Bois 1916; Banerjee and Sen 1995). Sometimes, segmental surface areas are estimated using Meeh constants (Faillie 1925), which have been derived from the assumption of cylindrical body segments.

The energy expenditure may be corrected to a reference bodyweight of a population and thereby to attain uniformity of expression. When the energy expenditure is expressed per unit of body weight, the uncertainty of body weight differences is partly obviated. The body weight is a combination of body fat and lean body mass (LBM), where body fat is considered as anhydrous. Body fat is metabolically inert, except under severe underfed conditions when the energy requirement of the body derives from the breakdown of body fat. Since the LBM has constant water content and cell solids within a range of age and body weight, LBM may also be preferred as a point of reference to express energy expenditure or O_2 consumption. Table 4.2 includes some equations for estimating different body parameters.

Table 4.2 Equations for some important body parameters

Parameter	Definition	Equation
Body Mass Index (BMI)	Defined as the ratio of the body weight to body height squared of a person (Keys et al. 2014; Roozbazar et al. 1979). It is also called as Quetelet Index	$BMI = \frac{Weight}{(Height)^2}$ where bodyweight is in kg and body height is in m
Ponderal Index	An index to classify body type and physique. The PI is also known as Rohrer index, after the name of Swiss physician Fritz Rohrer, who first proposed the index.	$PI = Weight/Height^3$ where body height is in m and body weight is in kg, i.e. PI is expressed as kg/m^3
Lean Body Mass (LBM)	It is the mass of the body minus the mass of fat. There are many methods for determining the LBM. The most commonly used way is to calculate body density from the sum of four skinfold thickness (Durnin and Womersley 1974) and then calculating the body fat using Siri (1956) equation	Calculation of body density (BD) using Durnin and Womersley (1974) equations: For Male : $BD =$ $1.1765 - (0.0744\,x \log \sum 4\,\text{Skinfolds})$ For Female : $BD =$ $1.1567 - (0.0717\,x \log \sum 4\,\text{Skinfolds})$; where BD is the body density in g/cc, and skinfolds (mm) measurement includes the sites of mid-biceps, mid-triceps, subscapular, and suprailiac **Calculation of body fat (Siri** 1956**):** Percentage of body fat $= \frac{495}{BD} - 450$ Absolute body fat $=$ Weight $\times \frac{\text{Percentage of body fat}}{100}$ LBM = Total body weight (kg)—Absolute body fat (kg)

Maximal Work Capacity

Usually, a person's maximum oxygen consumption rate ($VO_{2\,max}$) or aerobic capacity or maximal aerobic power is taken as the maximal work capacity (Shephard et al. 1968a, b). As the O_2 uptake is directly related to the intensity of work, beyond a specific limit, however, an increase in workload does not elicit any further increase in O_2 consumption. This level of O_2 consumption is called the $VO_{2\,max}$ and is a good measure of physical work capacity. When large groups of muscles are brought into continuous action, the O_2 demand exceeds the supply. The point at which the supply no longer keeps pace with the demand for O_2 is taken as the $VO_{2\,max}$. Age, gender, body build, and professional habituation to a type of work influence the $VO_{2\,max}$.

The workload while performing an activity is better expressed in terms of percentage of aerobic capacity. There are limitations about expertise and equipment availability for direct measurement of VO_{2max} of a large population. Step test is a simple technique to determine the VO_{2max} of a person. Queen's College Step Test (McArdle et al. 1972) is one of the commonly used step tests. The protocol involves stepping up and down a stool of 41.3 cm (16.25 inches) height for a total duration of 3 min at the stepping frequency of 24 cycles per minute. The stepping frequency of the subject is monitored with a metronome set at 96 beeps/min. The following equation may be used for the prediction of aerobic capacity.

$$\text{Predicted } Vo_{2max}(\text{l/min}) = 111.3 - 0.42 \times \text{pulse rate (heart rate) during first 15 seconds after cessation of the exercise}$$

While there are several approaches to predict VO_2 max based on sub-maximal workloads, and linear regression between heart rate and oxygen consumption or equivalent workload (Astrand and Ryhming 1954; Margaria et al. 1965), a linear extrapolation of age-related maximal heart rate (Maritz et al. 1961) is preferred. For collecting the data of heart rates at different workloads, either a treadmill or a bicycle ergometer is used. For testing with each of the equipment, various protocols (e.g. Bruce, Naughton, Modified Naughton protocols) are available, and one may select a protocol depending on the requirements and situation. Generally, the HRmax of a person is estimated as HR_{max} = [220-age (years)]. This formula gives an error of ±11 beats/min. Inbar et al. (1994) suggested HR_{max} = 205.8–0.685 age (years).

For a comparative analysis, the VO_{2max} of the population groups from the tropics is shown in Table 4.3. There is a difference in maximal VO_2 depending on the habituation of types of muscular work; for instance, the maximal VO_2 at arm work is about 70% of what may be attained in leg work. One may predict VO_{2max} concerning different individual characteristics, such as age, gender, and habitual physical activity.

Heart Rates—A Predictor of Energy Expenditure

As mentioned earlier, the heart rate (the number of ventricular beats per min) is usually counted from the records of the electrocardiogram, blood pressure curves, or simple auscultation with a stethoscope or palpation over the heart. The pressure waves that are propagated along the peripheral arteries, such as the carotid or radial arteries, are recorded as *pulse rate.* The heart rate is often taken as an indicator of cardiovascular stress. When the heart rate exceeds 80% of the age-related maximum (Table 4.4), undue stress occurs to a person.

The widespread use of heart rate to determine the load on the cardiovascular system is based on well-established physiological principles. Since the heart rates increase with the intensity of work, continuous recording of the heart rates throughout a working shift is useful in determining the strain imposed on the human body. Brouha

Table 4.3 $VO_{2\ max}$ of the population in the tropics

		$VO_{2\ max}$ (ml/kg. min)		
Country	Age (years)	Men	Women	References
Indian landless farmers	35	35.7	–	Nag et al. (1978)
Indian small farmers	25–33	42.4–46.6	36.3–36.4	Nag (1987)
Indian women farm workers	35	–	33.2	Singh et al. (2008)
Indian agricultural workers	30.6	44.1	–	CIAE (2012)
Indian Dock workers	33	46.6	–	Saha (1975)
Indian Steel mill workers	33.3	42.6	–	Saha (1978)
South African Bantu (rural)	32	37.6	–	Wyndham (1973)
South African Bantu (urban)	34	41.9	–	Wyndham (1973)
Trinidadian Negroes	25.5	38.3	–	Edwards et al. (1972)
East Indians	26	39.4	–	Edwards et al. (1972)
Ethiopian Adi Arkai	25	39.9	–	Baker and Baker (1978)
South African white miners	35	40	–	Wyndham and Sluis-Cremer (1969)
Easter Island Pascuans	25	42	–	Ekblom and Gjessing (1968)
Nigeria Yoruba inactive	25.8	45.9	–	Davies et al. (1972)
Nigeria Yoruba active	25.3	48.5	31.6	Davies et al. (1972)
Tanzanian inactive	21.8	47.2	–	Davies and Van Haaren (1973)
Tanzanian active	25.4	57.2	40.2	Davies and Van Haaren (1973)
Jamaican	26	47	27.3	Miller et al. (1972)
Venezuelan Warao Indian	22	51.2	–	Gardner (1971)
Israeli Yemenites	26.4	48.2	35.4	Samueloff et al. (1973)
Israeli Kurds	25.3	52.4	29	Samueloff et al. (1973)

Table 4.4 Maximum heart rates (beats/min) for men and women

Age range (years)	Men	Women
20–29	188	182
30–39	184	175
40–49	173	169
50–59	168	164
60–69	162	159

Table 4.5 Heart rate, VO_2 (l/min), and $VO_{2\,max}$ (%) in different age groups

	20–29 years		30–39 years		40–49 years	
Heart rates (beats/min)	VO_2 (l/min)	VO_2 max (%)	VO_2 (l/min)	VO_2 max (%)	VO_2 (l/min)	VO_2 max (%)
105–115	0.71	30	0.84	36	0.76	36
116–125	0.82	35	0.89	39	0.99	48
126–135	0.95	40	1.05	45	1.17	56
136–145	1.16	50	1.30	56	1.29	62
146–155	1.35	57	1.63	71	1.39	67
156–165	1.66	71	1.78	77	1.69	81
166–175	1.89	80	1.97	85	1.78	87
176–185	2.06	88	2.14	93	–	–

(1967) suggested that the first min recovery heart rate is maintained at about 110 beats/min or below, and when deceleration between the first and third min recovery heart rate is at least ten beats/min, no increasing cardiac strain occurs. The use of heart rate as a valid indicator of work severity has been greatly facilitated in recent years by the development of telemetric devices for the continuous recording of heart rates. The linear relationship between heart rate and energy expenditure has been validated, primarily in the case of dynamic muscular work (Astrand and Rodahl 1986). General estimates of the relative level of the work capacity against heart rates and VO_2, derived from studies on Indian male farm workers (Nag 1981), are shown in Table 4.5. This relationship may vary with the physical fitness of the person and differences in the stroke volume of the heart. Heart rate in maximum exercise may be elevated to about 2.2 to 2.5 times the resting level. For performance analysis of different jobs, the heart rates give a better evaluation of the strain imposed on the human body than the actual energy requirement. This is one of the bases for frequent use of continuous heart rate recording during occupational work.

There is a note of caution about the measures, since the heart rate may be significantly influenced by the factors other than the metabolic rate, such as ambient air temperature, emotional status, muscle groups used in work, and amount of static work involved. The VO_2 and the heart rates are useful when large muscle groups are in action. Static work drastically increases the heart rate. Much of the farming activity involves static work components in grasping and holding tools, and in postural control of trunk and head in varied work postures. Work with a mixture of static–dynamic components may yield disproportionate heart rate responses (Nag et al. 1986). The mental work does not show up on the relationship between heart rate and VO_2.

On the other hand, the heart rate response in mental work may be expressed as heart rate variability. Emotional and psychological stresses also increase the heart rate. The heart rate may be significantly influenced by environmental heat. Consequently, the heart rate in practical work should not be interpreted directly in terms of VO_2 or energy expenditure but should instead be explained as the strain on the cardiovascular system.

Energy Cost of Work in Farming

The resting level of energy output, including the increased metabolic rate resulting from the ingestion of food, is indeed biologically constant in all individuals. Apart from the period of childhood and adolescence when there are needs for growth and periods of pregnancy and lactation, the energy expenditure of a person depends on inter-related variables, such as level of physical activity, body size and composition, and age and environmental factors. The most crucial factor that influences the energy expenditure is usually the physical activity pattern, which may vary greatly depending on socio-cultural elements, occupation, ecological, and other environmental factors. Accordingly, the habitual physical activity may be differentiated between occupational and non-occupational activities. Since agriculture is seasonal, there are wide variations in workloads throughout the working seasons. As observed, farmers in rural India spend nearly 7.5 months in farm activities, and for the remaining 4.5 months, they do not have any specific work schedule (Nag et al. 1978; Nag and Chintharia 1985). The work period in the farming season may stretch to about 9 to 10 h. In the harvesting season, the work continues even at night.

The crop production encompasses many processes, including seedbed preparation, sowing, planting, weeding, harvesting, threshing, livestock rearing, materials handling, machinery operation and maintenance, fertilizer and pesticide application, water lifting and irrigation, crop processing, storage and transport, and other jobs. A multitude of farm machinery, including manually operated machines and hand tools, animal-drawn implements, tractors, and other powered machinery, are used for farm operations. Table 4.6 gives cardiorespiratory responses of men and women in different farm tasks.

Table 4.6 shows that the heart rates in farming work range from 76 beats/min during bird-scaring work to 172 beats/min while digging with a spade. The corresponding energy demand ranges from 3.1 to 47.6 kJ/min. Nag et al. (1980a) mentioned that there was no consistency in heart rate response with the severity of work. In agricultural activities, localized muscular fatigue is one of the main factors which makes the workload severe and compels the worker to stop the work.

Nag and Chatterjee (1981) observed that women had a whole-day energy expenditure of 10.61 MJ, in which about 52% of the total energy was expended for the workday and the remaining 48% was used for maintaining postural and other activities in day-to-day life. From the workday energy expenditure proportion, the average expenditure rate over the working hours was 11.11 kJ/min or about 28% of the $VO_{2\,max}$. This indicates that the work level of the women workers was within acceptable limits, with exceptions of some short spurts of activity of substantial severity. If the workers are allowed to work at this mean level or a slightly higher level of activity (up to 35% of $VO_{2\,max}$) for a prolonged period, it should not induce substantial amounts of physiological strain on the workers. When the workers have to perform more substantial jobs, they compensate by adjusting their rate of work in subsequent workloads. Consequently, they maintain their own pace of work without becoming unusually tired or weakened. This is a unique feature of body adjustment, which is possible in agricultural work, because of the unregulated pace of work.

An early report of Durnin and Passmore (1967) showed that manual milking of cows required 20.9 to 29.3 kJ/min. Modern farm milking machines significantly reduced the work content and severity, i.e. the work associated with handling the

Table 4.6 Cardiorespiratory responses of men and women in different farm operations and activities (data in parenthesis are of women workers)

Farm operations and activities	Heart rate (beats/min)	Energy demand (kJ/min)	References
Ploughing (Country plough, Mouldboard plough) in dryland and wetland	103–131	13.9–28	Kaul and Splinter (1964), Nag et al. (1980a), Nag and Chintharia (1985), De and Sen (1986), Pradhan et al. (1986), Gite (1991, 1993)
Harrowing with animal-drawn blade harrow	104	11.2	
Levelling soil by laddering or wooden rake		9.8–16	
Digging soil using a spade, hoe, pickaxe Bund trimming (wet and dry lands)	131–172	19.6–35.5 (13.5–25.5)	
Hoeing, low lift (41–70 stroke/min); High lift (21–35 stroke/min)		20.7–39	Pradhan et al. (1986), Nag and Pradhan (1992)
Tractor—Empty trailer on bitumen road (6–10 km/h)		9.9–13.6	TNAU (2002)
Tractor—Cultivating (3.0–6.0 km/h)		10.2–15	
Tractor—Disc ploughing (2.5–3.5 km/h)		11.7–16.7	
Power tiller—Transport on bitumen or farm road (3.5–5.0 km/h)		10.8–12.8	Pawar and Pathak (1980), Tiwari and Gite (2002)
Power tiller—rotapuddling (1.1 to 2.3 km/h)	101–119	11.7–18	
Power tiller—rototilling in the tilled and untilled field (1.5–2.4 km/h)	97–114	13.3–20.1	
Sowing		7.0–16.7	Nag and Dutt (1980)
Sowing with a seed drill	135		Singh et al. (2006)
Paddy seeder of various types, namely, 4, 6, or 8 rows	154	24.9—51.1	Nag and Dutt (1980), TNAU (2002)
Paddy transplanter (powered)	110	9.9 to 26.6	Gayatri and Sam (2017); Ojha and Kwatra (2014)

(continued)

Table 4.6 (continued)

Farm operations and activities	Heart rate (beats/min)	Energy demand (kJ/min)	References
Transplanting seedling (by hand) on puddled field	109	11.0–16.4	Nag and Dutt (1980), (Nag and Chatterjee 1981)
Weeding (using hand tool) in bent and squat posture	98–114	9.3–18.5	Nag and Dutt (1979), Nag and Chatterjee (1981), Yadav and Gite (1987), Tewari et al. (1991), Singh et al. (2006)
Weeding with a long-handled tool—single row sweep weeder, multiple sweeps weeder, blade and rake weeder, projection finger weeder	102–127	8.9–29.2	
Weeding—wheel hoe weeder	114–121	13.6–28.2	Nag and Dutt (1979), Nag and Chatterjee (1981), Tewari et al. (1991), Gite et al. (1993)
Weeding with Cono weeder	124	25.4–27.5	TNAU (2002)
Weeding with power weeder	123–134	23–25.8	CIAE (2007)
Fertilizer broadcasting by hand/with broadcaster	126–147	5.5–12.7	Nag et al. (1980a), Singh et al. (2004)
Spraying of pesticides (knapsack sprayer)	95–126	3.1–11.6	Nag et al. (1980a), Ghugare et al. (1991)
Watchkeeping to scare birds	76	4.5	Nag et al. (1980a)
Uprooting seedlings by hand (bent and squat posture)		6.8–19.7	Nag and Chatterjee (1981)
Harvesting paddy using a sickle	99–120	6.8–17.4	Nag et al. (1980a), Nag and Chatterjee (1981), Gite and Agrawal (2000)
Plucking vegetables		8.6–12.1	
Cutting sugarcane		10.6–13.9	
Self-propelled paddy harvester		23.7–27.4	TNAU (2002)
Load carrying on the head (20–25 kg)		10.1–22.4 (10.4–25.7)	Singh and Kaul (1972), Sen and Nag (1975), Ramanathan and Nag (1982)
Load carrying on the head (60–80 kg)		39.7–47.6	

(continued)

milking machinery required only 12.5 to 16.7 kJ/min. Farm tasks such as manual weeding, uprooting, and transplanting of seedlings are performed by hand while sitting with one or two legs flexed at the knee and in bent postures. The bent posture of work usually demands a higher physiological cost than in the sitting posture. Any sustained bent posture of work should be avoided, wherever possible, and thereby to minimize biomechanical stresses on the vertebral structures. In a puddled field, the workers perhaps can work only in a bent posture. Specific tasks such as harvesting,

Table 4.6 (continued)

Farm operations and activities	Heart rate (beats/min)	Energy demand (kJ/min)	References
Load carrying on yoke (60–75 kg)		32.2–42.2	
Threshing paddy by beating	136	17–21.5	Nag and Dutt (1980)
Threshing paddy by pedal thresher	141	21.0–34.2	
Winnowing (squat and stand)	125	4.5–20.1 (6.1–11.1)	Nag et al. (1980a), Nag and Chatterjee (1981)
Winnowing with winnower	112		Singh and Gite (2007)
Irrigation work (water channelling in the field)		7.5–14.1	Nag et al. (1980a), Tiwari and Gite (2002)
Irrigation work (water lifting using swing basket)		17.6–26.5	
Pounding grain		(13.2–25.2)	
Walking on the puddled field (1.1–2.3 km/h)	112	11.3–15.7 (8.5–14.5)	
Spreading grains and vegetables on the floor		(12.8–17.3)	

i.e. cutting paddy pinnacles by a sickle, may be higher in bent postures, about 50–55 cuts per min compared to sitting postures, i.e. 40–45 cuts per min. Harvesting in bent posture has an advantage of mobility compared to sitting with the leg bent at knee (Nag et al. 1988).

Estimating Energy Expenditure from Heart Rate

Many times, while carrying out physical activities in the field, the measurement of O_2 uptake is difficult. In such a case, the established relationship between heart rate and O_2 consumption rate is utilized to estimate the O_2 consumption rate for both men and women. Some such equations given herewith have relative accuracy, and therefore the readers may refer to the source for choosing an equation.

- Nag (1981) for Indian male farmworkers: $Y = 0.0183\ \mathrm{HR} - 1.28$;
- Verghese et al. (1994) for Indian women: $Y = (0.159\ \mathrm{HR} - 8.72)/20.9$;
- Singh et al. (2008) for Indian farm women: $Y = 0.0114\ \mathrm{HR} - 0.68$;
- Tiwari and Gite (2010) for Indian male farmworkers: $Y = 0.0156\ \mathrm{HR} - 0.88$;
- Gite (2017) for Indian male farmworkers: $Y = 0.014\ \mathrm{HR} - 0.80$; and Indian female farmworkers: $Y = 0.011\ \mathrm{HR} - 0.59$,
  where $Y = O_2$ consumption rate, l/min, and HR = Heart rate, beats/min.

Work Severity Classification

Agricultural work imposes a considerable degree of physical strains and fatigue, which may culminate in different nature of musculoskeletal injuries and accidents in the workplace. Minimizing the demands of work has a bearing in channelizing labour requirements at peak working seasons and enhancing work efficiency. Data given in Table 4.6 have various utilities in determining the job severity of farming tasks, assessing the impacts of tools and machinery, and formulating the basis to reorganize work.

Evaluation of workload imposed based on strain on the cardiorespiratory system in dynamic work is to determine the *relative load index* (RLI), i.e. the O_2 requirement of work expressed as a percentage of the $VO_{2\,max}$ (Andersen et al. 1971). By this approach, variations in age, sex, and work capacity may be taken into account. The expression of energy expenditure as a fraction of the human work capacity opens up its usability in work classification, as light, moderate, heavy, and extremely heavy work and the prediction procedures are handy for quick assessment of the workload of the farmers concerned. Various investigators (Astrand et al. 1973; Bonjer 1968; Nag et al. 1980a, b; Saha et al. 1979) emphasized that for long duration work, the activity levels should not exceed 35 to 50% of $VO_{2\,max}$. At an energy expenditure beyond 50% of $Vo_{2\,max}$, a substantial amount of anaerobiosis (accumulation of lactic acid in the blood) occurs in the working muscles. Nag (1983) examined blood lactate concentration in farmworkers, and after a 6-hour workday, the cumulative level reached 83 ± 27 mg%. During prolonged work, such accumulation disturbs the acid–base balance, affects respiratory regulation, and causes muscle fatigue. Based on lifting work, Petrofsky and Lind (1978) also suggested that above 50% of $VO_{2\,max}$, heart rate, and arterial lactate increased rapidly, and lifting could not be continued for periods more than 2 h. Often, therefore, the RLI is taken to categorize the severity of workload.

For instance, ploughing requires VO_2 of 1.5 l/min at STPD; for a regular farmer with a $VO_{2\,max}$ of 2.5 l/min, the RLI is, therefore, $(1.5/2.5) * 100 = 60\%$, which is considered as heavy work for the person. For the person who has relatively lower $VO_{2\,max}$, e.g. 2.0 l/min, the RLI is $(1.5/2) * 100 = 75\%$, which may be extremely heavy work for this person. Therefore, the assessment of work severity for classifying work is a debatable issue. Since the perception of physical effort is not the same to two different persons, and also age, sex, and nutritional status of a person influence on the perception of physical exertion, the work classification has been the subject of considerable debate and different guidelines proposed. For practical purposes, work classification may vary according to whether a person is trained or untrained, and athletic or nonathletic. Apart from physical strains, evaluation of occupational work begins with the account of the motion and time analysis of work elements and contents during a work shift. In addition to physiological and psychological approaches, the work time measurements make it possible to classify the aspects of work based on work severity.

Accordingly, different proposals have been given for grading human strains in occupational work, whereby physical activity can be graded in a simple, descriptive way like light activity or heavy work (Christensen 1953; Sen and Nag 1975). The classic attempt was of Christensen (1953) based on his studies in British ironworks. Edholm's scale (1966) is a standardized method for classifying daily activities into seven groups (Table 4.1), which are adjusted to multiples of basal metabolic rate (Mets). A linear relationship exists between the energy expenditure and cardiovascular indicator (heart rate), within certain limits and with exceptions of the influence of environmental heat. Accordingly, the intensity of work referred to as work severity may be classified as sedentary, light, moderate, heavy, and extremely heavy muscular work, as fractions of the human work capacity, i.e. up to 15%, 16–25%, 26–50%, 51–75%, and above 75% of $VO_{2\ max}$, respectively (Nag et al. 1980a, b). This contribution brings a compilation of different work severity classifications, as given in Table 4.7.

A perusal of work severity classification given in Table 4.7 and data of different farm tasks given in Table 4.6 indicates that the majority of the operations in farming are moderately heavy. Operations such as laddering (by two men) to level the ploughed ground and cutting crops are the light jobs. The operations such as water lifting, bund trimming in the wet and dry land, ploughing, and pedal threshing are the heavy jobs in farm work. Ramana Murthy and Belavady (1966) noted that puddling and bund trimming were heaviest agricultural work. According to Edholm et al. (1973), the heavy farm activities encountered by the Kurdish and Yemenite Jews were walking in mud, moving irrigation pipes, potato picking, forking, scything grass, and manure spreading. According to Nag et al. (1980a), men spend only 29% of the total working hours in light work, 64% in moderate work, and only 6% in heavy work. For only about 1% of the total working hours, the workers had to undertake extremely heavy work in farming. Thus, the physical activities in agriculture usually lie within a moderate level of activity, except for short periodical spurts of heavy activities. The situation is very similar in both men and women, only with a difference in the absolute level of energy expenditure.

Farm tasks for women are different from those performed by men. It may be obvious that women perform less strenuous tasks. None of the agricultural operations performed by women can be categorized as extremely heavy work. However, the task varieties are very large in case of women. For example, tasks like weeding and uprooting seedlings, grain pounding, and winnowing are usually women's work. Activities such as harvesting, transplanting, uprooting, and carrying loads are moderately heavy tasks.

Here, an important point to note that improvement in body status and work capacities of the farming population may bring improvement in work performance and productivity (Nag et al. 1978). Since the farm activities are predominantly moderately heavy, it may not be suitable to engage the farming population even in heavy or extremely heavy jobs unless their physical status is uplifted by providing better nutrition. Otherwise, severe health breakdown may result from the slightest epidemic and endemic outbursts, which are common in rural sectors in Asia and Africa. Various observations suggest that the pace of work in farming may be standardized by modifying the schedule of work and work methods, keeping the load of work, whole-day energy need, and relative demand of work capacity within allowable limits. In the

Table 4.7 Work severity classification based on heart rate and energy expenditure rate

Category of work	Energy expenditure, kJ/min (numbers in bracket are values of heart rates in beats/min)							
	Christensen (1953)	Malhotra et al. (1966)	Ramanathan et al. (1967)	Sen and Sarkar (1969)	Sen and Nag (1975)	Nag et al. (1980a)	Varghese et al. (1994)	Astrand and Rodahl (1986)
Sedentary/very light	11	4.5	4.2	<7.3 (<75)	–	–	<5.0 (<90)	–
Light	11–21	4.5–10.5	4.2–10.5	7.3–14.6 (75–100)	7.5–15.0	<9.1	5.1–7.5 (91–105)	<10.4 (<90)
Moderate	21–31	10.5–21	10.5–16.7	14.6–22.0 (101–125)	15.0–22.0	9.1–18.2	7.6–10.0 (106–120)	10.5–20.8 (90–110)
Heavy	31–41	21–31	16.7–25	22.0–29.3 (126–150)	22.0–29.0	18.2–27.2	10.1–12.5 (121–135)	20.9–31.2 (111–130)
Very heavy	41–51	31–41	25–33	29.3–36.6 (151–175)	29.0–36.0	>27.2	12.6–15.0 (136–150)	31.3–41.6 (131–150)
Extremely heavy	51	41	33	>36.6 (>175)	36.0	–	>15.0 (>150)	>41.7 (151–170)

labour-intensive countries, farm mechanization has obvious infrastructure limitations, primarily on the possibilities of labour displacement. Therefore, redesigning simple tools and machinery is necessary to minimize physical demands in the existing farm methods.

Physical Activity and Fitness During Ageing

An essential feature of the ageing process is a fall in the resting energy expenditure, either due to the economy in body movements or physiological lowering of basal metabolic rate. In men, the typical reference value for resting metabolism is about five kJ/min at the age of 25 years; it falls steadily with increasing age, reaching nearly 85% by the age of 65 years. Women have a resting metabolism of about four kJ/min at the age of 25 years. The resting metabolism of women is relatively constant in the first part of their life but declines with age after menopause at approximately the same rate as for men.

A comprehensive survey on the farmworkers indicates that ageing has no significant influence on the work strenuousness and hours of work. There might be a marginal slowing in the rate of work among the elderly. Due to job compulsions, it is hardly possible for the farmworkers to reduce the energy expenditure at work as one becomes older. Even in jobs that involve considerable physical activity, it would be difficult for most people to change their habitual occupations to a less strenuous work when one becomes older. The economic compulsions, unemployment, and competition for a job are the determining factors for the less likeliness of the possibility of changing a job. In dealing with the farming population, particularly in the situations in labour-intensive agricultural sector, the policy strategy towards developing large-scale ancillary rural industries is a positive step for employment generation for unskilled rural populations of different age groups.

Manual Material Handling Tasks

Inherently, the agricultural activities involve a variety of manual material handling (MMH) tasks, such as lifting, lowering, pulling, pushing, and carrying of heavy loads in the form of farm inputs, products or tools, and equipment. Substantial research has been undertaken on MMH tasks, showing potential risks of musculoskeletal strains and disorders, falls, spinal injuries, and so on. The *proposed limits of safe load lifting* are given in Table 4.8.

The fall injury rate increases dramatically when the fall height is more than 2 metres. The impact forces are reduced manifold if the victim falls on soft earth, hay, or sand. In rural areas, loads weighing 50 to 100 kg might be carried several miles daily (Sen and Nag 1975). In some countries, women and children have to fetch water in large quantities from a distance. These arduous tasks need to be minimized

Table 4.8 Safe load-lifting limit (in kg)

Source	Female	Male
Snook (1978)	14	23
Waters et al. (1993)	–	23
ISO 11228 Part I (ISO 2003)	15	25
Maiti and Ray (2004)	15	–
NIOSH (2007)	–	23
Sharma and Singh (2012)	15	–
Hunt (2017)	16	25

to the extent possible. Different methods of water carriage involve carrying on the head, on the hip, on the back, and on the shoulder. These have been associated with a variety of biomechanical effects and spinal disorders (Dufaut 1988). There are well-recognized good practices for MMH tasks. As a thumb rule, the load carried by a worker should not exceed 40% of one's body weight on a plain surface. In marketplaces, the grain bags of 100 kg have been gradually replaced by 30 kg or 50 kg capacity bags, in many countries. Load transportation using a transverse yoke and head load is more efficient than the frontal yoke. Load optimization is an essential requirement to safeguard the health of the workers. Attempts have been made for popularizing well-designed wheelbarrows and other efficient modes of transport in rural settings.

References

Andersen, K. L., Shephard, R. J., Denolin, H., Varnauskas, E., Masironi, R. (1971). *Fundamentals of exercise testing*. Geneva: World Health Organization.

Astrand, I., Astrand, P. O., Hallbäck, I., & Kilbom, A. (1973). Reduction in maximal oxygen uptake with age. *Journal of Applied Physiology, 35*(5), 649–654.

Astrand, P. O., & Ryhming, I. (1954). A nomogram for calculation of aerobic capacity (physical fitness) from pulse rate during submaximal work. *Journal of Applied Physiology, 7*(2), 218–221.

Astrand, P. O., & Rodahl, K. (1986). *Textbook of work physiology: Physiological bases of exercise*. New York: McGraw-Hill Book Company.

Baker, P. T., & Baker, P. T. (Eds.). (1978). *The biology of high-altitude peoples* (Vol. 14). Cambridge University Press.

Banerjee, S., & Sen, R. (1955). Determination of the surface area of the body of Indians. *Journal of Applied Physiology, 7*(6), 585–588.

Bonjer, F. H. (1968). Relationship between working time, physical working capacity and allowable calorie expenditure. In W. Rohmert (Ed.), *Muskelarbeit and Muskeltraining* (pp. 86–99). Stuttgart: Gentner Verlag.

Brouha, L. (1967). *Physiology in industry*. New York: Pergamon Press.

Christensen, E. H. (1953). Physiological evaluation of work in the Nykroppa iron works. In W. F. Floyd & A. T. Welford (Eds.), *Ergonomics society symposium on fatigue* (pp. 93–108). London: Lewis.

CIAE. (2007). Proceedings IV workshop. All India coordinated research project on Ergonomics and Safety in Agriculture. Bhopal, India: Central Institute of Agricultural Engineering.

CIAE. (2012). *Research highlights 2010–2012*. All India coordinated research project on ergonomics and safety in agriculture. Bhopal, India: Central Institute of Agricultural Engineering.
Consolazio, C. F. (1963). *Physiological measurements of metabolic functions in man*. McGraw-Hill Book Company.
Davies, C. T. M., Barnes, C., & Godfrey, S. (1972). Body composition and maximal exercise performance in children. *Human Biology*, 195–214.
Davies, C. T. M., & Van Haaren, J. P. M. (1973). Maximum aerobic power and body composition in healthy East African older male and female subjects. *American Journal of Physical Anthropology, 39*(3), 395–401.
De, A., & Sen, R. N. (1986). Ergonomic evaluation of ploughing process of paddy cultivation in India. *Journal of Human Ergology, 15*(2), 103–112.
Du Bois, D., & Du Bois, E. F. (1916). A formula to estimate the approximate surface area if height and weight be known. *Archives of Internal Medicine, 1916*(17), 863–871.
Dufaut, A. (1988). Women carrying water: how it affects their health. *Waterlines*, 23–25.
Durnin, J. V. G. A., & Passmore, R. (1967). *Energy*. London: Work and Leisure. Heinemann Educational Books Ltd.
Durnin, J. V., & Womersley, J. V. G. A. (1974). Body fat assessed from total body density and its estimation from skinfold thickness: measurements on 481 men and women aged from 16 to 72 years. *British Journal of Nutrition, 32*(1), 77–97.
Edholm, O. G. (1966). The assessment of habitual activity. In E. Evang & K. L. Anderson (Eds.), *Physical activity in health and disease*. Oslo: Universitets forlaget.
Edholm, O. G., Humphrey, S., Lourie, J. A., Tredre, B. E., & Brotherhood, J. (1973). VI. Energy expenditure and climatic exposure of Yemenite and Kurdish Jews in Israel. *Philosophical Transactions of the Royal Society of London. B, Biological Sciences, 266*(876), 127–140.
Edwards, R. H. T., Miller, G. J., Hearn, C. E. D., & Cotes, J. E. (1972). Pulmonary function and exercise responses in relation to body composition and ethnic origin in Trinidadian males. *Proceedings of the Royal Society of London. Series B. Biological Sciences, 181*(1065), 407–420.
Ekblom, B., & Gjessing, E. (1968). Maximal oxygen uptake of the Easter Island population. *Journal of Applied Physiology, 25*(2), 124–129.
Faillie, R. (1925). Measurement of the body surface in men and women. *Archives of Internal Medicine, 35*(5), 626–631.
Gardner, G. W. (1971). Physical fitness of primitive peoples: The Warao Indians of Venezuela. In V. Seliger (Ed.), *Physical fitness*. Prague: Charles University.
Gayathri, G., & Sam, B. (2017). Ergonomic evaluation of rice transplanters for assessing the suitability to women operators. *Agricultural Engineering Today, 41*(4), 58–66.
Ghugare, B. D., Adhaoo, S. H., Gite, L. P., Pandya, A. C., & Patel, S. L. (1991). Ergonomics evaluation of a lever-operated knapsack sprayer. *Applied Ergonomics, 22*(4), 241–250.
Gite, L. P. (1991). Optimum handle height for animal-drawn mould board plough. *Applied Ergonomics, 22*(1), 21–28.
Gite, L. P. (1993). Optimum handle height for an animal drawn blade harrow. *Journal of Human Ergology, 22*(1), 21–27.
Gite, L. P. (2017). Development of Ergonomical Design Guidelines for Agricultural Tools, Equipment and Work Places. Final report of Emeritus Scientist Project, ICAR-Central institute of Agricultural Engineering, Bhopal, India.
Gite, L. P., & Agrawal, N. (2000). Ergonomical comparison of local and improved sickles for wheat harvesting by women workers. *Agricultural Engineering Today, 24*(3), 7–12.
Gite, L. P., Bhardwaj, K. C., & Bohra, C. P. (1993). Ergonomics evaluation of wheeled type manual weeders. In *XXVIII Annual Convention of ISAE held at CIAE, Bhopal* (pp. 2–4).
Gite, L. P. & Ganeshsan, S. (2014). Energy from Animate Sources in Indian Agriculture: Potential, availability and Projections, Paper presented in the National Consultation on Energy in Agriculture organized by National Academy of Agricultural Sciences at New Delhi on September 15–16.
Gite, L. P., & Singh, G. (1997). *Ergonomics in agricultural and allied activities in India*. Bhopal: Central Institute of Agricultural Engineering.

Harris, J. A., & Benedict, F. G. (1918). A Biometric Study of Human Basal Metabolism. *Proceedings of the National Academy of Sciences of the United States of America, 4*, 370–373.

Harrison, M. H., Brown, G. A., & Belyavin, A. J. (1982). The 'Oxylog': an evaluation. *Ergonomics, 25*(9), 809–820.

Hildebrand, F. B. (1956). *Introduction to numerical analysis*. Bombay: Tata McGraw Hill.

Hunt, S. (2017). *A guide to manual handling and lifting techniques*. U.K.: Work Safety Advice Cheshire.

ICRP. (1975). *Reference man: anatomical, physiological and metabolic characteristics*. International Commision of Radiological Protection, ICRP 23, Pergamon Press, Oxford

Inbar, O., Oren, A., Scheinowitz, M., Rotstein, A., Dlin, R., & Casaburi, R. (1994). Normal cardiopulmonary responses during incremental exercise in 20-to 70-yr-old men. *Medicine and science in sports and exercise, 26*, 538–546.

ISO. (2003). International standard ISO 11228 (Part 1). Ergonomics - Manual handling - Part 1: Lifting and carrying. International Organization for Standardization, Geneva.

Kaul, R. N., & Splinter, W. E. (1964). *The man- machine concept as a guideline to the introduction and development of small implements in India*. Paper presented in the Annual General Meeting of ASAE held at Colorado, USA.

Keys, A., Fidanza, F., Karvonen, M. J., Kimura, N., & Taylor, H. L. (2014). Indices of relative weight and obesity. *International Journal of Epidemiology, 43*(3), 655–665.

Maiti, R., & Ray, G. G. (2004). Manual lifting load limit equation for adult Indian women workers based on physiological criteria. *Ergonomics, 47*(1), 59–74.

Malhotra, M. S., Ramaswamy, S. S., Dua, G. L., & Sengupta, J. (1966). Physical work capacity as influenced by age. *Ergonomics, 9*(4), 305–316.

Margaria, R., Aghemo, P., & Rovelli, E. (1965). Indirect determination of maximal O_2 consumption in man. *Journal of Applied Physiology, 20*(5), 1070–1073.

Maritz, J. S., Morrison, J. F., Peter, J., Strydom, N. B., & Wyndham, C. H. (1961). A practical method of estimating an individual's maximal oxygen intake. *Ergonomics, 4*(2), 97–122.

McArdle, W. D., Katch, F. I., Pechar, G. S., Jacobson, L., & Ruck, S. (1972). Reliability and interrelationships between maximal oxygen intake, physical work capacity and step-test scores in college women. *Medicine and Science in Sports, 4*(4), 182–186.

Miller, G. J., Cotes, J. E., Hall, A. M., Salvosa, C. B., & Ashworth, A. (1972). Lung function and exercise performance of healthy Caribbean men and women of African ethnic origin. *Quarterly Journal of Experimental Physiology and Cognate Medical Sciences: Translation and Integration, 57*(3), 325–341.

Nag, P. K. (1981). Predicting maximal oxygen uptake of workers engaged in agricultural tasks. *Journal of Human Ergology, 10*(1), 25–33.

Nag, P. K. (1983). Ergonomics: A new perspective of work organisation in traditional agriculture. In: *DSc thesis*. University of Calcutta.

Nag, P. K. (1984). Circulo-respiratory responses to different muscular exercises. *European Journal of Applied Physiology and Occupational Physiology, 52*(4), 393–399.

Nag, P. K. (1987). Maximal oxygen uptake of agricultural men and women in India. *American Journal of Physical Anthropology, 74*(2), 149–153.

Nag, P. K., & Chatterjee, S. K. (1981). Physiological reactions of female workers in Indian agricultural work. *Human Factors, 23*(5), 607–614.

Nag, A., & Chintharia, S. (1985). Physical activities of women in farming and non-farming seasons. *Journal of Human Ergology, 14*(2), 65–70.

Nag, P. K., & Datt, P. (1979). Effectiveness of some simple agricultural weeders with reference to physiological responses. *Journal of Human Ergology, 8*(1), 13–21.

Nag, P. K., & Dutt, P. (1980). Circulo-respiratory efficiency in some agricultural work. *Applied Ergonomics, 11*(2), 81–84.

Nag, P. K., Goswami, A., Ashtekar, S. P., & Pradhan, C. K. (1988). Ergonomics in sickle operation. *Applied Ergonomics, 19*(3), 233–239.

Nag, P. K., & Pradhan, C. K. (1992). Ergonomics in the hoeing operation. *International Journal of Industrial Ergonomics, 10*(4), 341–350.

Nag, P. K., Sebastian, N. C., Ramanathan, N. L., & Chatterjee, S. K. (1978). Maximal oxygen uptake of Indian agricultural workers with reference to age and sex. *Journal of Madurai Kamraj University (India), 7,* 69–74.

Nag, P. K., Sebastian, N. C., & Mavlankar, M. G. (1980a). Occupational workload of Indian agricultural workers. *Ergonomics, 23*(2), 91–102.

Nag, P. K., Sebastian, N. C., & Malvankar, M. G. (1980b). Effective heat load on agricultural workers during summer season. *Indian Journal of Medical Research, 72,* 408–415.

Nag, P. K., Pradhan, C. K., Goswami, A., & Ashtekar, S. P. (1986). Cardio-respiratory and muscle responses in static, dynamic and combined work. *Journal of Human Ergology, 15*(1), 73–77.

NIOSH. (2007). Ergonomics guidelines for manual material handling. Publication No. 2007-131. National Institute for Occupational Safety and Health, Cincinnati USA.

Ojha, P., & Kwatra, S. (2014). Analysis of different paddy transplanting methods in northern India: Ergo-economical study. *Journal of Applied and Natural Science, 6*(2), 654–658.

Pawar, J. G., & Pathak, B. S. (1980). Investigation of human energy requirements for power tiller operations. *Journal of Agricultural Engineering.*

Petrofsky, J. S., & Lind, A. R. (1978). Metabolic, cardiovascular, and respiratory factors in the development of fatigue in lifting tasks. *Journal of Applied Physiology, 45*(1), 64–68.

Pradhan, C. K., Goswami, A., Ghosh, S. K., & Nag, P. K. (1986). Evaluation of working with spade in agriculture. *Indian Journal of Medical Research, 84,* 424–429.

Ramana Murthy, P. S. U., & Belavady, B. (1966). Energy expenditure and requirement of in agricultural labourers. *Indian Journal of Medical Research, 54,* 977–979.

Ramanathan, N. L., Dutta, S. R., Roy, B. N., Chatterjee, A., & Mullick, L. N. (1967). Energy cost of different muscular exercise tests performed by Indian subjects. *Indian Journal Occupational Health, 10,* 253–261.

Ramanathan, N. L., & Nag, P. K. (1982). Energy costs of human labour in India. *ASCI Journal of Management, 11*(2), 1981–1982.

Roozbazar, A., Bosker, G. W., & Richerson, M. E. (1979). A theoretical model to estimate some ergonomic parameters from age, height and weight. *Ergonomics, 22*(1), 43–58.

Saha, P. N. (1975). Aerobic capacity of dock workers in Bombay. *American Industrial Hygiene Association Journal, 36*(4), 311–317.

Saha, P. N. (1978). Aerobic capacity of steel workers in India. *Ergonomics, 21*(12), 1021–1025.

Saha, P. N., Datta, S. R., Banerjee, P. K., & Narayane, G. G. (1979). An acceptable workload for Indian workers. *Ergonomics, 22*(9), 1059–1071.

Samueloff, S., Davies, C. T. M., & Shvartz, E. (1973). VII. The physical working capacity of Yemenite and Kurdish Jews in Israel. *Philosophical Transactions of the Royal Society of London. B, Biological Sciences, 266*(876), 141–147.

Seliger, V., Trefny, Z., Bartunkova, S., & Pauer, M. (1974). The habitual activity and physical fitness of 12 year old boys. *Acta Paediatrica Belgica, 28,* 54.

Sen, R. N., & Nag, P. K. (1975). Work organization of heavy load handling in India. *Journal of Human Ergology, 4*(2), 103–113.

Sen, R. N., & Sarkar, D. (1969). Maximal physical work capacity of Indian industrial workers in relation to age and heaviness of jobs. In *Proceedings of International Conference Human Factors Society, Philadelphia* (p. 9).

Sharma, R., & Singh, R. (2012). Determination of safe carrying load limit for women carrying water. *Journal of Ergonomics, 2*(103), 2.

Shephard, R. J., Allen, C., Benade, A. J. S., Davies, C. T. M., Di Prampero, P. E., Hedman, R., et al. (1968a). The maximum oxygen intake: An international reference standard of cardio-respiratory fitness. *Bulletin of the World Health Organization, 38*(5), 757.

Shephard, R. J., Allen, C., Benade, A. J. S., Davies, C. T. M., Di Prampero, P. E., Hedman, R., et al. (1968b). Standardization of submaximal exercise tests. *Bulletin of the World Health Organization, 38*(5), 765.

Singh, S. P., & Gite, L. P. (2007). Ergonomical evaluation of a hand operated paddy winnower by women workers. *Journal of Agricultural Engineering, 44*(4), 67–71.

Singh, S. P., Gite, L. P., & Agrawal, N. (2004). Ergonomical evaluation of manually-operated fertilizer broadcaster with farm women. *Journal of Agricultural Engineering, 41*(3), 22–25.

Singh, S. P., Gite, L. P., & Agarwal, N. (2006). Improved farm tools and equipment for women workers for increased productivity and reduced drudgery. *Gender, Technology and Development, 10*(2), 229–244.

Singh, S. P., Gite, L. P., Majumder, J., & Agarwal, N. (2008). Aerobic capacity of Indian farm women using sub-maximal exercise technique on tread mill. *Agricultural Engineering International: CIGR Journal*. Manuscript MES 08 001. Vol. X.

Singh, H., & Kaul, R. N. (1972). Human energy requirements of selected farm operations. *Journal of Agricultural Engineering., 9,* 44–52.

Siri, W. E. (1956). The gross composition of the body. In *Advances in biological and medical physics* (Vol. 4, pp. 239–280). Elsevier.

Snook, S. H. (1978). The Ergonomics Society the Society's Lecture 1978. THE design of manual handling tasks. *Ergonomics, 21*(12), 963–985.

Tewari, V. K., Datta, R. K., & Murthy, A. S. R. (1991). Evaluation of three manually operated weeding devices. *Applied Ergonomics, 22*(2), 111–116.

Tiwari, P. S., & Gite, L. P. (2002). Physiological responses during operation of a rotary power tiller. *Biosystems Engineering, 82*(2), 161–168.

Tiwari, P. S., & Gite, L. P. (2010). *Development of an ergonomically designed dynapod and its application for maize shelling operation*. Bhopal, India: Central Institute of Agricultural Engineering.

TNAU. (2002). Progress Report 2000–2002. *All India coordinated research project on human engineering and safety in agriculture*. Coimbatore: Tamil Nadu Agricultural University.

Varghese, M. A., Saha, P. N., & Atreya, N. (1994). A rapid appraisal of occupational workload from a modified scale of perceived exertion. *Ergonomics, 37*(3), 485–491.

Waters, T. R., Putz-Anderson, V., Garg, A., & Fine, L. J. (1993). Revised NIOSH equation for the design and evaluation of manual lifting tasks. *Ergonomics, 36*(7), 749–776.

Weir, J. D. V. (1949). New methods for calculating metabolic rate with special reference to protein metabolism. *The Journal of Physiology, 109*(1–2), 1–9.

Wyndham, C. H. (1973). The Work capacity of rural and urban Bantu in South Africa. *South African Medical Journal, 47*(29), 1239–1244.

Wyndham, C. H., & Sluis-Cremer, G. (1969). The capacity for physical work of white miners in South Africa. III. The maximum oxygen uptakes of normal miners and miners with cardiorespiratory diseases. *South African Medical Journal, 43*(32), 3–8.

Yadav, B. G., & Gite, L. P. (1987). *Manual weeder*. Central Institute of Agricultural Engineering, Bhopal, India: Final project report.

Chapter 5
Work Planning and Scheduling in Farming

Introduction

Worldwide, farming, to a great extent, is seasonal. The crop-specific farm operations are time-bound. Since individual families primarily maintain the small farm holdings, the work time scheduling for cropping is voluntary by the local farmers. In the industrial context, work time issues fall under the legislative and administrative policy decisions that are executed through the negotiated agreements between employer-employee unions. The primary objective of work planning and scheduling is to help workers to be productive without undue fatigue or any health issues. In other words, the work schedule follows the objective to standardize the minimum number of annual, weekly, or daily hours of work that should not regularly be exceeded (Nag 1991a). These conventional approaches of *work schedule* are that applicable in the organized industrial environment, which are out of place in farming due to its many necessary farming/crop growth requirements. This contribution elucidates the contemporary understanding of work schedules, working age, shift working system, work–rest allowance guidelines in forming strategies in reducing human fatigue in work.

The Rationale of Work Scheduling

Traditionally, work time practices in the industry have been stereotypic, reflecting social customs. For the protection of workers against excessive hours of work, different forms of work patterns have emerged, and policy guidelines formulated (ILO 2018; Cheong et al. 2013). The consideration includes (a) the legal and contractual obligations, (b) workers' social customs and preferences, (c) operational needs of the work processes, and (d) ergonomics concerns and human needs. These interactive factors allow making decisions on the work schedule, such as (i) reduction in

P. K. Nag and L. P. Gite, *Human-Centered Agriculture*, Design Science and Innovation,
https://doi.org/10.1007/978-981-15-7269-2_5

regular working time, (ii) rationalization of working age, (iii) work time and shift working system, and (iv) control of work overtime.

Reduction in Working Time

Reduction in working time is called to compensate for the impoverishment in the content of jobs. The workers must achieve better physical and mental relaxation, inside and outside of their occupational sphere. The ILO convention 47 on forty-hour week convention, (ILO 1935) Article 1(a) emphasizes the principle of reduction in working time in a manner that the standard of living is not reduced in consequence. The Reduction of Hours of Work Recommendation, as per ILO convention 116 (ILO 1962) and the Employment Policy Recommendation, ILO convention 122 (ILO 1964) re-emphasized the necessity of progressively reducing the hours of work, without a reduction in the wages.

The length of work and the pattern of working time vary with the national characteristics, demographic trends, energy availability, structural issues of the labour market, and technology deployment. Reduction in maximum or actual working time per day, per week, or year has several implications in the optimization of work content in occupational pursuits and extension of leisure time. It provides a stimulus for meaningful utilization of free time, such as sports, recreation, education, and cultural activities. Undoubtedly, there has been a consistent downward trend in weekly hours of work in the market economy countries, ranging from 35 to 40 h per week. Lee et al. (2007) examined the evolution of statutory limits on working hours in different countries, indicating a broad convergence towards a 40-hour limit on weekly working hours. The working hours tend to be lower for women than men. The ILO convention 184 (ILO 2001) specifies that the hours of work, night work, and rest periods for workers in agriculture shall be following national laws and regulations or collective agreements.

Demographic Trend

The average age of the working population in the industrialized countries is increasing by approximately one-half for each passing year during the 1980s and the 1990s (Rabbit 1991). The situation is very similar in industrially developing countries. Subramanian (2015) analyzed the data of the Indian agricultural workforce for the period across the four census periods from 1971 to 2001. Observation is beckoning that the male youth segments were moving out of agriculture, and more women remained in conventional activities, making *feminization of agriculture.*

It may be worth noting here, the age distribution of the world population and also in the populous nations, like China and India (Fig. 5.1). China has a total population of 1428 million, and the youth component (15 to 24 years) is relatively small (12.1%),

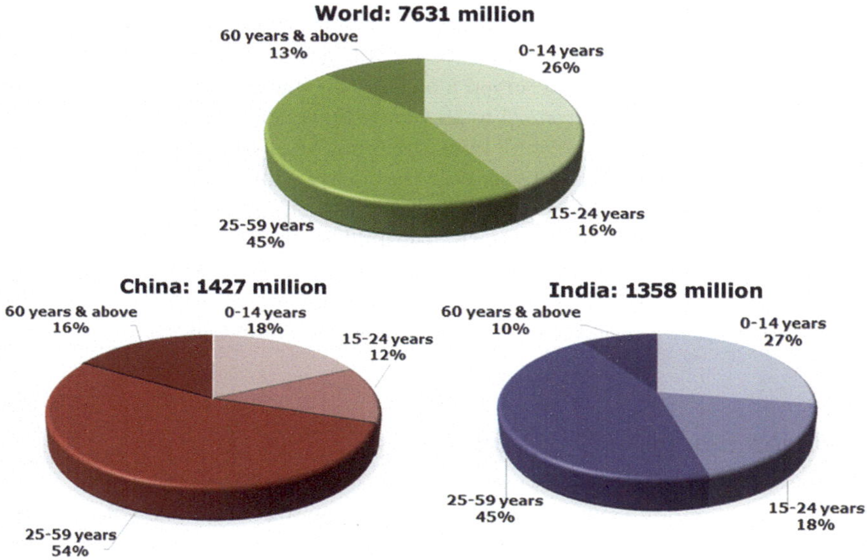

Fig. 5.1 Age-wise population distribution in 2018 (World, China, and India) (*Source* www.populationpyramid.net, 2019)

and about 16.2% of people are in the age group of 60 and above. The life expectancy in 2018 was 76.4 years. India, the 2nd largest population of the world, is growing older and faster. In 2018, the youth component of the population (15–24 years) was 18.2%. During the past two decades, the increase in the aged population (+60 years) was over 33%, when the rest of the people went up by only 18%. The life expectancy (as per WHO) in 2018 was 68.8 years, while it was 66 years in 2013, and 58 years in 1990 (www.worldlifeexpectancy.com) Out of total India's population of 1358 million (Populationpyramid.net), nearly 130 million people (9.5%) were above 60 years of age in 2018, as against 33 million (6%) in 1971. The question that looms frighteningly large is the *old elderly*. The country will face the increasing problem of integrating a massive number of older people into society, not merely as passive recipients of care but as active contributors to the economy.

In this context, one may distinguish the concept of disability from other related aspects, such as impairment and handicap. An impairment is primarily a disturbance of a healthy structure or function, which may lead to a disability, i.e. the inability to perform a task in the way that the majority of healthy people can perform. The disability or impairment may result in some psychosocial disadvantage (handicap) of the concerned, and the whole process may be referred to as a disablement. Undoubtedly, ageing is accompanied by increased risks of health complications, such as diabetes, blindness due to glaucoma, strokes, and loss of joint mobility due to arthritic conditions. As age advances, there is a steady decline in the performance of most measures of sensory, intellectual, and physical functions. Though the incidence of age-related pathologies is evident, most older people tend to complete their lives without suffering disability in strict medical and legal senses that entitle to exceptional assistance. Nag (1987) observed a shallow decline in the physical working capacity for the age ranging from 25 to 59 years, showing a reduction in $VO_{2\,max}$ at the rate of 0.41 ml/kg.min per year for Indian men and 0.32 ml/kg.min per year in case of Indian women for the 30- to 49-year old. The elderly witness a substantial decline in cognitive performance, such as simple memory tasks, information processing rate, scores on verbal and performance IQ tests. The mild hearing impairment in old age is well documented. Despite such changes, people growing older learn to compensate for the losses of capacity and skill, by finding alternative ways to deploy other skills and diversity of performance dimensions that they may still retain.

Rationalization of Working Age

There are two contrary views of fixing the age of retirement, i.e. one voice is loud that lowering of retirement age would result in the reduction of working time to the entire active life, and that would, in turn, contribute to a decrease in unemployment. The other view favours the old age as not necessarily a condition of disability for all practical purposes. Retirement age must not be seen as immutably fixed by laws of nature (Rabbit 1991). The concept of a fixed period for compulsory retirement leads to a noticeable loss of elderly qualities and skills. There is no uniformity in the legislation on the retirement age. The age of retirement on a full pension is still 65+ years to the vast majority of countries of both developed and the developing world.

Systems of progressive retirement combined with reduced working time, between 17 and 35 h per week, are in operation in some countries, such as Sweden, France, and the Netherlands. The demographic trend strengthens the case of adopting flexible arrangements in retirement, whereby human skills and capacities are effectively utilized. A new social approach is called for to adapt our policies to adjust occupational trajectories of the elderly, making them remain productive, until very late in their lives. However, the concept of retirement age in agriculture may not hold good in many developed countries as many of the farms are family farms. On the legislative count, agriculture is generally not considered as an industry.

Work Time Systems

The work time and schedules involve the length of the workday, the days of the week worked, the starting time, and the regularity of the plans. There is no uniform acceptance of work time systems, due to the nature of work practices in an occupation. The same organization with the horizontal differentiation of the enterprise structure may have variations in work patterns that operate simultaneously (Kogi 1991).

Flexible daily hours—While the regular base schedule is eight hours/day, five days/week, the *flexitime* varies from one week to one month, and carryover hours may vary up to 15 h. The minimum and maximum daily working hours, under the arrangement of *flexitime*, may be as short as 5 to 6 h, or as long as 10 to 12 h, thereby a worker may choose the working hours based on the seasonal requirements, as in farming. Flexibility in work time scheduling is possible by multiple combinations of options, such as full-time, part-time, early morning shift, night, and weekend work.

Compressed workweek/extended workday—Market economies adopt a variety of compressed workweek schedules, by squeezing the traditional 40 h workweek into working days less than 5 (e.g. four days of 9 h or more, three days of 12 h each, or a week of 4 working days alternating with a week of 5 working days). This would allow workers to reduce the number of consecutive workdays, and gain more consecutive days off. One of the available schedules is 4–40 (i.e. four straight days of work with 10-hour workday, followed by three straight days off). Such a system may not be suitable in agriculture as field operations associating to crops are done on the demand of the job, irrespective of whether it is a Sunday, holiday, or an off-day.

Overtime schedules—Workdays may extend from 12 to 16 h, with weekly working hours exceeding over 50 h. Considerable differences exist in different countries on the provisions of maximum hours of work. Since farming activities are grossly weather dependent and seasonal, the apparent over time is not uncommon in various activities, such as sowing, weeding, and harvesting. In the aberrant weather season, the farmers may not have work for days. It is recognized that a person's day is divided into blocks of sleep time, work time, personal time, and recreational time. Therefore, significant concerns regarding overtime are the amount and the schedule of work time. The overtime may be scheduled either before or after the regular work shift; however, the early work would mean sleep deprivation, while the later work affects recreational time. Working in the afternoon would mean using up of one's recreational time. Mostly the optimal work schedule would do away with all types of overtime to obviate excessive workload and its impact on health, comfort, and safety.

Shift Work Schedules—In the industrial scenario, the shift systems refer to the number of teams involved with or without weekend work, night work, and the regularity of the shift systems. However, the typical shift work schedules do not apply to tropical open-field farming.

Work System Analysis: Human Strain and Drudgery

In analyzing a working system, one look into the functions, such as effecting, controlling, and monitoring that may be carried out by a man or by technical equipment. In a manual work system, man fulfils all three above functions, but in a mechanized work system, man is replaced by machines to effect in the actual changes in a work. In a working arrangement with automatic performance, in addition to the functions of effecting, man is also replaced in monitoring the functional capacity of the system. Thus, the human interaction with the technical equipment leads to different kinds

Table 5.1 Type of work and affecting organ systems (Rohmert and Mainzer 1986)

Type of work	Work content	Organ system	Work type	Task example
Physical	Producing force	Muscles and cardio-respiratory systems	Muscular work	MMH tasks
	Motor and sensory coordination	Muscles and sensory organs	Sensorimotor work	Assembling task
Non-Physical	Converting information to reaction	Muscles and sensory organs	Non-muscular	Mainly controlling
	Transform input to output information	Sensory and mental ability	Mental work	Programming, process control
	Producing information	Mental ability	Mental work	Decision-making, designing

of human performance in the stated three types of man–machine work systems. Accordingly, five primary work contents in the physical and non-physical kind of work (Table 5.1) are ordered by using human capabilities and attributes (Rohmert and Mainzer 1986). Different physical and non-physical work is related both to the content of work and to its alternation. Besides, *physical* (e.g. noise, illumination, climate, radiation, vibration), *chemical* (e.g. dust, fumes, toxic exposures) as well as *social* environmental stressors (e.g. work team, plant hierarchy) have related effect on the human beings. Since the type of strain and many of the stressors are often alternating with time, the optimal work schedule and work–rest guidelines demand joint optimization of work content and work time patterns, as emphasized earlier.

Mainly, farming activities are drudgery prone, since the farmworkers in the labour-intensive countries apply a substantial amount of muscle power in carrying out farm operations. With low wages, they work hard to earn their living. Improvisation in the design of farm tools, machinery, workplaces, and work practices can help to reduce the drudgery in terms of reduction in cardiac cost and energy demand, and also help to increase the work output and enhance comfort and safety of workers. Table 5.2 presents the commonly used physiological methods of measuring strain about the type of work. Using telemetric devices, some of these methods (e.g. heart rates) offer an advantage in bringing practicality in agriculture. The heart rate is a sensitive and excellent discriminating measure in evaluating strains in muscular work. The *heart rate variability* is a useful indicator of human strain, particularly in non-physical work. Therefore, the heart rate is often taken as a measure in setting rest allowances in both physical and non-physical work. Brouha (1967) suggested that when the first minute recovery heart rate is maintained at about 110 beats/min or below, and deceleration between the first and third minute recovery heart rate is at least ten beats/min, no increasing cardiac strain occurs. Other indicators such as electrical potentials of the brain (EEG), variability in working capacity, reaction

Table 5.2 Physiological methods for assessment of strain and drudgery

Parameter	Type	Remark/observation	Source
Heart rate (beats/min), including heart rate variability	Objective	Index of cumulative stress	Sanchez and Monod (1979), Nag et al. (1980), Nag and Chatterjee (1981b), Legg and Mahanty (1985), Grandjean (1988)
O_2 consumption rate (l/min)	Objective	A reliable measure of physical workload. Commonly used	Srivastava et al. (1962), Andrews (1967), Saha et al. (1979)
Pulmonary Ventilation (l/min)	Objective	Used to estimate O_2 consumption rate. Easy to measure.	Sharkey et al. (1966), Legg and Mahanty (1985)
Respiratory rate (breathing/min)	Objective	Easy to measure. Weak correlation with energy expenditure. Unreliable	Sengupta et al. (1979)
Energy expenditure rate (kJ/min)	Objective	Calculated from O_2 consumption. Commonly used	Saha et al. (1979), Sengupta et al. (1979), Morrissey and Liou (1984), De and Sen (1986)
Body temperature (°C)	Objective	Low sensitivity to workload. Response time is relatively longer. Useful in thermal stress situations	Shoenfeld et al. (1977), Legg (1985)
Blood pressure (mm Hg)	Objective	Used for estimating static load. Difficult to use during work	Bedale (1924), Lind and McNicol (1968), Tuure (1992), Singal (1994)
Muscle activity (electromyography)	Objective	Difficult to use in an actual work situation. Expensive instrumentation	Nag and Chatterjee (1981a), Jager et al. (1984), Tanti and Masuda (1985), Nag et al. (1986a, b)
Intra-abdominal pressure (mm Hg)	Objective	Sophisticated instrumentation required. Difficult to use in an actual work situation	Mairiaux et al. (1984), Legg (1985)
Blood lactic acid and blood glucose	Objective	Difficult to use in an actual work situation	Shoenfeld et al. (1977)
Urine analysis	Objective	Useful for long-term effects and in limited situations	Dukes Dobos et al. (1967)
Eye movements and critical flicker fusion (Hz)	Objective	Rarely used because of high variability	Volley et al. (1979)

(continued)

Table 5.2 (continued)

Parameter	Type	Remark/observation	Source
Muscle strength and endurance	Objective	High variability	Nag PK (1991b), Hasalkar et al. (2007)
Borg's Rated Perceived Exertion Scale, 6-20 (RPE scale)	Subjective	Useful for assessing overall discomfort in a dynamic work situation	Ulmer et al. (1977), Legg and Myles (1981), Pepermans and Corlett (1983), Legg and Mahanty (1985), Hanger and Hagberg (1989)
Visual analogue discomfort scale, 0-10; Overall Discomfort Rating	Subjective	Useful for assessing overall discomfort. Simpler for workers	Legg and Myles (1981); Pepermans and Corlett (1983); Legg and Mahanty (1985); Gite (1996)
Corlett and Bishop scale—Overall discomfort 0–7 scale	Subjective	Useful for assessing postural/muscular discomfort.	Corlett and Bishop (1976); Yu and Kesarling (1989); Ghugare et al. (1991); Gite (1992)
Corlett and Bishop score for body part discomfort	Subjective	Useful for detailed analysis for locating specific problems	Corlett and Bishop (1976); Yu and Kesarling (1989); Ghugare et al. (1991); Gite (1992)

time, and recurrence of faults and accidents in workplaces have limited utility in human strain assessment.

Optimal Work and Rest Cycle

Proper guidelines on the work–rest period are useful for equitable distribution of workload over the whole work shift, and also in reducing fatigue and work hazards (Nag 1991a). While deciding the optimal work–rest cycle, one has to consider the type of work, climatic stress involved, chemical exposures, and physiological responses of workers, fatigue, and recovery aspects in various activities. The allied knowledge on the physiology of work and rest helps in devising work time schedules and optimization of rest allowances (RA). Usually, the percentage ratio between the length of an uninterrupted working period and the length of the following resting period is called fatigue allowance or rest allowance (RA) (Muller 1953). The RA may fall into the categories of either fixed or variable allowances, wherein the fixed RA takes account of the personal needs and underlying fatigue, and the variable RA supplements the fixed allowances for additional demands like excessive fatigue and environmental stresses. The allowances applied to time and motion analysis, and other work measurement systems are based on either the fixed percentage multipliers, which is often taken as 8% to all jobs regardless of the work or the work environment, or the fatigue factor set by environment and work factors. To compensate for the preceding fatigue, one may distinguish work interruptions in terms of

- Rest-pause, defined as tea or meal breaks, usually specified through a negotiated agreement between employers and employee unions;
- Organized break, the extensions of the provisions of work agreement between employers and employees (e.g. short breaks of 5 min each hour after 55 min of work);
- Unoccupied time, e.g. the period during machine or process-controlled time, and other idle times when a worker for some reason could not attend to work, interference time, and machine downtimes.
- Concealed break, where the worker deliberately or subconsciously prolongs ancillary times, which may or may not be necessary; this type of break seems to be the worst of all work breaks.

Static Work

The work carried out by a person may be static muscular work, dynamic muscular action, or a combination of both (Nag et al. 1986a). Task analysis helps in assessing the strain on workers. As explained earlier, in static (isometric) muscular contractions, the length of the muscle is unaltered with no movements of the joints. Each working period consists of either sustained isometric contraction or short-cycle intermittent isometric contraction. These are often expressed as a fraction of the duty cycle, i.e. the duration of static contraction in each period as part of the entire cycle. The static work hinders local blood flow, which rapidly leads to fatigue. The maximum tension produced by a muscle group and the length of time the tension is maintained (endurance) depend on the functional properties of the muscles (Ghosh and Nag 1986; Nag 1991b).

There is a certain generalization that no reduction in maximum strength occurs when the static muscular force is limited to 15% of maximum voluntary contraction (MVC) (*Endurance Limit*). The concept of *indefinite* endurance time at a static load below 15% MVC is out of place, and the strength and endurance vary greatly with the postural positions, suggesting the possible influence of unidentified dynamic components of the exerted postural force in specific cases (Nag 1982; Rohmert and Mainzer 1986). In a study on isometric quadriceps contractions, Sjogaard et al. (1988) observed that continuous contraction might be performed at a tension below 10% of MVC to maintain the muscle blood flow at a sufficient level. When the static work sustains longer with greater muscle tension exceeding 10–15% of MVC, the greater is the reduction in the available maximum strength. Compensating the decrease in maximal strength accounts for the degree of fatigue and requires the amount of recovery (i.e. regeneration of decreased maximal strength). Rohmert (1973a) expressed RA based on unlimited endurance and the heart rate remaining within certain limits. That is

$$RA = 18(t/T\text{max})^{1.4} * (f/F\text{max} - 0.15)^{0.5} * 100\%$$

for (f/F_{max}) is greater than 0.5. RA is expressed as percentage of t (holding time, min), T_{max}—(maximum holding time, min), f—(actual force exerted, kg), and F_{max}—(maximum force, kg). This expression was designed for static contractions but may also be used in intermittent contractions considering that each intermittent cycle is a separate static contraction. However, in repeated bouts of static work, a person is unlikely to return fully to the maximum starting strength in the subsequent work bouts. Therefore, the fraction f/F_{max} is likely to be altered. Also, if the recovery period is not enough, there is cumulative residual fatigue when the work is to be restarted. Further, the relationship between time and load is dependent on the type of muscle loading, muscle groups, and work posture (Deeb et al. 1992), relating different modes of isometric muscular work. Milner et al. (1986) developed a model that requires information about the length of the holding time and rest period to estimate RA, as below:

$$RA = (MHT - HT) + HT * e^{(-0.164\, HT/P)}$$

where MHT is maximal holding time, HT is observed holding time, and P is observed pause in each cycle. In this case, there is no factor of contraction intensity, apart from its implied component in MHT. Sheeley et al. (1991) proposed an algorithm for the prediction of time to fatigue in intermittent contractions, as

$$T = 1.99 + e^{(-17.53 * \%MVC)} + e^{(-0.88\, S)}$$

where T is time of fatigue (min) and S is rate of wrist flexion/extension per min.

Dynamic Work

In dynamic muscular work (such as cycling, cranking), the natural alternation of muscle contraction and relaxation facilitates blood flow, the supply of O_2, and removal of acidic metabolites from the muscles. The physiological responses in dynamic work are attributed to the differences in muscle mass used to perform the work, like in arm or leg work, and also the presence of static components in dynamic work (Nag et al. 1986b). Therefore, when the intensity of muscular contraction is too high and is above the endurance limit, and the relation between time for relaxation to contraction is too small, the additional rest pauses are required to postpone fatigue. Rohmert (1973b) expressed RA as follows:

$$RA = 1.9 * (t)^{0.145} * ((N_{eff}/N_{el}) - 1)^{1.4} * 100\%$$

where *t is* working time, N_{eff} is effective working capacity, and N_{el} (endurance limit). The energy delivery mechanism is better in dynamic work, and the extent of RA depends on the effective capacity due to the larger exponent than with the work time. This is in contrast to the equation for static work, where the holding time plays a decisive role in the design of work/rest schedules.

The inherent problem in occupational work, like in farming activities, is the physiological impurity of muscular action, referred to as *combined work.* In several farming operations, the local muscular fatigue due to repetitive arm or leg movements and awkward work posture is evident. The presence of either static or dynamic work in physical activity is understood in principle, but quantification in measurable terms is difficult (Sanchez and Monod 1979; Nag 1982). The difficulty lies in the method of identifying and classifying such combined work, e.g. the relative partition between arm and leg work, and muscle contraction behaviour in static and dynamic work. Analyzing combinations of different components of static and dynamic work, Nag et al. (1986a, b) observed that the combination of static and dynamic action caused the facilitation of the activity of some muscles based on the relative load of each component. The EMG changes support that static and dynamic work components are not additive in effect and that only the predominant part was primarily reflected in the responses.

Shreds of evidence suggest that sustained postural orientation in a given work results in local muscle fatigue, biomechanical load on the spinal structure, and increased risk of pathological reactions, such as varicose veins, thrombosis, and pulmonary embolism. In the case of spinal structure, the intervertebral disc compression is a visible mechanism for transporting fluids, nutrients, and wastes to and from the discs. However, when the compressive load becomes more significant than the interstitial osmotic pressure of the discs, these may undergo shrinkage due to loss of fluid through the disc wall, and the electrolyte concentration may increase inside the disc. Helander and Quance (1990) suggested that rest breaks when a person is either walking or standing, bring a temporary reduction in the compressive forces which perpetuate the pumping process; however, it does not eliminate spinal disc height shrinkage. Whereas, a productive work/rest schedule may alleviate problems arising from work stress and awkward work postures, the formulation of a work–rest sequence requires a considerable understanding of the interrelationship between physiological and biomechanical factors for a specific work pattern.

Environmental Stresses

Apart from the influence of nature of work on work time patterns and work–rest scheduling, attention is needed to the potential danger of the length of ones' exposure to physical environmental factors (e.g. climate, noise, and vibration) and also hazards associated to toxic chemicals (e.g. application of pesticides in the crops). These exposures for prolonged hours of work may have an adverse influence on health and safety. A reduction in working time that reduces exposure to risks below the statutory limit would increase the safety margin in practice. In the case of open-field farming work in a hot environment, there may be cumulative bodily strains due to combined metabolic and environmental heat load. Apart from the productivity loss, the exposed farmworkers might manifest symptoms of heat susceptibility and injury (Sen and Nag 2019). Hall (1968) used the P4SR (predicted four-hour sweat rate) in obtaining RA (% of shift).

$$\mathrm{RA} = [(3.0 - \mathrm{P4SR})/(\mathrm{B4SR} - \mathrm{P4SR})] * 100$$

This equation has a limitation when P4SR exceeds 3 l in combined stress of work and heat. From Wyndham's (1970) data on the productivity loss with an increase in ET (Effective temperature) among African miners, Axelson (1974) proposed a formula for predicting thermal reduction of work output.

$$R(t)\% = \left[k(t - t_o)^2\right]/(2t_o - t)$$

where $R(t)\%$ is the percentage reduction of work capacity at temperature (t) of ET and t_o of 24 °C ET and constant 'k' is equal to (9.8–1.6 V); V is air velocity in m/s.

In the case of chemical exposures during the workday, Brief and Scala (1986) proposed a simple linear equation to modify the existing threshold limit value (TLV) about the base schedule of 8 h/day, five days/week, as follows:

$$\text{TLV Reduction Factor} = (8/\text{hours worked}) * (\text{hours off work}/16)$$

However, the TLV adjustment for specific hazardous chemicals requires knowledge of metabolism and kinetics. Regarding the biological half-life of the chemical compounds exposed (including uptake and excretion kinetics), Hickey (1980) method yielded ~0.74 as the TLV adjustment factors for 12-hour shifts per week. For the substances with a very long half-life (exceeding 1000 h), the adjustment factor is merely proportional to the number of hours exposed. For such situations, establishing the adjustment factor based on the total weekly hours of exposure rather than the daily limits is useful. There is an obvious need for the judicious application of the work time in the farming sector since the farmers are exposed to a multitude of uncontrolled extraneous exposures.

Human Fatigue and Recovery

The basic premise is that the bodily strains and work fatigue culminate in different forms of musculoskeletal and other injuries, and accidents at workplaces. Despite conflicting hypotheses on *what is human fatigue,* all agree that *fatigue* is a state of disturbed homeostasis due to work and to the working environment. The condition denotes the feeling of physical and mental tiredness, a loss of efficiency and a disinclination for any effort. Knowing that not all functions of a human being become tired as a result of use, the primary and secondary sites of fatigue are difficult to determine. One may consider an organ and its system functions as the site of primary fatigue. The repercussion on the whole body is due to the remote effects of localized fatigue. That is, when *fatigue* is referred to as the reduction in the functional capacity of an organ or organism as a result of action, this broadly applies to muscles, sensory organs, nerve centres, and cells of the sweat glands.

Fatigue and recovery are the periodic processes, and conceived as the time functions that one may allocate as speed, that is, rapid or slow tiring, and early or delayed recovery. The state caused by tiring may be measured as a *degree of fatigue,* and an organism aims at balancing out the capacity lost by its activity that claims for equal recovery. One recognizes that the aspects of fatigue are the objective outcome of *stress (*sum of all parameters influencing man at work) *and strain* (summated effects of stress) of an organism. Two methodological approaches emerge in analyzing strain, i.e. the strain develops from the measures of stress (stressors), and (b) strain is developed by linking suitable measures of stress with the human capacities. Optimization of the stress–strain relationship is limited, since a level of stress may lead to different levels of strains due to variation in working capacities.

Kinds of Fatigue

The types of fatigue include muscular fatigue, mental fatigue, general fatigue, and the like. Muscular fatigue is characterized by reduced muscle power and slower movement, resulting in impaired muscle coordination, and increased liability to errors and accidents. In agriculture, farmers encounter muscular fatigue to a great extent due to their labour-intensive farming activities. Primarily, there are two theories of muscle fatigue, i.e. chemical phenomena and the central nervous system (CNS). In the former, the loss of performance of fatigued muscle results from the chemical processes (consumption of energy-producing phosphate compounds and accumulation of acidic metabolites) with the electrical phenomena in the muscles and nerves playing a secondary role. In CNS theory, however, the chemical processes are considered as the releasing stimulus for sensory impulses that travel along the nerves to the CNS. The afferent impulses inhibit the centres of the brain for motor control of movements, thereby brings about a reduction in the frequency of impulses along the motor neurons. Both processes combined contribute to muscular fatigue, and it manifests in a variety of human working capacity indicators. For example,

Aerobic capacity—The capacity of the human body to maintain the steady-state of O_2 supply to that of the needs at a high workload is termed as the aerobic capacity ($VO_{2\,max}$). Age, gender, body build, and professional habituation to a type of work influence one's aerobic capacity or physical work capacity. The occupational workload must remain at a permissible level, i.e. about 2/5th of one's capacity, beyond which there is a fast accumulation of fatigue (Nag 1981).

Anaerobic capacity—For short high-intensity work, more energy is required in comparison that is available from the oxygen-transporting system. The anaerobic power is derived from the high energy phosphate compounds through the breakdown of glycogen to lactic acid. The anaerobic component is present in any work when the energy cost of work reaches about 25 kJ/min, depending upon one's habituation to work.

Endurance capacity—The endurance expresses the ability of a person to withstand stress over a prolonged period. The capacity of endurance depends on the muscular strength and the ability to use O_2 and dispose of metabolites from muscles and blood. A person with high endurance may carry on exhausting work for an extended period, establishing a steady state of action. Also, his recovery from fatigue is faster than a person with poor endurance. The strength of the working muscles is a limiting factor in endurance. The muscle strength can be maximally used by the equated distribution of the workload, whereby the muscle fibres are alternately contracted and relaxed in the course of force exertion. Besides, the endurance of discomfort includes the measures of the compulsive drives that motivate a person. Human skill develops with training and experience, thus benefitting a person to make efficient movements and recuperate faster from bodily strains.

Besides, mental fatigue may be induced by both mental overloading (e.g. excessive stress on the psychomotor system due to long hours of mentally demanding tasks) and underloading (e.g. when the work is dull and uninteresting). Both the overload and underload increase stress, and if these are prolonged, both reduce the level of one's arousal. The physiological mechanisms of muscular and mental fatigue are not directly related; however, heavy muscular work causes some degree of CNS fatigue. There are possibilities of improving CNS fatigue by interrupting the mental work with recreational activities.

Commonly understood that general fatigue results from the stress of performing tedious activities. While the onset of fatigue is seldom noticed, a significant symptom is a general sensation of weariness, and persons feel inhibited to work. As fatigue advances, attention on the job begins to fade away, and eventually human skill in work tends to deteriorate. This may be overcome by recuperative processes like physical and mental rest. The chronic form of the fatigue is symptomatic of the long-term effects of tiredness, loss of interest in work, increased irritability, and tremor of outstretched hands. The problems involved in chronic fatigue are emotional and situational, not purely physiological.

Prolonged exposure to noise and stress on the auditory organs results in the auditory fatigue. The recovery of the acoustical threshold may be accelerated by rest breaks in noiseless surroundings. Visual fatigue results from over-straining the eyes. In long-time inspection work at inadequate illumination, the eye fixation time and the corrective movements increase, but the rate of movement decreases. In severe stress, when requiring the eyes to move at the maximum pace, profuse tearing of the eyes may be interpreted as the spread of innervation of other effectors. Circadian or nychthemeral fatigue is a part of the day–night rhythm. The after-effects of work are superimposed over the day–night cycle of sleep and wakefulness. The downswings are sometimes called nychthemeral fatigue. Sensory deprivation is visible among the maladjusted shift workers, due to sleep disturbances.

Physiology of Work and Rest

There have been several classical physiological studies on work–rest guidelines (Astrand et al. 1960a, b; Christensen et al. 1960a, b; Lehmann 1958; Margaria et al. 1969; Muller and Karrasch 1955). Studies evolved certain general principles about the patterns of work–rest. By splitting the work into short periods of work and rest (0.5 or 1 min), a person might be able to tolerate heavy work for about an hour. With more extended periods of work and rest (2 or 3 min), the work output approaches the upper limit of performance, with utmost strain. The transport of O_2 from blood to the muscles depends upon the amount of O_2, which the tissue disposes at the initiation of work. The bound O_2 to myohemoglobin and haemoglobin during shorter work bouts (0.5 min) is the prerequisite in aerobic work. In shorter pauses (0.5 min), the myohemoglobin is reloaded with O_2 before the next work begins.

On the other hand, in the case of longer work spells, i.e. 2 or 3 min, the O_2 transport to the muscle is relatively higher but never becomes adequate. Therefore, the O_2 bound to myohemoglobin is of lesser importance. The observations evolved that the length of work periods was the deciding factor, while the rest duration was of secondary importance.

Edwards et al. (1973) noted that for the same total amount of work, the physiological responses to intermittent exercise were higher than in continuous activity. A more significant proportion of the physiologic costs occurred only during recovery in intermittent work than in continuous operation. Saltin et al. (1976) observed no difference in mechanical efficiency between continuous and intermittent work, with varied length and frequency of breaks. Margaria et al. (1969) viewed that the high energy phosphate sources are a possible O_2 store for intermittent exercise, and the duration of the rest period is of importance.

Simonson (1971) detailed the concept of *micro pauses*, proposing sub-divisions of rest pauses of (a) 0.5 to 2 or 4 s, (b) short breaks from 10 s to a few minutes, and (c) longer pauses, exceeding five min. The concept of micro pause has not been received much attention because of apparent misconception about the inability to manage short work periods in the occupational activity. The shortness of the work period is very much critical than the length of each rest-pause. Also, shorter periods of work and rest are more advantageous than the long periods of work and rest. Rest pauses increase one's endurance to work to no small extent. The probable mechanism of micro pause is better muscular pumping action, favouring muscle blood flow, and utilization of myoglobin O_2 store. Many old workers have likely learned this skill to sustaining heavy jobs. There appears to be a lesson that one must give adequate consideration to the beneficial effects of training, skill, and motivation of workers towards the effectiveness of micro pauses.

Individuals differ in their working capacity, and therefore, Astrand and Rodahl (1986) viewed that a workload taxing 30 to 40% of the ones' $VO_{2\,max}$ seems to be a reasonable upper limit for physical work performed over an eight-hour working day. Saha et al. (1979) reported that 35% of one's $VO_{2\,max}$ capacity could be considered as an acceptable workload for Indian workers, with the energy expenditure and the corresponding heart rate at a level of 18 kJ/min and 110 beats/min, respectively, along with two tea breaks of 15 min and a lunch break of 1 h.

Nag et al. (1980) and Nag and Chatterjee (1981a, b) suggested that the work levels for eight hourly activities for men and women farmworkers should not exceed beyond 35 and 28% of one's $VO_{2\,max}$, respectively. Nag and Pradhan (1992) brought out an example of the work–rest ratio in hoeing tasks. The O_2 deficit of a person (e.g. 5.8 l) during maximal work was taken as the indicator of energy reserve. The reserve remains undisturbed as long as the work is aerobic (e.g. VO_2 of 1 to 1.25 l/min). If the VO_2 demand in hoeing tasks varies from 1.5 to 1.9 l/min, for the excess VO_2 needs of 0.5 to 0.7 l/min, the energy reserve will last about 10 min. At this point, the fatigue will ensue, unless work breaks replenish the reserve. The duration of rest can be determined by subtracting the resting VO_2 from the assumed aerobic level of 1.0 to 1.25 l/min. This leaves the remaining VO_2 requirement (0.8 to 0.9 l/min) to replace the energy reserve, requiring about 6 to 7 min. The derivation evolves that a cycle of work at a VO_2 of 1.5 to 1.9 l/min may be organized into periods of 10 min work and 7 min rest, and thereby, the energy reserve is replenished. This approach can be utilized to arrive at a work–rest ratio in diverse farming activities.

Energy Cost of Work as a Guideline for Rest

It is generally emphasized that the work content is a primary step to define the acceptable limit of energy output over the work shift. The energy demand of a task must meet the constraints, as stated in the following examples. Limitation for short-term work refers to the energy expenditure level for about 4 to 5 min. For this short duration, a person can sustain a high workload, even at the level of ones' $VO_{2\,max}$. For a comparative analysis, the $VO_{2\,max}$ of the population groups from the tropics is given in Chap. 4. Taking the average of the $VO_{2\,max}$ of different population groups, the limit of energy expenditure rate arrives at 50 kJ/min for men and 40 kJ/min for women.

Eight-hour energy expenditure limitation refers to a constant rate of work that may be maintained for the working day. As stated earlier that for the eight hourly work levels, the O_2 uptake should not exceed beyond 30 to 40% of ones' $VO_{2\,max}$. If one must exceed this limit, time on the job should be shortened, and additional work breaks need to be scheduled to allow for recovery. Assumed that a workload may be maintained at 35% of one's $VO_{2\,max}$, i.e. 2.5 l/min, the worker should not exceed an energy expenditure rate of (0.35 * 2.5 * 20.31) 17.8 kJ/min. If the $VO_{2\,max}$, is 2.0 l/min, the worker should not exceed an energy expenditure rate of (0.35 * 2.0 * 20.31) 14.2 kJ/min. Taking the daily maximum energy expenditure for German men of 19.5 MJ (4800 kcal), and 8.1 MJ/shift of 8 h, Spitzer (1951) assigned RA guidelines, as

$$RA = ((M/4) - 1) * 100\%$$

where RA is expressed as % of work time and M is the metabolic rate of work (kcal/min). It would imply that for a task with the energy demand of 6 and 8 kcal/min, the RA will be 50 and 100%, respectively. Spitzer's formula was further modified (Murrell 1965) as

$$R = T\,(K - S)/(K - 1.5)$$

where R is the rest period (min), T is working time (min), K is the work-energy cost (kcal/min). The standard energy cost of work (S) is taken as 4 kcal/min and 1.5 kcal/min in the denominator is an approximation of the resting level. The age range and gender may factor the standard and resting energy costs. That is, for young men, an acceptable rate is 5 kcal/min for an eight-hour work shift, and for young women, the maximum rate is factored by about 70%. Further, Karger and Hancock (1982) computed fatigue allowance (min) for eight hour work period, as

$$\mathrm{FA} = [\mathrm{A}(480 + \mathrm{MT}) - \mathrm{M}(\mathrm{W}) - \mathrm{RM}(\mathrm{RB} + \mathrm{MT} + \mathrm{PFD} + \mathrm{CU})]/\mathrm{RM}$$

where A = acceptable energy expenditure (kcal/min); MT = the meal time (min); M = energy expenditure of the job (kcal/min); W = the length of work time (min); RM = metabolic rate (kcal/min) during rest and habitual tasks; RB = rest break time (min) during the shift; PFD = the personal, fatigue, and delay time (min), and CU = clean up time (min).

Tiwari and Gite (2006) carried out a study on the operation of a power tiller (two-wheel walking type tractor) for the preparation of seedbed in drylands. The power tiller was operated (for rotatilling) by five subjects with four work–rest schedules. The work involved controlling a power tiller and walking at a speed of 1.63 km/h. The first work–rest schedule included four work bouts of 90 min each with 15 min rest in between two bouts and a 1-hour lunch break between the second and third bouts. The second schedule involved four work bouts of 75 min each and a final session of 60 min with 15 min rest in between two rounds and a 45 min lunch break between the third and fourth bouts. The third schedule had six work bouts of 60 min duration with 15 min rest between two rounds and a 30 min lunch break between the fourth and fifth bouts. The fourth schedule involved eight work bouts of 45 min each with 10 min rest in between the two rounds and a 30 min lunch break between the fifth and sixth bouts. The total work duration in each schedule was six hours. The overall analysis of the average working heart rates and discomfort rating in different work–rest plans yielded that the work rest schedules did influence the physiological and postural workload. The investigators suggested work schedules for power tilling operation as 60–75 min work bouts with 15 min rest between two work bouts, and a lunch break of 45 min.

In situations like open-field farming activity, a farmworker might sustain a certain level of physical work over the days and weeks. The long-term workload may preferably be compensated by food intake. With an average food intake of about 10.5 MJ/day for a person of 60 kg body weight, his energy expenditure rate may be limited to (10,500 kJ/(60 kg BW * 24 h * 60 min) 7.3 kJ/min. Sen and Nag (1975) suggested that 5.0 MJ/8 h should be the net optimal energy output for heavy MMH jobs in farming and other informal sectors.

The above expressions have necessary implications in the area of work design, whereby to determine how long and how intense individual workers should perform the activity levels. Also, these are helpful where changes in work content are called for because of changes in shift duration. The policy demands *equal rest for equal effort* for a particular segment of time, rather than for *equal work done*. Such recognition seems a must for a society that aspires for the humane treatment of workers, regardless of age, sex, and body status.

Perspectives of Work Time Planning in Farming

The approach to work design in a system of production, like in agriculture, requires intertwining of both the technical and social systems. In labour-intensive agriculture, arguments favoured that the controlled development and effectiveness of the sector may be achieved through appropriate technological innovation, and not at the cost of accelerated work pace that aggravates occupational hazards. Since the work time patterns are structural interventions in working life, the work time and work–rest scheduling are often presented as the elements of economic and employment policies in the production sector. The nature of the agricultural system is unstructured and primarily comprised of small and marginal farm holdings across the world. In contrast to industrial and service organizations, the farming sector requires different forms of work time systems.

Since most farming activities are weather and season dependent, one distinct notion is that shortening working hours does not by itself change job content. In such situations, the small farm holdings would require to involve more hired workforce for completion of jobs within the stipulated time and protect crops. On the reverse, there is a possibility of more substantial machinery dependence owing to speeding up of the cropping activity. Since machinery incorporation is not uniform across the cropping activities, there is a likelihood of more significant nervous fatigue among the farmers. In general, the farmers are deprived of enough leisure, which in turn creates conditions of workers' discontent and ill health, and incur workplace accidents.

For the open-field agriculture in labour-intensive cropping areas, there exists undefined shorter spells of shift work (morning and afternoon stretching over to evening hours). A critical review of literature on shift work systems indicates two schools of thought that—(i) a person can adjust to any schedule and that long blocks of time on a fixed schedule are preferable, and (ii) adjustment is not ever complete. One prefers a shorter spell of work before rotating. Multiple criteria have been suggested for optimal work time practices, whereby to (i) link work time options with the job content, and (ii) consider the possibilities together with the human limits of body clocks adjustment, and risks to health, comfort, and safety. The optimal time planning in the farming sector would encompass much consideration. These include, for example, (a) the workers' preferences and options on work time and adjustment of the sleep-wakefulness cycle, and (b) long-term measures to optimize work content and work time, and secure restful recreations for better health, family, and social life.

For work schedule construction, one may follow specific criteria. People seem to approve that short cycles and conventional systems are convenient to plan and participate in a healthy social life. Night shift work is non-existent in farming, other than in infrequent crop handling, processing, and storing activities, or in some organized farming (e.g. horticulture). A single shift of night work does not markedly disturb circadian rhythms, and also several days of consecutive night shifts are required for re-entrainment of the rhythms (Rutenfranz and Knauth 1983). An extent of free time on the following day may be allowed after each night shift, to compensate for any sleep deficit.

For farming activities that are unstructured and self-paced by the farmworkers, flexible working times have advantages in executing heavy jobs in shorter shifts. An extended workday may be considered for a range of farm activities that have a severity of light to moderate. Any flexibility in work schedules may reflect on factors, such as attaining longer operating hours and that allows flexibility in adjusting the operating capacity to demands and labour productivity. The workers may change in their attitudes towards work and leisure. The multi-skilled groups of workers may cover one or more workplaces simultaneously, and women farmers can adjust work time to personal and family needs.

A unique problem prevails in large farmhouses that primarily depend on the temporary migratory workers from one region of the country to another. The social and ethnic customs of work time and patterns may conflict with those of the community in which they come to live. For example, Muslims, during the annual month-long Ramazan festival, must fast between sunrise and sunset. Therefore, any suggestion of a work time system requires adjustment in the local context (Wedderburn 1989), and this may be one of the reasons that, despite a myriad of work time schedules available, each of them has its limitation of applicability to a specific context.

The question looming significant is how best to facilitate the adjustment of individual workers and to (re)design work time schedules with minimal disruption of the socio-demographic characteristics of the farmworkers. Since the introduction of night shift in open-field agriculture is not a feasible option due to safety concerns, other options, and compensatory packages, such as longer work breaks, fewer hours per shift, compensatory rest periods are the other alternatives. The work schedule in farming requires situation-specific planning to the kind of activities to be performed in the field, and no uniform guidelines may be applied to all situations. The farmworkers and people involved in agro-product handling, processing, trade, and commerce, are over-taxed by substantial physical activity, particularly in peak crop harvest seasons. Recognizing that discrepancies exist in RA guidelines, as described earlier in the chapter, the optimization of the energy demand for the workday is a likely feasible approach for application in farming. For the workers involved in long-term physical activity for weeks and months together, there is a likelihood that the work may be organized in a manner to allow workers to get time for restful recreation during work. The knowledge of total energy cost for the work shift and the available aerobic power of the worker would help in rationally restructuring the work–rest mix for farmworkers. One recognizes that the human fatigue builds up exponentially. There is a rule of thumb that *repeated short breaks* ensures short working periods with a small degree of fatigue and also a high rate of recovery. More systematic research would evolve guidelines for the nutritionally deficient farmworkers to reduce their drudgery and plan workday with equitable distribution of work and rest.

References

Andrews, R. B. (1967). Estimation of values of energy expenditure rate from observed values of heart rate. *Human Factors, 9*(6), 581–586.

Åstrand, I., Åstrand, P. O., Christensen, E. H., & Hedman, R. (1960a). Intermittent muscular work. *Acta Physiologica Scandinavica, 48*(3–4), 448–453.

Åstrand, I., Åstrand, P. O., Christensen, E. H., & Hedman, R. (1960b). Myohemoglobin as an oxygen-store in man. *Acta Physiologica Scandinavica, 48*(3–4), 454–460.

Åstrand, P. O., & Rodahl, K. (1986). *Textbook of work physiology, physiological bases of exercise*. New York: McGraw-Hill.

Axelson, O. (1974). *Heat stress in forest work: an attempt to evaluate the physical work capacity of forest workers as influenced by a hot climate* (No. Folleto 9309). FAO. Rome.

Bedale, M. (1924). Comparison of the energy expenditures of a woman carrying loads in eight different positions. Industrial Hygiene Research Board Report No.29. Medical Research Council, London.

Brief, R. S., & Scala, R. A. (1986). Occupational health aspects of unusual work schedules: a review of Exxon's experiences. *American Industrial Hygiene Association Journal, 47*(4), 199–202.

Brouha, L. (1967). *Physiology in industry*. New York: Pergamon.

Cheong, D., Jansen, M., & Peters, R. (2013). *Shared harvests: agriculture, trade and employment*. International Labour Office and United Nations Conference on Trade Development, Geneva.

Christensen, E. H., Hedman, R., & Holmdahl, I. (1960a). The influence of rest pauses on mechanical efficiency. *Acta Physiologica Scandinavica, 48*(3–4), 443–447.

Christensen, E. H., Hedman, R., & Saltin, B. (1960b). Intermittent and continuous running (A further contribution to the physiology of intermittent work.). *Acta Physiologica Scandinavica, 50*(3–4), 269–286.

Corlett, E. N., & Bishop, R. P. (1976). A technique for assessing postural discomfort. *Ergonomics, 19*(2), 175–182.

De, A., & Sen, R. N. (1986). Ergonomic evaluation of ploughing process of paddy cultivation in India. *Journal of Human Ergology, 15*(2), 103–112.

Deeb, J. M., Drury, C. G., & Pendergast, D. R. (1992). An exponential model of isometric muscular fatigue as a function of age and muscle groups. *Ergonomics, 35*(7–8), 899–918.

Dukes-Dobos, F. N., Keenan, R. G., & Carlow, T. J. (1967). Urinary mucoprotein excretion in physical exercise. *Human Factors, 9*(6), 595–601.

Edwards, R. H. T., Ekelund, L. G., Harris, R. C., Hesser, C. M., Hultman, E., Melcher, A., et al. (1973). Cardiorespiratory and metabolic costs of continuous and intermittent exercise in man. *The Journal of Physiology, 234*(2), 481–497.

Ghosh, S. N., & Nag, P. K. (1986). Muscular strains in different modes of load handling. *Clinical Biomechanics, 1*(2), 64–70.

Ghugare, B. D., Adhaoo, S. H., Gite, L. P., Pandya, A. C., & Patel, S. L. (1991). Ergonomics evaluation of a lever-operated knapsack sprayer. *Applied Ergonomics, 22*(4), 241–250.

Gite, L. P. (1996). *Some investigations on a side lever operated knapsack sprayer from mechanical and ergonomical considerations* (Doctoral dissertation, IIT Kharagpur).

Gite, L. P. (1992). Ergonomics aspects of lever-operated knapsack sprayers. Proceedings—'Small sprayer standards, safety and future directions for Asia'. International Rice Research Institute, Manila, Philipines, 25–38.

Grandjean, E. (1988). *Fitting the task to the man - A textbook of occupational ergonomics*. London: Taylor and Francis.

Hall, S. A. (1968). Heat stress amongst outdoor manual workers in East Africa. *East African Medical Journal, 45*(5), 351–357.

Hanger, I. M., & Hagberg, M. (1989). Evaluation of two floor-mopping work methods by measurement of load. *Ergonomics, 32*(4), 401–408.

Hasalkar, S., Shivalli, R., & Budihal, R. (2007). Musculo-skeletal disorders of the farm women while performing the top dressing of fertilizer activity. *Journal of Human Ecology, 21*(2), 109–112.

Helander, M. G., & Quance, L. A. (1990). Effect of work—rest schedules on spinal shrinkage in the sedentary worker. *Applied Ergonomics, 21*(4), 279–284.

Hickey, J. L. (1980). Adjustment of occupational exposure limits for seasonal occupations. *American Industrial Hygiene Association Journal, 41*(4), 261–263.

ILO. (1935). Convention No. 47 - Forty-Hour Week Convention. International Labour Office, Geneva.

ILO. (1962). *Recommendation 116—Reduction of Hours of Work Recommendation*. Geneva: International Labour Office.

ILO. (1964). *Convention No. 122—Employment policy convention*. Geneva: International Labour Office.

ILO. (2001). *Convention No. 184—Safety and health in agriculture*. Geneva: International Labour Office.

ILO. (2018). *Ensuring decent working time for the future*. Geneva: International Labour Office.

Jäger, M., Luttmann, A., & Laurig, W. (1984). The load on the spine during the transport of dustbins. *Applied Ergonomics, 15*(2), 91–98.

Karger, D. W., & Hancock, W. M. (1982). *Advanced work measurement*. Industrial Press.

Kogi, K. (1991). Job content and working time: The scope for joint change. *Ergonomics, 34*(6), 757–773.

Lee, S., McCann, D., & Messenger, J. C. (2007). *Working time around the world-trends in working hours, laws and policies in a global comparative perspective*. Geneva: International Labour Organization.

Legg, S. J. (1985). Metabolic, cardiovascular and spinal strain of a representative fuel replenishment task. *Ergonomics, 28*(5), 773–780.

Legg, S. J., & Mahanty, A. (1985). Comparison of five modes of carrying a load close to the trunk. *Ergonomics, 28*(12), 1653–1660.

Legg, S. J., & Myles, W. S. (1981). Maximum acceptable repetitive lifting workloads for an 8-hour work-day using psychophysical and subjective rating methods. *Ergonomics, 24*(12), 907–916.

Lehmann, G. (1958). Physiological measurements as a basis of work organization in industry. *Ergonomics, 1*(4), 328–344.https://doi.org/10.1080/00140135808964608

Lind, A. R., & McNicol, G. W. (1968). Cardiovascular responses to holding and carrying weights by hand and by shoulder harness. *Journal of Applied Physiology, 25*(3), 261–267.

Mairiaux, P., Davis, P. R., Stubbs, D. A., & Baty, D. (1984). Relation between intra-abdominal pressure and lumbar moments when lifting weights in the erect posture. *Ergonomics, 27*(8), 883–894.

Margaria, R., Oliva, R. D., Di Prampero, P. E., & Cerretelli, P. (1969). Energy utilization in intermittent exercise of supramaximal intensity. *Journal of Applied Physiology, 26*(6), 752–756.

Milner, N. P., Corlett, E. N., & O'Brien, C. (1986). A model to predict recovery from maximal and submaximal isometric exercise. In E. N. Corlett, J. Wilson, & I. Manenica (Eds.), *The ergonomics of working postures* (pp. 126–135). London: Taylor & Francis.

Morrissey, S. J., & Liou, Y. H. (1984). Metabolic costs of load carriage with different container sizes. *Ergonomics, 27*(8), 847–853.

Müller, E. A. (1953). The physiological basis of rest pauses in heavy work. *Quarterly Journal of Experimental Physiology and Cognate Medical Sciences: Translation and Integration, 38*(4), 205–215.

Müller, E. A., & Karrasch, K. (1955). Der Einfluß der Pausenanordnung auf die Ermüdung bei Schwerarbeit. *Internationale Zeitschrift für angewandte Physiologie einschließlich Arbeitsphysiologie, 16*(1), 45–51.

Murrell, K. F. H. (1965). *Human performance in industry.* New York: Reinhold Publishing Corporation.

Nag, P. K. (1981). Predicting maximal oxygen uptake of workers engaged in agricultural tasks. *Journal of Human Ergology, 10*(1), 25–33.

Nag, P. K. (1982). Influence of posture and speed of arm and leg work on physiological responses. *The Journal of Sports Medicine and Physical Fitness, 22*(4), 426.

Nag, P. K. (1987). Maximal oxygen uptake of agricultural men and women in India. *American Journal of Physical Anthropology, 74*(2), 149–153.

Nag, P. K. (1991a). Planning Rest Allowances for Better HRM in Industry: A critical appraisal. *Indian Journal of Social Work, 52*(4), 515–526.

Nag, P. K., & Chatterjee, S. K. (1981a). Voluntary muscular contraction with reference to agricultural tasks. *Indian Journal of Medical Research, 74,* 111–119.

Nag, P. K., & Chatterjee, S. K. (1981b). Physiological reactions of female workers in Indian agricultural work. *Human Factors, 23*(5), 607–614.

Nag, P. K., & Pradhan, C. K. (1992). Ergonomics in the hoeing operation. *International Journal of Industrial Ergonomics, 10*(4), 341–350.

Nag, P. K., Chintharia, S., Saiyed, S., & Nag, A. (1986a). EMG analysis of sitting work postures in women. *Applied Ergonomics, 17*(3), 195–197.

Nag, P. K., Pradhan, C. K., Goswami, A., & Ashtekar, S. P. (1986b). Cardio-respiratory and muscle responses in static, dynamic and combined work. *Journal of Human Ergology, 15*(1), 73–77.

Nag, P. K., Sebastian, N. C., & Mavlankar, M. G. (1980). Occupational workload of Indian agricultural workers. *Ergonomics, 23*(2), 91–102.

Nag, P. K. (1991b). Endurance limits in different modes of load holding. *Applied Ergonomics, 22*(3), 185–188.

Pepermans, R. G., & Corlett, E. N. (1983). Cross-modality matching as a subjective assessment technique. *Applied Ergonomics, 14*(3), 169–176.

Rabbitt, P. (1991). Management of the working population. *Ergonomics, 34*(6), 775–790.

Rohmert, W. (1973a). Problems in determining rest allowances: Part 1: Use of modern methods to evaluate stress and strain in static muscular work. *Applied Ergonomics, 4*(2), 91–95.

Rohmert, W. (1973b). Problems of determination of rest allowances Part 2: Determining rest allowances in different human tasks. *Applied Ergonomics, 4*(3), 158–162.

Rohmert, W., & Mainzer, J. (1986). *Influence and assessment methods for evaluating body postures* (pp. 183–217). The ergonomics of working postures. London: Taylor and Francis.

Rutenfranz, J., & Knauth, P. (1983). Criteria for the evaluation of shift system. In T. O. Kvalseth (Ed.), *Ergonomics of Workstation Design* (pp. 142–151). London: Butterworths.

Saha, P. N., Datta, S. R., Banerjee, P. K., & Narayane, G. G. (1979). An acceptable workload for Indian workers. *Ergonomics, 22*(9), 1059–1071.

Saltin, B., Essén, B., & Pedersen, P. K. (1976). Intermittent exercise: its physiology and some practical applications. In *Advances in exercise physiology* (Vol. 9, pp. 23–51). Karger Publishers.

Sanchez, J., & Monod, H. (1979). Physiological effects of dynamic work on a bicycle ergometer combined with different types of static contraction. *European Journal of Applied Physiology and Occupational Physiology, 41*(4), 259–266.

Sen, R. N., & Nag, P. K. (1975). Work organization of heavy load handling in India. *Journal of Human Ergology, 4*(2), 103–113.

Sen, J., & Nag, P. K. (2019). Human susceptibility to outdoor hot environment. *Science of the Total Environment, 649*, 866–875.

Sengupta, A. K., Sarkar, D. N., Mukhopadhyay, S., & Goswami, D. C. (1979). Relationship between pulse rate and energy expenditure during graded work at different temperatures. *Ergonomics, 22*(11), 1207–1215.

Sharkey, B. J., McDonald, J. F., & Corbridge, L. G. (1966). Pulse rate and pulmonary ventilation as predictors of human energy cost. *Ergonomics, 9*(3), 223–227.

Sheeley, G. A., Hallbeck, M. S., & Bishu, R. R. (1991). Wrist fatigue in flexion and extension. In *Proceedings of the International Safety and Industrial Ergonomics Conference* (pp. 55–60). Taylor & Francis, London.

Shoenfeld, Y., Shapiro, Y., Portugeeze, D., Modan, M., & Sohar, E. (1977). Maximal backpack load for long distance hiking. *Journal of Sports Medicine, 17,* 147–151.

Simonson, E. (Ed.). (1971). *Physiology of work capacity and fatigue.* Thomas.

Singal, S. (1994). Changes in blood pressure during manual grinding of dry spices in different body postures by female workers. *Research and Development Reporter, 11*(1&2), 71–78.

Sjøgaard, G., Savard, G., & Juel, C. (1988). Muscle blood flow during isometric activity and its relation to muscle fatigue. *European Journal of Applied Physiology and Occupational Physiology, 57*(3), 327–335.

Spitzer, H. (1951). Physiologische grundlagen für den erholungszuschlag bei schwerarbeit. *REFA-Nachrichten, 2,* 37–42.

Srivatava, K., Johnson, A. C., & Zahradnik, J. W. (1962). *Energy requirements, harvest rates and efficiency compared for several hand grain harvesting implements.* Washington, DC: Annual Meeting of ASAE.

Subramanian, S. (2015). *Emerging trends and patterns of India's agricultural workforce: Evidence from the census.* Bangalore, India: Institute for Social and Economic Change.

Tanti, K., & Masuda, T. (1985). A kinesiologic study of erector spinae activity during trunk flexion and extension. *Ergonomics, 28*(6), 883–893.

Tiwari, P. S., & Gite, L. P. (2006). Evaluation of work-rest schedules during operation of a rotary power tiller. *International Journal of Industrial Ergonomics, 36*(3), 203–210.

Tuure, V. M. (1992). Determination of physical stress in agricultural work. *International Journal of Industrial Ergonomics, 10*(4), 275–284.

Ulmer, H. V., Janz, U., & Löllgen, H. (1977). Aspects of the validity of Borg's scale. Is it measuring stress or strain?. In G. Borg, (Ed.), *Physical work and effort* (pp. 181–196). Pergamon. 181–185.

Volle, M., Brisson, G. R., Perusse, M., Tanaka, M., & Doyon, Y. (1979). Compressed work-week: Psychophysiological and physiological repercussions. *Ergonomics, 22*(9), 1001–1010.

Wedderburn, A. (1989). *Negotiating shorter working hours in the European Community.* Dublin: European Foundation for the Improvement of Living and Working Condition.

www.populationpyramid.net (2019).

www.worldlifeexpectancy.com (data retrieved in May 2020).

Wyndham, C. H. (1970). Adaptation to heat and cold. In D. H. K. Lee & D. Minard (Eds.), *Physiology, environment and man.* New York: Academic Press.

Yu, C. Y., & Keyserling, W. M. (1989). Evaluation of a new work seat for industrial sewing operations. *Applied Ergonomics, 20,* 17–25.

Chapter 6
Engineering Anthropometry of Farmworkers

Introduction

Today, more and more manually operated and powered farm implements and machinery are brought into farm operations. Many times, due to poor design of devices or flawed process or lack of safety awareness, the farm activities are drudgery and accident-prone. It has been imperative to design the farm equipment as well as workplaces with due account of the man–machine functional compatibility. The equipment and machinery should always be a good match between the operator's biological limits and an operational range of movements, and the physical requirements of the machinery/ equipment (Kroemer 1983; Das and Grady 1983; Grandjean 1988; Das and Sengupta 1996; ISO 2013). The matching of machinery requirements with human capabilities is necessary for the optimum performance of the man–machine system. In this regard, the essential information required is the body dimensions of the users of tools and equipment. Neglect of applying anthropometric data in the design of farm tools and equipment might cause reduced efficiency of operation, problems of safety, operational discomfort and failure to the operator (Gite and Yadav 1989; Rogan and O'Neill 1993).

Anthropometry is the technology of measuring human physical traits, primarily such factors as size, mobility, and strength. However, the body dimensions vary with age, sex, and different ethnic groups (Roebuck et al. 1975; Sanders and McCormick 1992). In many countries efforts have been made to establish an anthropometric database for different population groups, such as civilians, military personnel, students, and industrial workers (NASA 1978a, b; Haslegrave 1980; Marras and Kim 1993; Lee 2000; Bolstad et al. 2001; Wang et al. 2002). ISO 15535 (ISO 2013) specifies general requirements for establishing an anthropometric database. In contrast, ISO 7250 (Part 1, 2, and 3) (ISO 1996, 2010, 2015) deals with the definition of the anthropometric parameters, and the data from some member countries along with other related information.

P. K. Nag and L. P. Gite, *Human-Centered Agriculture*, Design Science and Innovation,
https://doi.org/10.1007/978-981-15-7269-2_6

The use of anthropometric data in the design and application of farm tools and equipment brings better functional ease, reduces drudgery, and enhances the productivity and safety of the farmworkers. The design of tools, equipment, and workplace requires data of (a) typical measurements of the static human body, (b) dynamic anthropometry including the range of movement of body joints, (c) pictorial forms for anthropometric parameters, including the dimensions and devices of a workplace; and (d) direct recommendations for designing objects.

Anthropometric Data

The Aerospace Medical Research Laboratories, Dayton, Ohio (USA) assembles and maintains the largest single repository of raw anthropometric data in the world (NASA 1978a,b). It contains data of the U.S. army and air force personnel as well as civilians. Some foreign population data are also available in the NASA data bank. ERGODATA is another data bank located at the Anthropology Laboratory of Paris, University of France. It mostly contains European anthropometric data. An All India Anthropometric Survey initiated by the Anthropological Survey of India (ASI) in the period 1961–1969, covered nearly 60,000 male population from about 300 different castes/tribes and communities. The body dimensions included in the survey were body stature, sitting height, body weight, and a few others. During the period 1972–1980, an All India Bio-Anthropological Survey by the ASI gathered baseline information, such as physique, bodily disabilities, diseases, anomalies, demography, and food habits of the population. Only three body measurements, viz. body stature, body weight, and chest circumference, were included in the survey. Anthropometric data on people of armed forces are available from the Defence Institute of Physiology and Allied Sciences (DIPAS), Delhi. The National Institute of Design (NID, Ahmedabad) monogram on anthropometric data of Indians provided data on 1,000 volunteers of student community and other occupational groups (Chakrabarti 1997). Some anthropometric surveys carried out in different countries (Sen et al.,1977; Gupta et al. 1983; Moustafa et al. 1987; Gite and Yadav 1989; Yadav et al. 1997; Liu et al. 1999; Philip and Tewari 2000; Tewari and Ailavadi 2002; Victor et al. 2002; Lin et al. 2004; Dewangan et al. 2005, 2008) provide a glimpse of the case studies and mix of dimensions covered in the surveys.

The all India Coordinated Research Project (AICRP) on Ergonomics and Safety in Agriculture (ICAR-CIAE Bhopal) gathered anthropometric data for 14618 (8970 male and 5648 female) agricultural workers (Gite et al. 2009; Gite 2017), referring to the requirements of farm equipment design. The study covered 79 body dimensions. The standard terminologies and definitions, as given in the Anthropometric Source Book (NASA 1978a, b) and ISO 7250 (part 1), were used, as provided in Table 6.1 and Fig. 6.1a–d.

Table 6.1 Definitions of body dimensions

Sr. No.	Dimension	Definition
1	Body Weight	It is measured on a calibrated weighing scale
2	Stature	The vertical distance from the standing surface to the top of the head. The person stands erect and looks straight forward
3	Vertical reach	The vertical distance from the standing surface to the height of the middle finger when arm, hand, and fingers extend vertically
4	Vertical grip reach	The vertical distance from the standing surface to the height of the pointer held horizontal to the person's s fist when the arm extends upward. The person stands erect and looks straight forward
5	Eye height	The vertical distance from the standing surface to the inner corner of the eye. The person stands erect and looks straight forward
6	Acromial height	The vertical distance from the standing surface to the acromion process. The person stands erect and looks straight forward
7	Elbow height	The vertical distance from the standing surface to the top of the radiale bone. The person stands erect and looks straight forward
8	Olecranon height	The vertical distance from the standing surface to the height of the undersurface of the elbow, measured with the arm flexed 90° and the upper arm vertical. The person stands erect and looks straight forward
9	Iliocristale height	The vertical distance from the standing surface to the top of the ilium in the midaxillary plane. The person stands erect and looks straight forward
10	Iliospinale height	The vertical distance from the standing surface to the height of the iliospinale (the inferior aspect of the tip of the anterior superior iliac spine). The person stands erect and looks straight forward
11	Trochanteric height	The vertical distance from the standing surface to the height of the trochanter. The person stands erect and looks straight forward
12	Metacarpal- III height	The vertical distance from the standing surface to the height of the knuckle where the middle finger joins the palm. The person stands erect and looks straight forward

(continued)

Table 6.1 (continued)

Sr. No.	Dimension	Definition
13	Knee height	The vertical distance from the standing surface to the mid-point of the knee cap. The person stands erect and looks straight forward
14	Medial malleolus height	The height of the most medially projecting point of the medial ankle bone
15	Lateral malleolus Height	The height of the most lateral projecting point of the lateral ankle bone
16	Menton to top of the head	The distance from the lower edge of the tip of the chin to the top of the head
17	Waist back length	The vertical distance along the spine from the waist level to the top of the 7th cervical Vertebra. The person stands erect and looks straight forward
18	Scapula to waist back length	The surface distance from the superior angle of scapula to the back at waist level.
19	Span	The distance between the tips of right and left middle fingers when the person's arms are maximally extended laterally
20	Span akimbo	The distance between the elbow points measured with the arms flexed, held horizontally, palms down, fingers straight and together and palm and thumbs touching the chest
21	Arms reach from the wall	The distance from the wall to the tip of the middle finger measured with the person's shoulder against the wall, his hand and arm extended forward
22	Thumb tip reach	The distance from the wall to the tip of the thumb measured with the person's shoulder against the wall, arm extended forward, and index finger touching the tip of the thumb
23	Shoulder grip length	The horizontal distance from a pointer held in the person's fist to a wall against which he sits, measured with the arms extended forward and horizontal
24	Wall to acromion distance	The horizontal perpendicular distance from the wall to the acromion
25	Wall to lumbosacral joint	The transverse distance from the wall to the posterior iliac spine. The person stands erect against the wall and looks straight forward

(continued)

Table 6.1 (continued)

Sr. No.	Dimension	Definition
26	Abdominal extension to the wall	The vertical distance from the most laterally protruding point of the abdomen to a wall against which the person stands with minimal wall-buttock contact. The person stands erect and looks straight forward
27	Chest depth	The maximum depth of the torso measured at the nipple level. The person stands erect and looks straight forward.
28	Biacromial breadth	The transverse distance across the shoulder from right to the left acromion. The person stands erect and looks straight forward
29	Bideltoid breadth	The horizontal distance across the maximum lateral protrusion of the right and left deltoid muscles. The person stands erect and looks straight forward
30	Chest breadth	The maximum depth of the torso measured at the nipple level. The person stands erect and looks straight forward
31	Inter scye breadth	The tape distance across the back between the posterior axillary folds at a lower level of the armpit
32	Waist breadth	The breadth of the torso at the waist level. The person stands erect and looks straight forward
33	Hip breadth	The maximum breadth of the lower torso. The person stands erect and looks straight forward.
34	Heel breadth	The maximum breadth of the heel as measured behind the projection of the ankle bones
35	Bimalleolar Breadth	The distance across the protrusion of the medial and lateral malleolus bones
36	Chest circumference	The circumference of the torso measured at the nipple level. The person stands erect and looks straight forward
37	Waist circumference	The circumference of the torso measured at the waist level. The person stands erect and looks straight forward
38	Thigh circumference	The circumference of the upper leg measured as high in the crotch as possible
39	Calf circumference	The maximum circumference of the lower leg. The person stands erect and looks straight forward
40	Wrist circumstance	The circumference of the wrist at the level of the tip of the styloid process of the radius

(continued)

Table 6.1 (continued)

Sr. No.	Dimension	Definition
41	Grip diameter (inside)	The width of the broadest level of a cone that the person can grasp with his thumb and middle finger touching
42	Grip diameter (outside)	The distance between the joint of 1st and 2nd phalanges of the thumb and knuckles of the middle finger, with the hand held to the grip
43	Middle finger-palm grip diameter	The diameter of the broadest level of the cylinder, which the person can grasp with his palm and middle finger touching.
44	Vertical grip reach height sitting	The height above the sitting surface of a pointer held horizontal in the person's fist when the arm is extended maximally upward
45	Sitting height	The height, from the sitting surface to the top of the head. The person sits erect and looks straight forward
46	Sitting eye height	The height from the sitting surface to the external canthus. The person sits erect and looks straight forward
47	Sitting acromial height	The height from the sitting surface to the top of the acromion. The person sits erect and looks straight forward
48	Elbow rest height	The height of the bottom of the tip of the elbow above the sitting surface
49	Thigh clearance height sitting	The height of the highest point of the thigh above the sitting surface. The person sits erect and looks straight forward
50	Knee height sitting	The height from the footrest surface of the musculature just above the knee. The person sits erect and looks straight forward
51	Popliteal height sitting	The height of the underside of the upper leg above the footrest surface. The person sits erect and looks straight forward
52	Hand thickness at metacarpal - III	The width of the metacarpal-phalangeal joint of the middle finger
53	First phalanx digit III length	The length of the first segment of the middle finger across the 3rd metacarpal and 2nd phalanx while the hand is held in a fist
54	Grip span	The maximum distance between the palm and the fingers in grip position
55	Maximum grip length	The maximum length between the tip of the middle finger and the tip of the thumb while the palm, thumb, and finger are in grip position

(continued)

Table 6.1 (continued)

Sr. No.	Dimension	Definition
56	Index finger diameter	The diameter of the index finger of the right hand as determined by the smallest hole in which the finger can be inserted
57	Head length	The maximum length of the head as measured from glabella to the back of the head
58	Head breadth	The maximum breadth of the head
59	Coronoid fossa to hand length	The distance from the tip of the middle finger with arms bent at 90° to the coronoid fossa (anterior arm joint of the humerus)
60	Forearm hand length	The distance from the base of the hand to the top of the middle finger measured along the axis of the hand
61	Elbow grip length	The distance from the tip of the bent elbow to the centre of the clenched fist
62	Hand length	The distance from the base of the hand to the top of the middle finger measured along the long axis of the hand
63	Palm length	The distance from the base of the hand to the furrow where the middle finger folds upon the palm
64	Handbreadth across thumb	The breadth of the hand measured at the level of the distal end of the 1st metacarpal of the thumb
65	Handbreadth at metacarpal—III	The breadth of the hand as measured across the distal ends of the metacarpal bones
66	Buttock knee Length	The horizontal distance from the rearmost surface of the buttocks to the front of the knee caps
67	Buttock-popliteal length	The horizontal distance from the rearmost surface of the buttock to the back of the lower leg
68	Abdominal depth sitting	The distance from the rearmost surface of the waist to the front of the umbilicus
69	Hip breadth sitting	The breadth of the body as measured across the widest portion of the hips. The person sits erect, looking straight forward
70	Elbow-elbow breadth	The distance across the lateral surfaces of the elbows flexed and resting lightly against the body
71	Knee-knee breadth	The maximum horizontal distance across the lateral surfaces of the knees measured when the knees were gently touching

(continued)

Table 6.1 (continued)

Sr. No.	Dimension	Definition
72	Foot length	The length of the foot measured parallel to its long axis
73	Instep length	The distance from the plane of the heel to the point of maximum medial protuberance of the foot
74	Foot breadth (Ball of foot breadth)	The maximum width of the foot as measured at the right angle to its long axis
75	Functional leg length	The distance from the back at the waist level to the heel, measured along the long axis of the leg, when the person sits erect on the edge of a chair, his leg extended forward with knee straightened
76	Biceps skinfold	The skinfold thickness parallel to the arm on the mid-biceps between the acromion and the elbow
77	Triceps skinfold	The skinfold thickness on the back of the arm halfway between the acromion and the tip of the elbow parallel to the long axis of the upper arm
78	Subscapular skinfold	The skinfold thickness just at the inferior angle of the right scapula and parallel to the tension line of the skin
79	Supra-iliac skinfold	The skinfold thickness along the midaxillary line at the level of the crest of the ilium

Presentation of Anthropometric Data

Human beings vary in size. Each person is different from the other in terms of body dimensions. This variability in anthropometric dimensions follows a normal or Gaussian distribution curve. This curve is symmetrical about its highest point, i.e. mean value. For any distribution, there are two main parameters, namely, the mean (M) and standard deviation (SD). Another parameter used for describing the variation in anthropometric data is the coefficient of variation, that is, the SD expressed as a percentage of the mean. Variability in a set of anthropometric data is expressed as a range, that is, the difference between maximum and minimum values in a data set. It is a common practice to express anthropometric design limits in terms of percentiles. A particular percentile indicates the percentage of people within the population who have a body dimension of a specific size or smaller. The median is the 50th percentile value, that is, the point in the data where 50% of the data falls below that point, and the other 50% are above it.

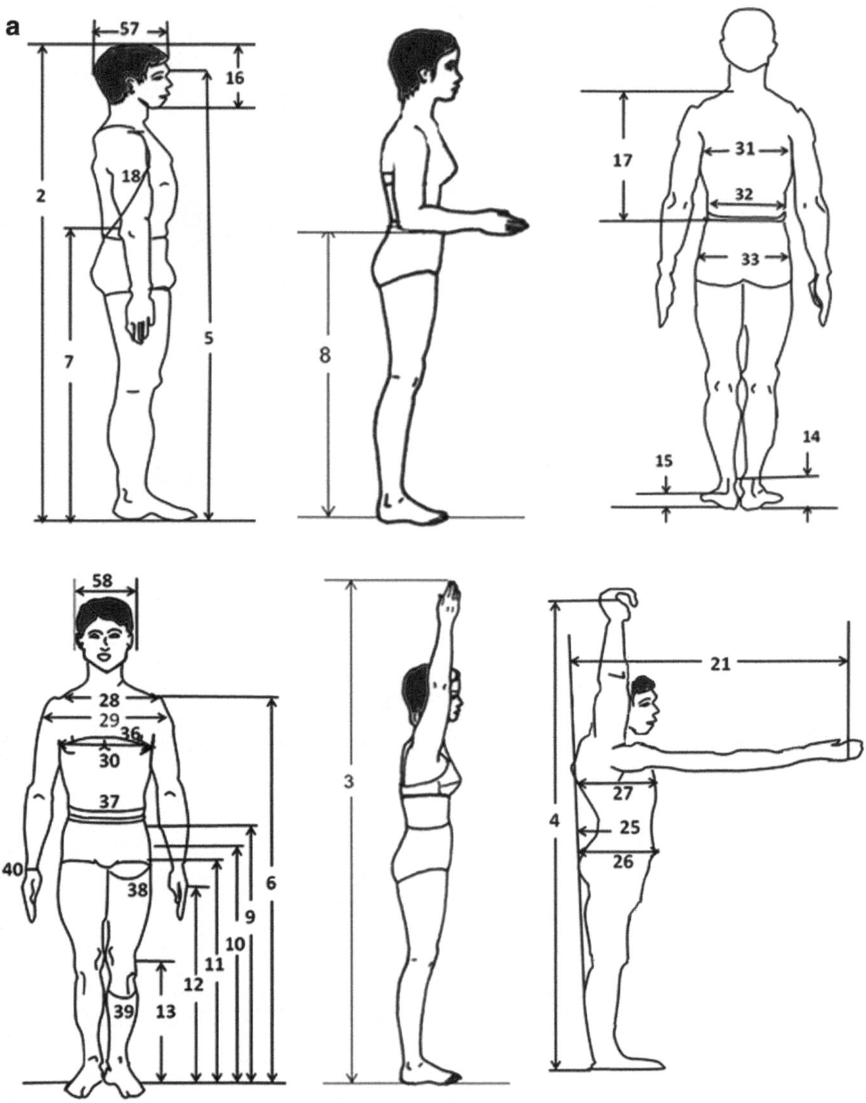

Fig. 6.1 **a** Body dimensions for farm machinery design (Dimension numbers refer to the serial number of Table 6.1). **b** Body dimensions for farm machinery design (refer to the serial number of Table 6.1). **c** Body dimensions for farm machinery design (refer to the serial number of Table 6.1). **d** Body dimensions for farm machinery design (refer to the serial number of Table 6.1)

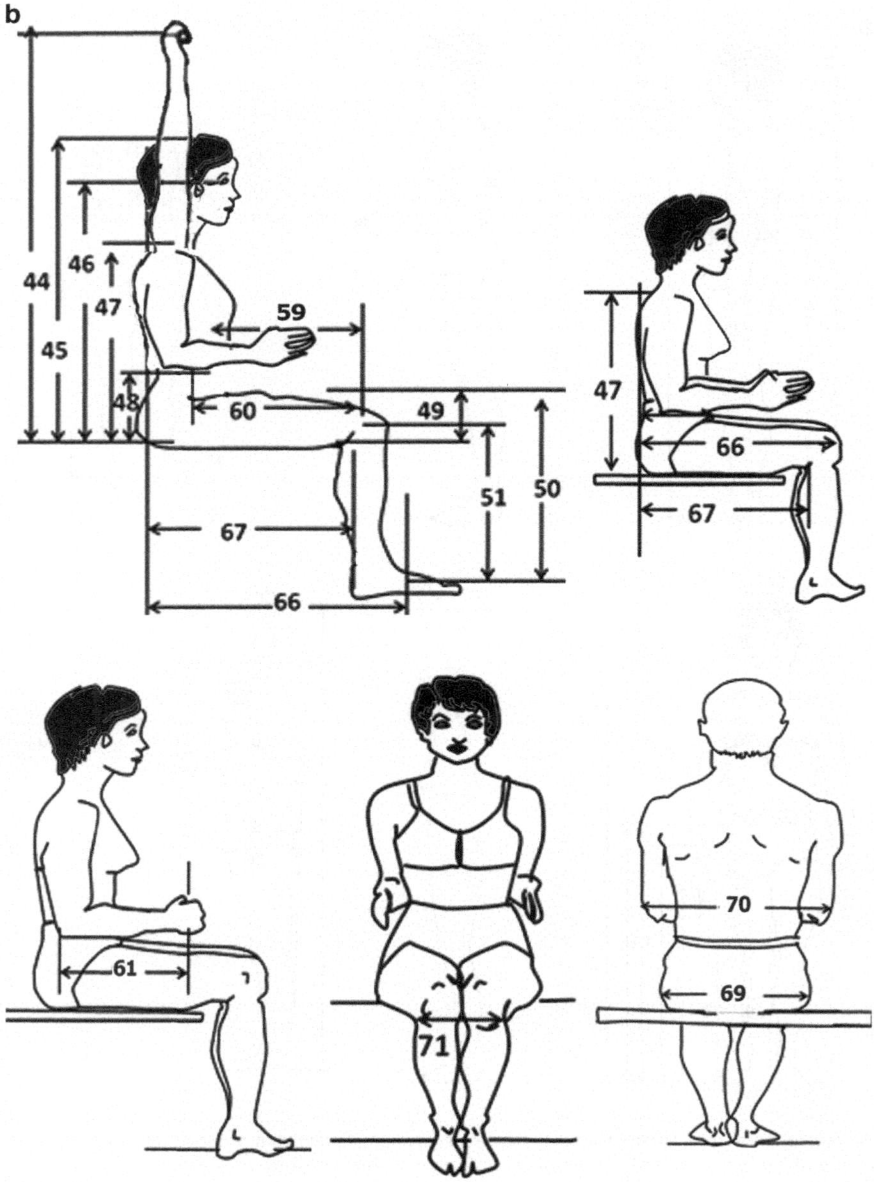

Fig. 6.1 (continued)

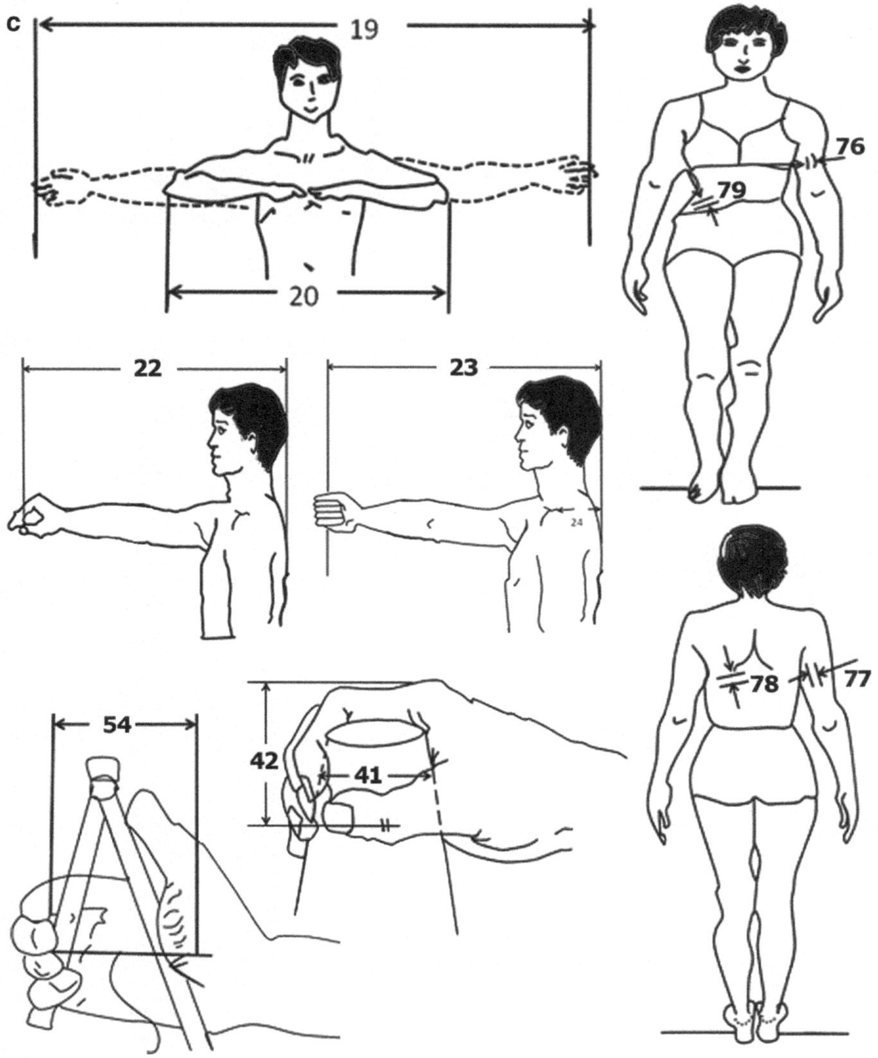

Fig. 6.1 (continued)

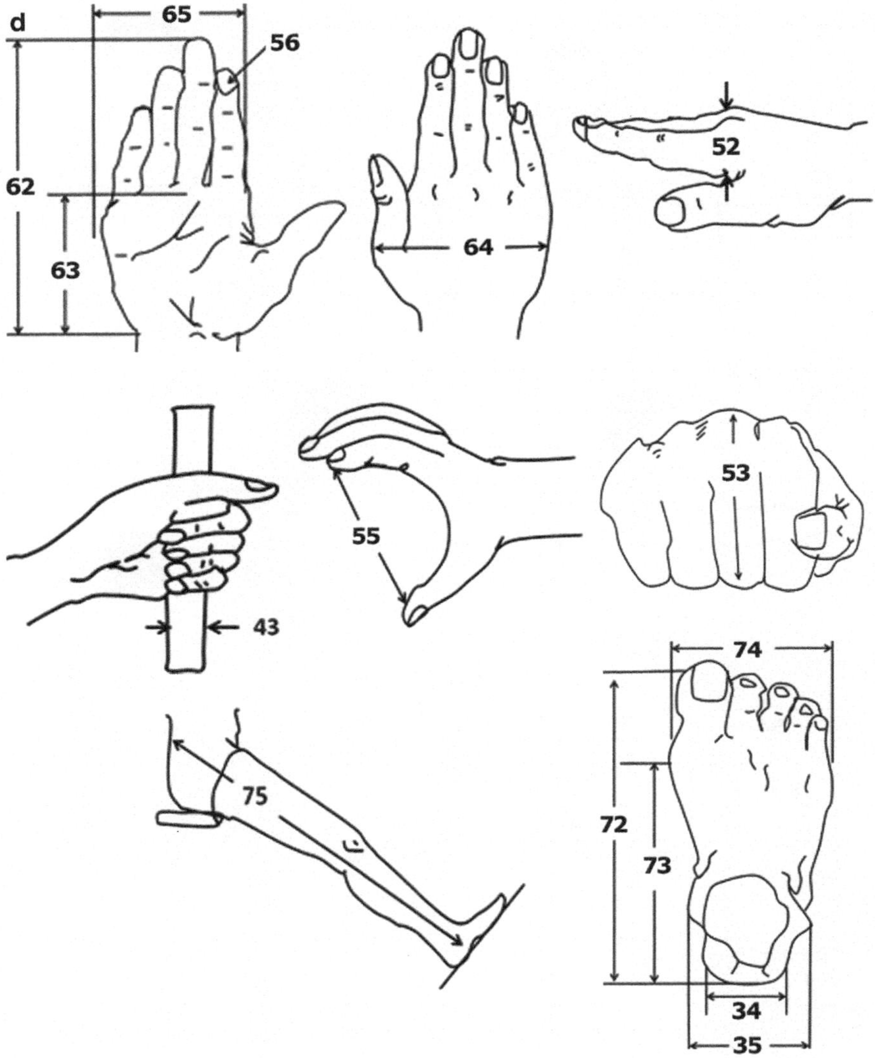

Fig. 6.1 (continued)

Anthropometric Data of Population

Table 6.2 gives the mean, SD, as well as the 5th and 95th percentile values of anthropometric data for 79 dimensions of 8,970 Indian men and 5,648 women farmworkers, as reported by Gite (2017). Figures 6.2 and 6.3 show the 5th and 95th percentile values of men and women, respectively. Such data are not available from other countries where human beings are involved as workers in large numbers. However, to have a comparison of anthropometric data from different countries, average values of 17 body dimensions of men and women are given in Tables 6.3 and 6.4. There is a considerable variation in body dimensions of people in different countries due to race, ethnicity, nutrition, and other characteristics. For example, the body height of male workers varies from 1620 mm in Indonesia to 1779 mm in Sweden.

In contrast, the bodyweight varies from 55.1 kg for Indian workers to 83 kg for American workers. The differences in the anthropometric data of men and women workers emphasize the need for keeping gender consideration in the design of equipment and workplaces. The ethnic and regional variations in the data show the necessity of having country-wise anthropometric data and then applying those data in the design of farm tools, equipment, and workplaces to reduce drudgery, improve work efficiency, and enhance safety in various agricultural activities.

Muscular Strength Data

Muscular strength measurement is of two types—static (isometric) and dynamic (isotonic). Static muscular force is usually larger than dynamic force. The dynamic force, depending on the velocity of the shortening muscles, amounts to about 50 to 90% of the maximal static force. Generally, it is a common practice to measure the static strength to compare different populations or to set the limits for force application. Kroemer (1970) defines strength as the maximal force muscles can exert isometrically in a single voluntary effort. It is a vector quantity and therefore, needs to be described by magnitude and direction. As per Mital and Kumar (1998), static muscle strengths are the capacities of muscles to produce force or torque by a single maximal voluntary isometric exertion (the body segment involved remains stationary and no displacement). Here, the internal muscular effort, amplified by the mechanical advantage of the body members involved, is measured as the external static effect in the form of force or torque. The standard protocols for the measurement

Table 6.2 Anthropometric data of Indian agricultural workers (Values in mm unless otherwise stated)

Sr. No.	Dimension	Men				Women			
		Mean	SD	Percentile		Mean	SD	Percentile	
				5th	95th			5th	95th
1	Body weight, kg	55.1	9	42	71.5	47.8	8.9	35	65
2	Stature	1636	71	1523	1756	1521	64	1420	1634
3	Vertical reach	2082	100	1920	2245	1923	89	1783	2080
4	Vertical grip reach	1986	99	1827	2150	1832	88	1690	1985
5	Eye height	1525	72	1410	1646	1408	64	1310	1521
6	Acromial height	1364	68	1255	1477	1265	60	1173	1370
7	Elbow height	1029	56	941	1121	963	48	890	1045
8	Olecranon height	1001	53	916	1090	938	46	865	1018
9	Iliocristale height	954	56	861	1050	900	50	820	984
10	Iliospinale height	908	59	810	1005	856	52	780	945
11	Trochanteric height	818	69	700	922	784	65	680	894
12	Metacarpal III height	692	47	620	770	653	44	587	729
13	Knee height	473	35	419	530	439	31	390	490
14	Medial Malleolus height	80	10	64	96	74	11	55	89
15	Lateral Malleolus height	69	9	54	85	62	9	46	76
16	Menton to top of the head	213	20	180	248	198	23	164	240
17	Waist back length	445	43	380	520	388	47	315	470
18	Scapula to waist back length‘	561	104	390	730	547	69	445	670
19	Span	1698	85	1560	1842	1555	80	1435	1700
20	Span akimbo	871	58	780	965	790	54	705	884
21	Arm reach from the wall	840	53	754	925	775	48	701	855
22	Thumb tip reach	771	55	681	862	712	50	630	796
23	Shoulder grip length	720	58	625	814	667	54	570	750
24	Wall to acromion distance	101	18	72	132	99	17	70	128

(continued)

Table 6.2 (continued)

Sr. No.	Dimension	Men				Women			
		Mean	SD	Percentile		Mean	SD	Percentile	
				5th	95th			5th	95th
25	Wall to lumbo-sacral joint distance	44	14	24	70	44	13	25	68
26	Abdominal extension to wall	229	35	176	292	216	39	159	290
27	Chest depth	208	23	174	245	209	35	155	275
28	Biacromial breadth	332	45	273	420	296	36	246	360
29	Bideltoid breadth	417	34	365	475	375	36	325	440
30	Chest breadth	277	35	220	337	256	31	212	313
31	Inter scye breadth	304	33	251	360	276	35	224	336
32	Waist breadth	268	32	220	325	244	36	193	308
33	Hip breadth	303	29	259	352	291	32	240	342
34	Heel breadth	60	8	48	73	55	9	42	70
35	Bimalleolar breadth	68	7	55	79	62	9	50	78
36	Chest circumference	847	63	750	960	820	80	700	962
37	Waist circumference	772	88	650	940	736	95	605	911
38	Thigh circumference	443	49	362	520	416	58	325	510
39	Calf circumference	312	34	260	370	295	41	230	360
40	Wrist circumference	160	11	140	180	150	12	130	170
41	Grip diameter (inside)	48	6	39	58	45	6	35	57
42	Grip diameter(outside)	88	10	74	106	78	11	62	95
43	Middle finger-palm grip diameter	28	6	18	38	25	4	20	32
44	Vertical grip reach sitting	1178	69	1062	1287	1092	62	996	1193
45	Sitting height	831	53	745	910	776	45	706	851
46	Sitting eye height	728	52	645	803	673	45	605	750
47	Sitting acromial height	569	46	500	641	532	41	470	599
48	Elbow rest height	216	34	167	276	211	30	161	261

(continued)

Table 6.2 (continued)

Sr. No.	Dimension	Men				Women			
		Mean	SD	Percentile		Mean	SD	Percentile	
				5th	95th			5th	95th
49	Thigh clearance height sitting	130	23	98	173	126	24	90	170
50	Knee height sitting	505	36	450	566	469	32	416	520
51	Popliteal height sitting	418	31	369	470	393	32	341	447
52	Hand thickness at metacarpal-III	28	6	20	40	27	6	19	38
53	First phalanx digit III length	62	5	54	70	56	6	45	65
54	Grip span	61	11	45	83	59	7	45	70
55	Maximum grip length	121	18	93	150	108	17	84	138
56	Index finger diameter	17	3	12	23	16	3	12	20
57	Head length	186	12	167	208	181	16	156	210
58	Head breadth	148	14	130	175	143	15	121	170
59	Coronoid fossa to hand length	393	30	346	440	360	28	320	410
60	Fore arm hand length	454	31	401	503	419	30	371	467
61	Elbow grip length	355	32	305	405	324	29	280	373
62	Hand length	179	12	160	198	167	10	150	185
63	Palm length	101	8	89	113	94	8	80	106
64	Hand breadth across thumb	97	8	85	110	89	8	76	102
65	Hand breadth at metacarpal-III	80	7	69	91	73	7	61	85
66	Buttock-knee Length	536	37	474	597	510	38	440	567
67	Buttock-popliteal length	443	37	382	502	427	37	365	487
68	Abdominal depth sitting	220	35	170	283	219	36	165	285
69	Hip breadth sitting	314	35	257	372	309	37	250	371
70	Elbow-elbow breadth sitting	379	49	305	465	358	45	288	440
71	Knee-knee breadth	184	22	154	222	176	24	143	220
72	Foot length	245	15	220	270	224	17	199	250

(continued)

Table 6.2 (continued)

Sr. No.	Dimension	Men				Women			
		Mean	SD	Percentile		Mean	SD	Percentile	
				5th	95th			5th	95th
73	Instep length	185	15	160	210	169	16	143	195
74	Foot breadth (Ball of the foot)	94	10	78	110	89	8	75	102
75	Functional leg length	977	63	880	1085	931	60	834	1027
76	Biceps skinfold	5	3	2	10	4	2	1	7
77	Triceps skinfold	7	3	3	14	5	3	2	11
78	Subscapular skinfold	11	5	5	20	9	4	3	16
79	Supra iliac skinfold	11	5	3	21	10	4	3	20

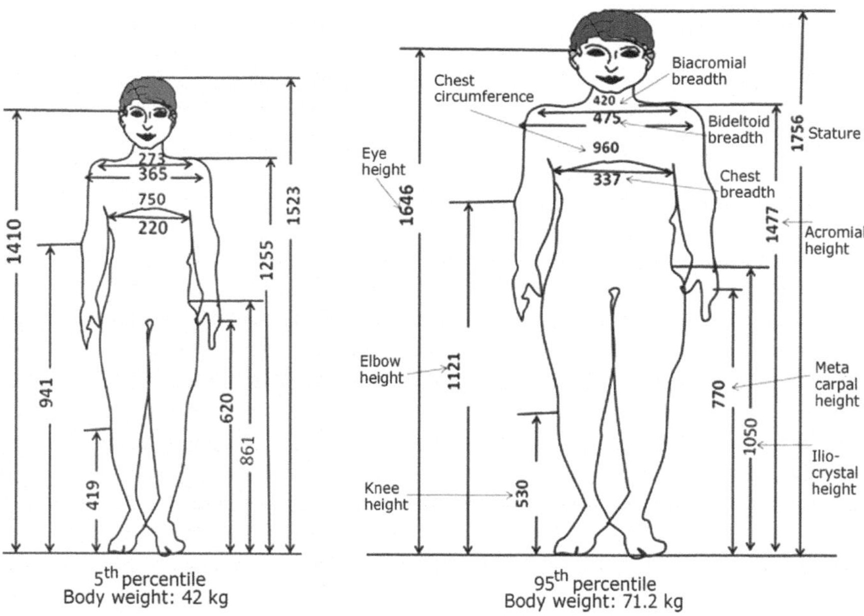

Fig. 6.2 The 5th and 95th percentile values for body dimensions of Indian men farm workers (All sizes in mm unless otherwise stated)

of isometric strength are available (Kroemer 1970; Mital and Kumar 1998; Gite et al. 2009). It involves a strength measurement setup and a procedure to be followed for measurement. A person is asked to build up the strength in the first 2 to 3 s and then maintain the peak strength for the next 3 s. Strength depends not only on the muscular

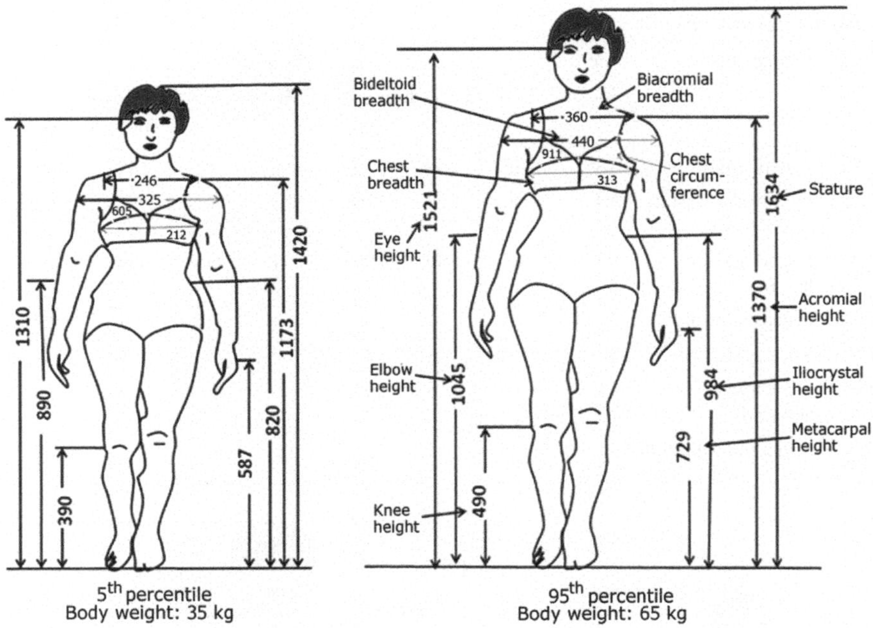

Fig. 6.3 The 5th and 95th percentile values for body dimensions of Indian women farmworkers (All sizes in mm unless otherwise stated)

capability, but also on health and training, technique, experience, and motivation. The personal and task factors that influence human strength are age, gender, posture, and reach distance, arm and wrist orientations, speed of exertion, and duration and frequency of efforts. The muscle strength is maximum at a young age, and it goes on decreasing with age. Mital and Kumar (1998) reported that at the age of 24 years, the strength is maximum, and at the age of 65, it decreases nearly 30% of the maximum.

Anthropometric data are available to an extent for different populations, but that is not the case for muscular strength data. From some countries of the West, strength data are available for reference and use (e.g. Davis and Stubbs 1977a, b, 1978, 1980; Laubach 1978; Snook 1978; Snook and Ciriello 1991; Mital and Kumar 1998). However, in the case of developing countries, very few data are available (e.g. Mamansari and Salokhe 1996). The muscle strength data for 16 parameters of 9515 Indian farm workers (5570 men and 3945 women) (Gite 2017) are described herewith. These strength parameters might be useful for the design of farm tools and equipment, as described in Chapters 12 and 13, concerning the ergo-design of farm tools and devices. The definitions of the strength parameters are given in Table 6.5, and the same is illustrated in Fig. 6.4a, b. The mean values for the parameters are given in Table 6.6.

Table 6.3 Anthropometric data of the male population of different countries (mean values in mm, unless otherwise stated)

Sr. No.	Dimension	India (Gite 2017)	Thailand (ISO 7250-2, ISO 2010)	Indonesia (Syuaib 2015)	Malaysia (Hassan et al. 2015)	China (ISO 2010)	Japan (ISO 2010)	Kenya (ISO 2010)	Nigeria (Onuoha et al. 2012)	Sweden (Hanson et al. 2009)	Britain (Pheasant 2003)	America (ISO 2010)	Germany (ISO 2010)	Portugal (Barroso et al. 2005)
1	Body weight, kg	55.1	64.3	57.1	–	60.0	67.5	70.6	56.7	75.0	75.0	83.2	79.0	74.0
2	Stature	1636	1666	1620	1696	1678	1696	1717	1634	1779	1740	1767	1750	1690
3	Acromial height	1364	1370	1358	1407	1367	1370	1440	1384	1459	1425	1445	1450	1395
4	Elbow height	1029	1025	1011	1045	1024	1031	1052	1012	1108	1090	–	1100	1050
5	Iliac spine height	908	918	–	–	–	912	922	–	–	–	–	–	–
6	Arm reach from the wall	840	–	814	–	–	–	–	841	–	–	–	–	–
7	Grip reach-forward reach (shoulder grip length)	720	719	699	–	–	693	764	728	–	665	–	740	730
8	Abdominal extension to wall	229	–	290	–	–	296	286	–	241	270	278[a]	285	265
9	Biacromial breadth	332	372	–	327	374	402	363	–	395	400	419	405	335
10	Bideltoid breadth	417	446	417	–	432	459	449	431	478	–	490	480	475
11	Grip diameter (inside)	48	–	43	–	–	–	–	43	–	–	–	–	–
12	Sitting height	831	870	827	856	908	910	842	837	946	910	921	910	920

(continued)

Table 6.3 (continued)

Sr. No.	Dimension	India (Gite 2017)	Thailand (ISO 7250-2, ISO 2010)	Indonesia (Syuaib 2015)	Malaysia (Hassan et al. 2015)	China (ISO 2010)	Japan (ISO 2010)	Kenya (ISO 2010)	Nigeria (Onuoha et al. 2012)	Sweden (Hanson et al. 2009)	Britain (Pheasant 2003)	America (ISO 2010)	Germany (ISO 2010)	Portugal (Barroso et al. 2005)
13	Popliteal height sitting	418	414	–	–	414	405	439	434	491	440	447[a]	450	400
14	Buttock-popliteal length	443	471	467	478	457	448	497	482	493	495	521[a]	495	485
15	Hip breadth sitting	314	345	300	–	323	361	361	290	387	360	376	375	380
16	Foot length	245	247	–	–	247	251	264	249	266	265	271[a]	265	–
17	Foot breadth	–	100	101	–	96	101	100	–	101	95	101[a]	101	–

[a] *Source* Hsiao et al. (2005)

Table 6.4 Anthropometric data of the female population of different countries (mean values in mm, unless otherwise stated)

Sr. No.	Dimension	India (Gite 2017)	Thailand (ISO 7250-2, ISO 2010)	Indonesia (Syuaib 2015)	Malaysia (Hassan et al. 2015)	China (ISO 2010)	Japan (ISO 2010)	Kenya (ISO 2010)	Nigeria (Onuoha et al. 2012)	Sweden (Hanson et al. 2009)	Britain (Pheasant 2003)	America (ISO 2010)	Germany (ISO 2010)	Portugal (Barroso et al. 2005)
1	Body weight, kg	47.8	53.5	52.3	58.7	52	51.9	67.9	51.3	65.0	63.0	69.6	66.0	64.0
2	Stature	1521	1547	1525	1557	1571	1570	1629	1568	1674	1610	1638	1625	1565
3	Acromial height	1265	1170	1250	1276	1272	1263	1370	1291	1359	1310	1343	1345	1295
4	Elbow height	963	951	952	962	961	953	996	984	1042	1005		1020	965
5	Iliac spine height	856	872	–	–	–	837	885	–	–	–	–	–	–
6	Arm reach from the wall	775	–	761	–	–	–	–	772	–	–	–	–	–
7	Grip reach-forward reach (shoulder grip length)	667	678	660	–	–	635	724	677	–	705	692[a]	690	675
8	Abdominal extn. to wall	216	–	–	–	–	248	280	–	226	255	266[a]	290	260
9	Biacromial breadth	296	344	–	–	350	358	345	–	357	355	376	370	300
10	Bideltoid breadth	375	400	404	379	398	407	437	417	425	395	431	435	445
11	Grip diameter (inside)	45	–	38	–	–	–	–	36	–	–	–	–	–
12	Sitting height	776	824	774	815	855	849	797	748	892	850	864	860	865

(continued)

Table 6.4 (continued)

Sr. No.	Dimension	India (Gite 2017)	Thailand (ISO 7250-2, ISO 2010)	Indonesia (Syuaib 2015)	Malaysia (Hassan et al. 2015)	China (ISO 2010)	Japan (ISO 2010)	Kenya (ISO 2010)	Nigeria (Onuoha et al. 2012)	Sweden (Hanson et al. 2009)	Britain (Pheasant 2003)	America (ISO 2010)	Germany (ISO 2010)	Portugal (Barroso et al. 2005)
13	Popliteal height sitting	393	389	–	400	379	371	412	395	447	400	413[a]	415	365
14	Buttock-popliteal length	427	474	450	431	434	448	509	432	478	480	510[a]	485	470
15	Hip breadth sitting	309	352	299	353	345	361	402	296	414	370	410	390	400
16	Foot length	224	227	223	227	229	230	246	214	243	235	247[a]	245	–
17	Foot breadth	89	94	93	89	88	93	92	–	91	90	92[a]	92	–

[a] *Source* Hsiao et al. (2005)

Table 6.5 Definitions of muscular strength parameters

Sr. No.	Parameter	Definition
1	Handgrip strength (right)	The grip strength of the right hand measured with a handgrip dynamometer when the person stands erect with his arms hanging downwards
2	Handgrip strength (left)	The grip strength of the left hand measured with a handgrip dynamometer when the person stands erect with his arms hanging downwards
3	Push strength (both hands), standing	The maximum push force applied with both hands on a load cell in a standing position with one leg forward
4	Pull strength (both hands), standing	The maximum pull force applied with both hands on the load cell in a standing position with one leg backward
5	Push strength (left hand), sitting	Maximum push strength applied with the left hand on the load cell at seat height level in the horizontal plane while sitting on a chair
6	Pull strength (left hand), sitting	The maximum pull force applied with left hand on the load cell at seat height in the horizontal plane while sitting in a chair
7	Push strength (Right hand), sitting	The maximum push force applied with the right hand on the load cell at seat height level in the horizontal plane while sitting on a chair
8	Pull strength (Right hand), sitting	The maximum pull force applied with the right hand on the load cell at seat height level in the horizontal plane while sitting in a chair
9	Leg strength (right), sitting	The maximum force applied by the right leg on the load cell while the person sits erect on a chair with his knee joint flexed at an angle between 90° and 110°
10	Foot strength (right), sitting	The maximum force applied by the right foot on the load cell while keeping the heel as the fulcrum and the person sits straight on a chair with his knee flexed at an angle 90^0 to 110^0
11	Leg strength (left), sitting	The maximum force applied by the left leg on the load cell while the person sits erect on a chair with his knee flexed at an angle between 90° and 110°

(continued)

Table 6.5 (continued)

Sr. No.	Parameter	Definition
12	Foot strength (left), sitting	The maximum force applied by the left foot on the load cell while keeping the heel as the fulcrum and the person sits erect on a chair with his knee flexed at an angle between 90^0 and 110^0
13	Torque strength (preferred hand), standing	Maximum torque applied with the preferred hand in a vertical plane on the crank in standing posture (length of the crank to be specified)
14	Torque strength (both hands), standing	Maximum torque applied by both hands in a vertical plane on the crank in standing posture (length of the crank to be specified)
15	Torque strength (both hands), sitting	Maximum torque applied by both hands in the clockwise direction in an inclined plane in a sitting position. (length of the crank to be specified)
16	Handgrip torque (preferred hand), sitting	Maximum grip torque applied by preferred hand in the anti-clockwise direction

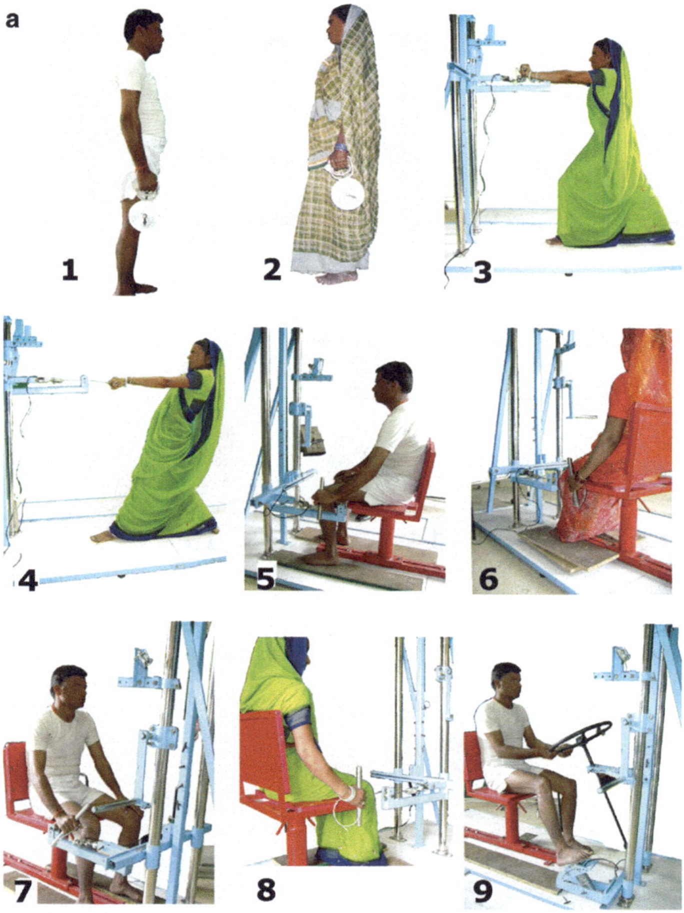

Fig. 6.4 **a** Muscle strength parameters for agricultural equipment design (parameter numbers refer to the serial number in Table 6.5). **b** Muscle strength parameters for agricultural equipment design (refer to the serial number in Table 6.5)

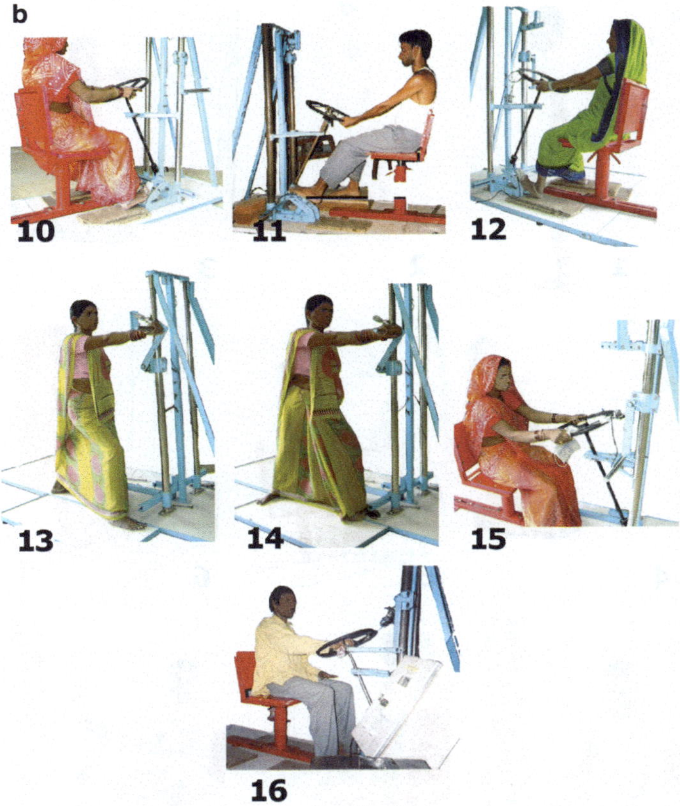

Fig. 6.4 (continued)

Variation in Strength Capacity

Most of the published literature indicates that the strength of women workers ranges from 60 to 70% that of men. Therefore, the muscular strength of women is generally taken as 2/3rd that of the men. Also, the dynamic strength amounts to about 50 to 90% of the maximal static force, depending on some factors. Handedness has not much effect on muscle strength. In right-handed people, the right hand is slightly stronger than left, though, the differences may not be significant. The maximum strength values of the persons are related to their body mass. The strength values vary from region to region owing to differences in body mass and several other factors. A comparison of North American data and Indian data are shown in Table 6.7.

Some studies have shown that many measures of strength correlate positively with each other and with anthropometric data and body weight, but the correlations are weak to have much predictive value for the designers (Laubach 1978). Therefore, it is preferable to measure the muscular strength data instead of calculating it from

Table 6.6 Muscular strength data of Indian agricultural workers

		Men				Women			
Sr. No.	Dimension	Mean	SD	Percentile		Mean	SD	Percentile	
				5th	95th			5th	95th
1	Handgrip strength (right)	347	91	206	510	220	72	118	353
2	Handgrip strength (left)	331	92	195	496	207	75	101	349
3	Push strength (both hands), standing	220	59	133	323	143	45	84	230
4	Pull strength (both hands),standing	218	54	140	320	159	44	101	240
5	Push strength (left hand) sitting	81	24	46	124	62	25	29	112
6	Pull strength (left hand), sitting	98	28	58	152	74	24	42	122
7	Push strength (right hand), sitting	84	24	47	125	64	24	30	109
8	Pull strength (right hand), sitting	101	30	54	156	76	23	42	122
9	Leg strength (right), sitting	426	116	251	636	304	97	160	486
10	Foot strength (right), sitting	317	110	155	516	222	88	98	390

(continued)

Table 6.6 (continued)

Sr. No.	Dimension	Men				Women			
		Mean	SD	Percentile		Mean	SD	Percentile	
				5th	95th			5th	95th
11	Leg strength (left), sitting	419	121	231	626	291	96	152	470
12	Foot strength (left), sitting	299	103	149	496	209	85	94	372
13	Torque strength (preferred hand), standing, Nm	46	15	26	78	35	11	21	55
		(175)*	(57)*	(98)*	(290)*	(135)*	(42)*	(79)*	(211)*
14	Torque strength (both hand), standing, Nm	56	19	31	95	41	12	25	62
		(217)*	(73)*	(120)*	(364)*	(158)*	(46)*	(97)*	(239)*
15	Torque strength both hand (sitting), Nm	62	17	36	91	41	12	23	63
		(293)**	(80)**	(173)**	(434)**	(194)**	(58)**	(110)**	(301)**
16	Hand grip torque (preferred hand), sitting, Nm	7 (31)**	3 (15)**	3 (13)**	12 (58)**	4 (21)**	2 (10)**	2 (10)**	9 (42)**

Values are in Newton (N) unless otherwise stated; Newton-meter (Nm)
*Data in parentheses are force levels in N; lever arm length is 0.26 m
**Data in parentheses are force levels in N; lever arm length is 0.21 m

Table 6.7 Muscle strength values of Indian and North American population

Sr. No.	Strength parameter	India	Source	North America	Source
1	Push force with both hands, N	220 (M) 143 (F)	Gite (2017)	372(M) 180 (F)	Chaffin et al. (1983)
				393 (M) 223 (F)	Kumar (1995)
2	Pull force with both hands, N	218 (M) 159 (F)	Gite (2017)	267 (M) 166(F)	Chaffin et al. (1983)
				518 (M) 346(F)	Kumar (1995)
3	Strength in elbow flexion (right), N	195 (M)	Shrimali (2005)	287 (M)	Pytel and Kamon (1981)
4	Hand grip strength (right), N	347 (M) 220 (F)	Gite (2017)	459 (M)	Karwoski and Mital (1986)
				502 (M) 319 (F)	Hsiao et al. (2005)

regression equations and correlation factors. Further, it is necessary to have the muscle strength data of each country and to use those in the design of tools/equipment for better human performance and safety.

References

Barroso, M. P., Arezes, P. M., da Costa, L. G., & Miguel, A. S. (2005). Anthropometric study of Portuguese workers. *International Journal of Industrial Ergonomics, 35*(5), 401–410.

Bolstad, G., Benum, B., & Rokne, A. (2001). Anthropometry of Norwegian light industry and office workers. *Applied Ergonomics, 32*(3), 239–246.

Chaffin, D. B., Andres, R. O., & Garg, A. (1983). Volitional postures during maximal push/pull exertions in the sagittal plane. *Human Factors, 25*(5), 541–550.

Chakrabarti, D. (1997). *Indian anthropometric dimensions for ergonomic design practice.* Ahmedabad, India: National institute of design.

Das, B., & Grady, R. M. (1983). Industrial workplace layout design an application of engineering anthropometry. *Ergonomics, 26*(5), 433–447.

Das, B., & Sengupta, A. K. (1996). Industrial workstation design: a systematic ergonomics approach. *Applied Ergonomics, 27*(3), 157–163.

Davis, P. R., & Stubbs, D. A. (1977a). Safe levels of manual forces for young males (1). *Applied Ergonomics, 8*(3), 141–150.

Davis, P. R., & Stubbs, D. A. (1977b). Safe levels of manual forces for young males (2). *Applied Ergonomics, 8*(4), 219–228.

Davis, P. R., & Stubbs, D. A. (1978). Safe levels of manual forces for young males (3): *Applied Ergonomics, 9*(1), 33–37.

Davis, P. R., & Stubbs, D. A. (1980). Force limits in manual work. Guilford: IPC Science and Technology Press.

Dewangan, K. N., Kumar, G. P., Suja, P. L., & Choudhury, M. D. (2005). Anthropometric dimensions of farm youth of the north eastern region of India. *International Journal of Industrial Ergonomics, 35*(11), 979–989.

Dewangan, K. N., Owary, C., & Datta, R. K. (2008). Anthropometric data of female farm workers from north eastern India and design of hand tools of the hilly region. *International Journal of Industrial Ergonomics, 38*(1), 90–100.

Gite, L. P. (2017). Development of Ergonomical Design Guidelines for Agricultural Tools, Equipment and Work Places, Final Report of the Emeritus Scientist Project, Central Institute of Agricultural Engineering, Bhopal, India.

Gite, L. P., Majumder, J., Mehta C. R., & Khadatkar, A. (2009). Anthropometric and strength data of Indian agricultural workers for farm equipment design. CIAE/2009/4, ICAR-Central Institute of Agricultural Engineering, Bhopal.

Gite, L. P., & Yadav, B. G. (1989). Anthropometric survey for agricultural machinery design: an Indian case study. *Applied Ergonomics, 20*(3), 191–196.

Grandjean, E. (1988). *Fitting the task to the man.* New York: Taylor & Francis.

Gupta, P. K., Gupta, M. L., & Sharma, A. P. (1983). Anthropometric survey of Indian farm workers. *AMA, Agricultural mechanization in Asia, Africa and Latin America., 16*(1), 27–30.

Hanson, L., Sperling, L., Gard, G., Ipsen, S., & Vergara, C. O. (2009). Swedish anthropometrics for product and workplace design. *Applied Ergonomics, 40*(4), 797–806.

Haslegrave, C. M. (1980). Anthropometric profile of the British car driver. *Ergonomics, 23*(5), 437–467.

Hassan, S. N., Yusuff, R. M., Zein, R. M., Hussain, M. R., &Tamil Selvan, H. K. (2015). Anthropometric Data of Malaysian Workers. In: New ergonomics Perspective, Yamamoto (Ed.),Taylor and Francis Group, London.

Hsiao, H., Whitestone, J., Bradtmiller, B., Whisler, R., Zwiener, J., Lafferty, C., et al. (2005). Anthropometric criteria for the design of tractor cabs and protection frames. *Ergonomics, 48*(4), 323–353.

ISO. (1996). ISO 7250-Part 1. Basic Human Body Measurements for Technological Design - Body measurement definitions and landmarks. International Standard Organization, Genava.

ISO. (2003). *ISO 15535.* International Standard Organization, Genava: General requirements for establishing an anthropometric database.

ISO. (2010). ISO 7250-Part 2. Basic Human Body Measurements for Technological Design - Statistical summaries of body measurements from national populations. International Standard Organization, Genava.

ISO. (2013). *ISO 14738-Safety of machinery—Anthropometric requirements for the design of workstations at machinery.* Genava: International Standard Organization.

ISO. (2015). ISO 7250-Part 3. Basic Human Body Measurements for Technological Design - Worldwide and regional design ranges for use in product standards. International Standard Organization, Genava.

Karwowski, W., & Mital, A. (1986). Isometric and isokinetic testing of lifting strength of males in teamwork. *Ergonomics, 29*(7), 869–878.

Kroemer, K. E. (1970). Human strength: terminology, measurement, and interpretation of data. *Human Factors, 12*(3), 297–313.

Kroemer, K. H. E. (1983). Engineering anthropometry: work, space and equipment to fit the user. In D. Oborne & M. Gruneberg (Eds.), *The physical environment at work* (pp. 39–68). London: Wisley.

Kumar, S. (1995). Upper body push-pull strength of normal young adults in sagittal plane at three heights. *International Journal of Industrial Ergonomics, 15*(6), 427–436.

Laubach, L. L. (1978). Human muscular strength. (in Anthropometric Source Book, Volume I: Anthropometry for Designers, NASA reference publication 1024). National Aeronautics and Space Administration, Washington, USA, pp-VII-1 to 55.

Lee, Y. S. (2000). *Applied Korean anthropometric database for product design: clothing design.* MOCIE, Korea: Agency for Technology and Standards.

Lin, Y. C., Wang, M. J. J., & Wang, E. M. (2004). The comparisons of anthropometric characteristics among four peoples in East Asia. *Applied Ergonomics, 35*(2), 173–178.

Liu, W. C. V., Sanchez-Monroy, D., & Parga, G. (1999). Anthropometry of female maquiladora workers. *International Journal of Industrial Ergonomics, 24*(3), 273–280.

Mamansari, D. U., & Salokhe, V. M. (1996). Static strength and physical work capacity of agricultural labourers in the central plain of Thailand. *Applied Ergonomics, 27*(1), 53–60.

Marras, W. S., & Kim, J. Y. (1993). Anthropometry of industrial populations. *Ergonomics, 36*(4), 371–378.

Mital, A., & Kumar, S. (1998). Human muscle strength definitions, measurement, and usage: Part I-Guidelines for the practitioner. *International Journal of Industrial Ergonomics, 22*(1–2), 101–121.

Moustafa, A. W., Davies, B. T., Darwich, M. S., & Ibraheem, M. A. (1987). Anthropometric study of Egyptian women. *Ergonomics, 30*(7), 1089–1098.

NASA. (1978a). Anthropometric Source Book, Vol. I- Anthropometry for Designers, NASA Reference Publication 1024. National Aeronautics and Space Administration, Washington.

NASA. (1978b). Anthropometric Source Book, vol. II – A Handbook for Anthropometric Data, NASA Reference Publication 1024. National Aeronautics and Space Administration, Washington.

Onuoha, S. N., Idike, F. I., & Oduma, O. (2012). Anthropometry of south eastern Nigeria agricultural workers. *International Journal of Engineering and Technology, 2*(6), 962–968.

Pheasant, S. (2003). *Body Space, Anthropometry, Ergonomics and the Design Work*. London: Taylor & Francys.

Philip, G. S., & Tewari, V. K. (2000). Anthropometry of Indian female agricultural workers and implication on tool design. *AMA, Agricultural Mechanization in Asia, Africa and Latin America, 31*(1), 63–66.

Pytel, J. L., & Kamon, E. (1981). Dynamic strength test as a predictor for maximal and acceptable lifting. *Ergonomics, 24*(9), 663–672.

Roebuck, J. A., Kroemer, K. H., & Thomson, W. G. (1975). *Engineering anthropometry methods* (Vol. 3). John Wiley & Sons.

Rogan, A., & O'Neill, D. (1993). Ergonomics aspects of crop production in tropical developing countries: a literature review. *Applied Ergonomics, 24*(6), 371–386.

Sanders, M. S., & McCormick, E. J. (1992). Human factors in engineering and design. McGraw-Hill International ed. McGraw-Hill Inc., Singapore.

Sen, R. N., Nag, P. K., & Ray, G. G. (1977). Some anthropometry of the people of eastern India. *Journal of Indian Anthropological Society, 12,* 201–208.

Shrimali, H. (2005). Ergonomical study on an agricultural foot sprayer. Ph.D. thesis, Maharana Pratap University of Agriculture and Technology, Udaipur, India.

Snook, S. H., & Ciriello, V. M. (1991). The design of manual handling tasks: revised tables of maximum acceptable weights and forces. *Ergonomics, 34*(9), 1197–1213.

Snook, S. H. (1978). The ergonomics society the society's lecture. (1978). THE design of manual handling tasks. *Ergonomics, 21*(12), 963–985.

Syuaib, M. F. (2015). Anthropometric study of farm workers on Java Island, Indonesia, and its implications for the design of farm tools and equipment. *Applied Ergonomics, 51,* 222–235.

Tewari, V. K. & Ailavadi, R. (2002). Ergonomic database for engineering design of agricultural machines. Final Report, Department of Agricultural and Food Engineering, Indian Institute of Technology, Kharagpur, West Bengal, India.

Victor, V. M., Nath, S., & Verma, A. (2002). Anthropometric survey of Indian farm workers to approach ergonomics in agricultural machinery design. *Applied Ergonomics, 33*(6), 579–581.

Wang, M. J. J., Wang, E., & Lin, Y. C. (2002). The anthropometric database for children and young adults in Taiwan. *Applied Ergonomics, 33*(6), 583–585.

Yadav, R., Tewari, V. K., & Prasad, N. (1997). Anthropometric data of Indian farm workers—a module analysis. *Applied Ergonomics, 28*(1), 69–71.

Part III
Farm Development

Chapter 7
Farm Mechanization: Nature of Development

Nature of Farm Mechanization

Agriculture is in transition today, with the coexistence of traditional farming and infusion of large machinery. Labour-intensive traditional practices are fast overtaken by the newer adopted manually and power-operated machinery in various farm operations. Equipping people with machinery and devices in doing their daily chores is usually referred to as mechanization. Generally speaking, mechanization connotes the application of machinery, tools, and devices to carry out farm activities as an alternative to those traditionally performed by bullocks, horses, and other draught animals or by human labour. Farm activities cover, for example, seedbed preparation, sowing, weeding, inter-cultivation, spraying, fertilizer application, harvesting, combining, baling, transportation, water-lifting for irrigation, threshing, winnowing, fodder cutting, and feed grinding. Design, development, and application of mechanical aids, motorized equipment and devices in crop production, irrigation control, material handling, storing and processing, dairy applications, oil pressing, cotton ginning, and the like are broadly recognized as the indicators of farm mechanization.

The machinery, such as power tillers, tractors, combines, electric motors, harvesters, threshers, and hauling equipment, are the assembles of gears, shafts, chains, belts, knives, and shakers, to perform the desired action either in a stationary position or while moving across a field. The mobile machinery is usually powered by an integral petrol/diesel engine or by a separate driver. There may be partial mechanization when only a part of the farm work is done by machine. Complete mechanization would indicate that the animal or human labour is wholly dispensed with power operated machines.

The first commercial tractor powered by petrol was sold in 1902 in the United States. Henry Ford introduced the Fordson tractor during World War I. A tractor provides machine power to perform farm tasks, like pulling of farm implements of various complexities for land tilling, planting, cultivating, spraying of agrochemicals, and harvesting crops, and also for hauling materials and personal transportation.

P. K. Nag and L. P. Gite, *Human-Centered Agriculture*, Design Science and Innovation,
https://doi.org/10.1007/978-981-15-7269-2_7

Often the terms agricultural mechanization and farm mechanization are used synonymously. The agricultural mechanization covers the entire gamut of activities in crop production and processing on or off the farms. On the other hand, farm mechanization focuses on the machinery application within the boundaries of farms, primarily including production machinery for raising crops. For example, the extent of tractor power applied to carry out agricultural operations is often described as the level of tractorization. A whole range of farm machinery is available today that are primarily observed in large farms and corporate farming. Some examples of gigantic farm machines are shown in Fig. 7.1. However, the significant categories of machinery are:

Fig. 7.1 Large format farm machinery; **a** Narrow track tractor, **b, c** Land tilling machines, **d** Precision seed drill, **e** Planter, **f** Inter-cultivator, **g, h** Weeders, **i** Sprayer, **j** Combine harvester, **k** Potato harvester, **l** Bale lifter

Machinery for soil tillage—ploughs, tillers, subsoilers, harrows, rollers, levellers, graders, and the like. The machines are designed to turn, agitate, and level the soil for planting. These may be available in different sizes and power sources (e.g. rototillers used in paddy cultivation are small machines; combined subsoilers, drills, and harrows are large machines).

*Machinery for p*lanting—planters, drills, and broadcast seeders are the machines to put seeds and seedlings at defined depth and spacing on the ground. Planters are designed for a single-row seeding mechanism, or as the multi-row planters with attachments for applying fertilizer, pesticides, and herbicides simultaneously.

Machinery for cultivation—rotary hoes, cultivators, and weeders. These machines are available in different sizes and formats for use in removing unwanted weeds from the soil and improving the soil tillage.

Machinery for forage harvesting—mowers, choppers, and balers, and the like. These machines are used to cut stems of roughage crops and to bundle those for storing. A simple mower merely cuts the harvest, and the chopper separates the stalk from the root, chop the entire plant into pieces, and put the yield to a wagon for transportation. Some fodder crops undergo the field-drying process and then baled. Pelleting machines and balers compress the fodder crops. Bales may be of different sizes, from 20 to 40 kg, to as large as 400 to 500 kg, which are handled by mechanical handling systems.

Machinery for grain and fibre harvesting—reapers, combines, threshers, cotton, and corn pickers. These machines collect the ripe grain or fibre from the plant, and thresh grains and fill in a bin or sack for storage and transportation. For instance, the moving combine harvester cuts the standing ripe grain crop plants; separates grain from the stalk; cleans, collects in a big container, and loads the contents to a wagon.

Machinery for vegetable harvesting—diggers and lifters are used to dig and separate the crops or lift the plant from the earth. A potato combine harvester, for example, comprises of potato digging, sorting, grading, polishing, and putting in bags.

Machinery for fruit and nut harvesting—machines used, for example, to harvest fruit, berries, and coconuts. It may be a tractor-mounted, vibrating tree shaker to separate ripe fruits from the tree, or cut the fruits, catch the falling fruits directly to the storage container.

Machinery for transport and elevation—These machines range from a simple wagon attached to a platform on wheels to a self-loading and stacking transport unit. Moving chain, belt conveyors, or pneumatic conveyors are used to move materials horizontally or vertically.

Machinery for chemical applications—These machines apply agrochemicals, such as fertilizers, herbicides, and pesticides, either in the form of liquid, powder, or granules. The applicators may be portable manually or power-operated, or vehicle-mounted that distributes chemicals either by pressure through a nozzle or by centrifugal force. At places, there is a growing use of aircraft for chemical applications.

Machinery for sorting and packaging—These machines include fanning mill, seed mill, graders that grade and clean grains and seeds. Machinery for packaging primarily grades fruit and vegetables, and wrap the produce into a bag or any other plastic container.

Scopes and Challenges of Farm Mechanization

Mechanization is essentially about deploying multiple power sources, including powered machinery, electric or diesel operated engines/pump sets for irrigation efficiency, animal (bullocks, buffaloes, camels, donkey, and mules) drawn implements and manually operated equipment and devices. Mechanization is also unravelling newer avenues of intelligent, lean, and efficient engineering solutions. These newer technological avenues facilitate, to increase cropping intensity, to get high yield of crop varieties, and also to minimize impacts of heavy machinery on the soil and water resources. Modern technology in precision farming makes efficient use of seeds and fertilizer and also crop protection chemicals.

There is a general recognition that mechanization is the avenue to increase production in all scales of farms. In conventional farming, a farmer is estimated to walk nearly 65 km to till about one hectare of land once by bullocks with a country plough having 15 cm furrow width. To transplant the one-hectare area, a woman would be inserting her hand fingers in mud for about 0.15 million times. A worker makes approximately 40 semi-circular movements of sickle per minute in manual harvesting of wheat or rice. Farm operations, such as bund trimming, manual carrying of load, are substantial energy demanding, causing workers more fatigue. The mechanical power available through tractors and stationary engines is efficient in producing useful work, in terms of yield per hectare of land, and also earnings per capita. The efficiency of diesel engines and petrol engines varies between 25 and 40%, depending on the size of the tractors in use. Electricity and renewable energy sources like solar and wind are used in irrigation.

Farm mechanization is capital intensive. However, efficient use of mechanized inputs and resources has a direct bearing on optimized utilization of human power, reduction in drudgery, improvement in the working environment, reduction in crops and food losses, timeliness and precision of agricultural operations, and increase in farm production and productivity. A direct relationship exists between farm power availability and input savings through improvement in the quality of cultivation, tilling techniques, use of seeds and fertilizers, and reduction of production costs.

On consideration of the multitude of aspects, such as technological, economic, social, environmental, and cultural aspects, sustainable mechanization in a sector or a country contributes to the sustainability of the food production and supply (Kienzle et al. 2013). The economically weaker countries view that the extensive farm mechanization is crucial in shaping the future of small and marginal farm holdings, and the avenue for the freedom of their backwardness. Therefore, the performance of agriculture in a country or region might be adjudged from the extent of farm mechanization.

There is a continuing debate, however, about the relevance of mechanization development in abundant labour economies. Some inherent drawbacks are evident, like fragmented land size, constraints of small and marginal landholdings, and general lack of awareness of the farmers about efficient farm machines. Undoubtedly, tractor power supplements crop intensification, leading to an increase in production and productivity and a reduction in the cost of production in many regions of the world. However, the concerns are vivid in the developing world of massive displacement of rural labour, particularly in harvesting seasons. In the countries that face constraints of significant scale of unemployment, displacing of human labour from agriculture bears consequences of livelihood. There has been a recognizable shortfall in the availability of the workforce in farming seasons in many countries. The migration of people from agriculture to other occupational pursuits is a definite trend. For example, in populous India, in 2011, the agricultural workforce was about 56% of the total workforce (Mehta et al. 2014a, b). In 2016, it went down to 44% (FAO 2018), and would likely go down to 25% in another three decades. There is a deterring concern that mechanization would eliminate draught animals mainly from farming.

Global Scenario of Farm Mechanization

United States of America

Farm mechanization scenario varies from region to region, and country to country. The United States (US), with 2.1 million farms covering an area of about 370 million hectares and the average farm size of 175 ha, is a net exporter of food (corn, soybean, wheat, meat, milk) to the global market (USDA 2014). About 50% of the US population was employed in agriculture in 1870 (Wikipedia.org 2019), and the number today has fallen to a mere 1.7% of the total population (FAO 2018).

Mechanization in the US farming started long back, with several significant milestones, including the introduction of John Deere's steel plough, Cyrus McCormick's mechanical reaper, Eli Whitney's cotton gin, and the Fordson tractor and combine harvester. Now, modern agriculture in the US, however, reached a gigantic proportion all over the croplands and rangelands. The total number of tractors and other self-propelled machines (combines and harvesters) is estimated at 5.3 million in the US (USDA 2014). Most farming activities are mechanized, barring a few sectors like an apple orchard, berry picking. The machines used on the US farms include ploughs, harrows, cultivators, seed drills, fertilizer applicators, sprayers, rain guns, sprinklers, grain and bean combines, cotton pickers, sugarcane harvesters, potato harvesters, forage harvesters, and hay balers, and other crop-specific implements and devices.

Advances in biological and chemical research, coupled with massive farm mechanization in the US had a revolutionary impact on its food production and economy, and also employment generation in other allied sectors. The stories of mechanization and its implications on economic growth in different countries of the western world are very much similar. Despite a high magnitude of farm mechanization, agriculture ranks among the most hazardous industries. The farmworkers encounter risks of injuries, noise-induced hearing loss, lung and skin diseases, and cancers due to exposure to pesticide and solar exposures. In 2017, 416 farmers and farm workers died from work-related injuries resulting in a fatality rate of 20.4 deaths per 100,000 workers. The transport accidents and overturn of the tractor were the leading cause of death. Also, every day, about 100 agricultural workers suffer a lost-work-time injury in agriculture (NIOSH 2019).

Eastern Europe and Central Asia

The region of eastern Europe and central Asia includes the countries of the former Soviet Union (Armenia, Azerbaijan, Belarus, Estonia, Georgia, Kazakhstan, Kyrgyzstan, Latvia, Lithuania, Moldova, Russia, Tajikistan, Turkmenistan, Ukraine, and Uzbekistan) and also the countries of Eastern Europe (Albania, Bulgaria, Czechoslovakia, Hungary, Poland, Romania, and Yugoslavia). In most of these countries, agriculture was organized either as collectives (cooperatives) or as state farms. Today, many of these countries have become members of the European Union (EU). The policy restructuring of forming large farm sizes had a significant impact on agricultural development, efficient use of farm technologies, and the development of mechanization in the region. The Russian Federation, however, witnessed a consistent decline in the total sown area between 1991 and 2005 from 115 to 77 million hectares, and gross grain output fell from 89 to 78 million tonnes (Clarke 2013). The recovery trend has begun in recent years with the improvement in investment in farm technologies and machinery.

China

New agricultural technology and mechanization are the vital driving force towards agrarian development. With these technologies, China has made a remarkable advance in transforming the country's traditional farming into modern agriculture. During the last 60 years, China's population has increased from 450 million to 1.44 billion (FAO 2018). The arable land is approximately 1.4 million sq.km, representing 10% of the total cultivable land in the world. About half of the arable farmland and irrigated areas permanently support crops, to feed 20% of the global population. The rural economic system reform (Stavis 1978) has achieved the nation's ability to feed the projected 1.6 billion people by 2030.

China comprises about 30 million hectares of paddy fields. In the 70s, China practised agricultural mechanization (tractorization) in state-owned farms. In 2013, the mechanization levels were reported as 42% for tillage, 10% for sowing, 0.7% for paddy transplanting, and 3% for rice harvesting (Wang 2013). The Government

State Law on Promotion of Agricultural Mechanization, 2004 provides for all government bodies to include the promotion of agricultural mechanization to the national and social development plans, and to increase financial support for its development. The government adopted the principles to (a) simultaneously develop full and partial mechanization, including improvisation in hand tools technology, (b) equally emphasize on manual, animal, and electro-mechanical power, and (c) combine R&D for engineering, agricultural, and biological technologies. The agricultural workforce has been reduced from 49% in 1995 to 18% in 2016 (FAO 2018). The state farm mechanization level is estimated to reach 70% by 2020 and 85% by 2030, with more considerable attention to production and post-harvest processing of cash crops, fruits, and vegetables, livestock, particularly in hilly and mountainous areas.

India

India, with a population of 1.34 billion in 2016, is very close to the doorstep of China. About 2/3rd of its people live in rural areas and are dependent on agriculture, and nearly, 44% of the country's total workforce is employed in agriculture (FAO 2018). Of India's arable land area of 160 million hectares, its gross irrigated crop area is the largest in the world, with about 83 million hectares. About 35% of the cultivated area is irrigated, and the remaining is the rain-fed (Singh 2013, 2015). The climatic conditions are full of extremes, with the temperature conditions varying from arctic cold to equatorial, hot, and rainfall from extreme aridity with less than 100 mm in the Thar Desert of Western India to the world's maximum rainfall zone of 11,200 mm at Mowsinram in the northeast.

The food grain production during 2017–18 was about 285 million tonnes (GOI 2019). As per 2011 census, there were 263 million agricultural workers, of which about 37% were women. About 48% is the cultivated area distributed among 120 million landholdings. Nearly 85% of the total farm holdings are the small and marginal category, accounting for 44% of the total cultivated area, and the average size of landholding is 1.16 hectares (Mehta et al. 2014a, b; Mondal 2019). The average farm size in India is infinitely small in comparison to the regions of the European Union (14 hectares) and the US (170 hectares). The situation of land sizes and holdings has an inherent limitation to suit *economies of scale* and an impact on the penetration of machinery, overall efficiency, and farm output (Mondal 2019). Large farm machinery is challenging to operate on such landholdings, which, in some cases, are entirely unsuitable (Tewari et al. 2012; Singh et al. 2015). The capital intensive farm machinery is not affordable by the majority of the farmers. At places, there is a local arrangement of custom hiring of farm machinery.

In 2013, the total power availability on Indian farms was about 1.84 kW/ha. Of this, 46% was from tractors, 27% from electric motors, 16% from diesel engines, 5% each from human labour and draught animals, and 1% from power tillers. The cropping intensity in Indian agriculture increased with the increase in power availability. The power availability of 0.36 kW/ha during 1975–76 increased fivefold during 2012–13.

Farm power availability and food grain productivity have a direct relationship, and the Indian states with higher farm power availability have, in general, more productivity. The farm power availability during the year 2020 will reach about 2.2 kW/ha.

Among the Asia-Pacific countries, India is one of the first nations that fuelled the growth of the agricultural equipment market during the recent decades, with a surge in demand for machinery, like tractors, power tillers, combine harvesters, rotavators, threshers, and rice transplanters. The sale of tractors has grown from 0.22 million in 2001–02 to 0.80 million in 2017–18. The total tractor population in the country is about 4.8 million, and the density is 33 tractors per 1000 ha of cultivated area. The power tillers are mainly used in the eastern and southern states of India, mostly for rice crop cultivation. Overall, power tiller density is 2.21 per 1000 ha of cultivated area (Mehta et al. 2014a, b).

The extent of penetration of mechanization technologies varies with farm sizes, such as the large farms (>10 hectares) followed by medium farms (4 to 10 hectares). The level of mechanization ranges from 42% in soil working and seedbed preparation, 29% in seeding and planting, 34% in plant protection, and 37% in irrigation. The level of mechanization is about 60–70% for wheat and rice harvesting and threshing (Mehta et al. 2014a, b). The machines that are getting popular are ploughs, harrows, cultivators, seed drills, seed cum fertilizer drills, planters, zero till drills, paddy transplanters, reapers, reaper binders, threshers, combine harvesters, straw reapers, and balers. A significant geospatial variation has been observed in tractor density. Northern India has the highest concentration of tractors, followed by southern and western India, and eastern India.

In the overall analysis, farm mechanization in India stands at about 40% (Mehta et al. 2014b), which is still low as compared to countries such as the US (95%), Brazil (75%), and China (57%) (Renpu 2014). India follows a policy of appropriate mechanization to achieve optimum utilization of the available farm power sources. For small and marginal landholdings, improved hand tools and animal-drawn equipment are used whereas for mechanizing medium and large landholdings, tractor-drawn and self-propelled farm machinery suit to the needs of the farmers. Hill agriculture covers about 1/5th of the cultivated area, and suitable machines are needed to promote mechanization in these areas.

India's food processing mainly involves primary processing. Three important agro-processing industries are wheat milling, rice milling, and pulse milling. To produce edible oils, mechanical oil expellers, oil mills, and solvent extraction plants are used. The post-production losses in agriculture are 5–15% in durables, 20–30% in semi-perishables, and 30–40% in perishables. About half of the production losses could be prevented using appropriate post-harvest approaches (Kulkarni 2009).

Bangladesh

The cultivated land area in Bangladesh is 4.62 million hectares. The average size of landholding is less than 2 ha, close to the scenario of land holdings in India (Justice and Biggs 2013). About 3/4th of all land preparation and other tillage operations are mechanized, performed by 0.4 million small 2-wheel tractors, with rotary cultivators (power tiller) and about 15,000 4-wheel tractors (Roy and Singh 2008). Importantly, about 40% of the trailers (of 1.5 to 2 tonnes capacity) used with 2-wheel tractors are locally manufactured. The 4-wheel tractors are mainly used in tea estates and sugarcane farming. Bangladesh is a river dependent country. Nearly 60% of the cultivated land is irrigated using almost 1 million small diesel-powered pump sets. Manually operated weeders (e.g. push-type inter-row weeders) are used widely in puddled rice cultivation. The use of reapers and combine harvesters are limited; however, the small belt-driven threshers and maize shellers are becoming popular.

The farming scenario in Bangladesh has transformed during the last two decades. Currently, entire maize shelling activity gets done by powered maize shellers, and over 80% of wheat and a large percentage of rice is being threshed by machines (Islam and Shirazul 2009). There is extensive use of low-cost open drum pedal-operated threshers in paddy threshing. The government of Bangladesh has taken the initiative to promote closed drum threshers for use in rice, wheat, and pulses (Roy and Singh 2008). For grain cleaning, women predominantly winnow the rice grain by manual method (Kabir 2004). A small number of pedal and power-operated winnowers are also used. The traditional way of foot-powered *dheki* of rice processing has disappeared from rural households, with the introduction of milling machinery. Perhaps this may be recognized as one of the significant technological changes in the farming sector. At present, the entire milling activity has become economically viable with the availability of mobile door-to-door rice hullers, and other small and medium-sized rubber roller mills, and fully automatic rice mills.

Nepal

The agro-climatic conditions of Nepal have a great diversity, covering the flatlands of the Terai (Gangetic plains), valleys between hills, and the lands of the Himalayan mountains. Due to diversity, Nepal's mechanization processes show wide variations. The prevailing inequalities in ownership of property and other economic conditions also reflect in the spread of farm machinery and other devices in farming activity. Only about 15 to 20% of tillage is mechanized in Nepal. The total number of 4-wheel tractors in Nepal, with an average of over 30 hp is in the range of 30,000 (Biggs et al. 2011). The fragmentation of landholdings and the low average holding size (0.98 hectares) make the growth of 4-wheel tractors slow (Justice and Biggs 2013). However, one may notice in Nepal that all tractors and trailers are multifunctional, including its predominant use as transport vehicles in both rural and peri-urban areas. Even the majority of the two-wheel tractors have trailers for use in other activities. Besides, the rapidly growing horticulture, poultry, dairy, and animal feed industries use engine-powered equipment in the production, processing, transport, and marketing activities.

Machines thresh most of the rice and wheat crops in the Terai region. Currently, there are over 90,000 wheat and rice threshers, and the pump set engines operate many of them. The spike-tooth threshers provide the benefit of cutting and processing the wheat straw for animal fodder. The open drum, wire loop type pedal-powered threshers are in abundant use in hills. A trend is visible that the small pedal threshers are powered via a belt and pulley attached to 1–2 hp electric motors. One may commonly observe the hand-cranked winnowing fans in nearly every terai village. These fans can clean up about 100 kg of grain per hour per person.

The analysis of farm machinery used in many countries, such as in Bangladesh and Nepal, would indicate the use of 2-wheel tractors for significant farm operations. According to the FAO definition, a tractor is a 4-wheel tractor. Therefore, the unified accounting of the extent of farm mechanization across countries has inherent limitations.

East Africa

Africa has abundant land resources, and about 2/3rd of the farm power is manual labour. The population in high proportion lives in rural areas and is engaged in agriculture. These workers are mostly the elderly, women, and children. In the East African nations, agriculture contributes to about 28% of the GDP (Hatibu 2013). Farm mechanization is at a superficial level and is coupled with low utilization of other productivity-enhancing inputs, such as improved seed and fertilizers, and water management. Generally, the small- and medium-scale farming sectors are low yielding. In Sudan, Tanzania, and Kenya, various farm operations such as planting, spraying, fertilizer application, weeding, and harvesting are primarily carried out manually. Farm mechanization levels vary with crops. Sudan has higher mechanization in irrigation (i.e. about 1/3rd of the cultivated land is irrigated). In Kenya and Tanzania, significant operations in sugarcane production, namely, land preparation, irrigation, cane loading, and transport are highly mechanized. In Tanzania, about 3/4th of commercial wheat production comes from mechanized farms. The majority of farmers in the hilly topography in Burundi, Ethiopia, Rwanda, and Uganda depend exclusively on hand tools/hoe for land preparation. Engine-powered mechanization is grossly restricted in these countries. The political desire to modernize agriculture and the development of large irrigated farms would drive agricultural mechanization to a tangible level.

Southern Africa

The countries of Southern Africa, namely Zimbabwe, Zambia, South Africa, Namibia, Swaziland, and Mozambique, are characterized by high population density, and about two-thirds of the population resides in rural areas. Agriculture in these sub-Saharan countries predominantly depends on age-old hand tool technology (Rukuni 1994). The emphasis is required on mechanization technologies in farming and other agri-processing activities to enhance agricultural production and productivity. Agriculture contributes about 18% of the GDP to the Zimbabwean economy. In Zimbabwe, the large commercial farms practice mechanized, high-input high-output farming.

In contrast, family-based farming is followed by nearly 1.8 million households, having cultivated land size of 3 hectares. Rain-fed agriculture practices have been followed using low-input low-output technologies. The hand hoes, axes, rakes, sickles, diggers, and watering cans are the most common age-old hand tools and implements used in these small landholdings. Some tractor-drawn implements, such as ploughs, disc harrows, ridgers, and rippers manufactured locally, are also used.

Zambia has a total area of 75 million hectares, out of which only 9 million hectares are arable, and only about 1.3 million hectares are used for crop production (Simalenga 2013). The agriculture sector employs about 2/3rd of the labour force, who are still dependent on age-old tools and devices for farming operations. In Zambia, 95% of the farmers have small- and medium-sized farm holdings, and the presence of the commercial farmers is meagre. In Namibia and also in Swaziland, agriculture plays a significant role in the economy of the countries. Nearly 2/3rd of the population live in subsistence farming and allied industrial activities. The arable land in Namibia and Swaziland amounts to only about 10% of the total land area, and the vast remaining areas are utilized in the settlement, urban dwellings, forestry, grazing, and other uses. In South Africa, the commercial sector with mechanized farming covers 82 million hectares of land, contributing nearly entire formally marketed agrarian output of the country. On the other hand, the small landholdings (14 million hectares spread over in 4 million households) have low production efficiency, and the production merely covers their subsistence requirements.

West and Central Africa

Agriculture is the backbone of the economy of most countries in West and Central Africa. The regions of West Africa comprise 16 countries, including Nigeria, Ghana, Senegal, Gambia, and others. Central Africa consists of nine countries, including Angola, Congo, Cameroon, and others. Central Africa employs about 3/5th of the workforce in the agricultural sector, whereas West Africa employs nearly half of the total workforce in the sector (Fonteh 2013). On a variety of yardstick, growth in agriculture in the region is perceived as a prerequisite for economic development. With the trend of rural exodus due to the low productivity of agriculture, the obvious consequence would be a consistent decrease in employment in agriculture. In both regions, the urban population is increasing at a faster rate, and hence before long, the urban population will supersede the rural community.

In spite that tractors have been introduced in many countries in West and Central Africa (Ghana and Sierra Leone) in the 1950s, the tractor intensities in the regions (0.6 tractor per 1000 ha of arable land) are nearly half to that of other areas of Africa. Agriculture in these regions sustains on the meagre mechanical power input and vastly depends on human labour, draught animal power, and hand tool technologies.

South America

The critical features of the South American sub-continent are the long chain of uninterrupted Andes Mountain from North to South, and the vast catchment basin of the river Amazon, including its rainforest. Brazil is a vast country, covering nearly half of the total surface area of the sub-continent. Agriculture in Brazil occupies about 264 million hectares, including 61 million hectares of arable land, and 196 million hectares of permanent crops and pastures (Pinto et al. 2013). The farmers cultivate temperate crops, such as cereals, potatoes, and tubers on steep slopes of the Andean regions. In the lowland, tropical areas, cultivation is still managed with manual techniques to grow wide varieties of subsistence crops, such as maize, cassava, and fruits.

The family agriculture covers 85% of the Brazilian farms. It accounts for the nearly entire national production of cassava, 70% of beans, 46% of corn, 38% of coffee, 34% of rice, and 21% of wheat in about 1/4th of the total land area. The average farm size of family agriculture was 18.4 hectares (IBGE 2006a, b). Vastly advanced mechanization has been beneficial to make better use of other agricultural technologies, such as proper fertilizer application, improved seeds of crop varieties having a short growing period of crop, as well as efficient pest and disease control. The production of sugar cane traditionally involved arduous labour inputs. However, in recent times, mechanical sugar cane harvesting has become popular. The development of mechanized harvesting technology has facilitated bean production. Manual harvesting of coffee was always a challenging endeavour due to the non-availability of skilled labour. Mechanization in coffee harvesting was strategic in Brazil's economy, and therefore, different types of harvesting machines and equipment have been developed.

Brazil is the world's second-leading country in the area with the no-tillage system covering about 25 million hectares. This conservation technique is beneficial, not only to large flat farms but also to medium as well as small farms in hilly regions. Suitable technologies for no-tillage planting are emerging with the introduction of mechanical multi-row tractor-drawn planters.

Chile stretches from the Atacama Desert down to the most southern regions. There is a broad diversity in the crops grown in many small farm holdings that comprise over 2/3rd of the farm households. Argentina has about 32 million hectares of arable land. The country has an excellent farm equipment manufacturing industry for tractors, harvesters, and other equipment for handling, primary processing, and agro-industries.

Approaches in the Evaluation of Farm Mechanization

There have been efforts in both the developed and developing countries of using various strategies in arriving at the farm mechanization index about the totality or crop-specific operation. Knowledge and methodology to determine the level of mechanization are available from research studies on large farms of the developed world (Pimentel et al. 2002). These farms primarily depend on co-operative farming and large-scale utilization of mechanical power. In mechanized farming, a positive correlation exists between farm power utilization and productivity. Accordingly, this basis has been taken to understand the state of mechanization in a sector, crop, or region. Also, contemporary research emphasizes the quantification of farm power utilization with reference to mechanical power estimates to farm mechanization (Basu and Nandi 2014).

Various studies indicate different input variables and their cumulative effect, referring to the state of mechanization in a farming sector, with a general premise that total input given is compared with the total output of the farm. This presumption is contrary to examining the effective utilization of human labour, taking into account the efficiency of human energy expenditure to that of the work produced. Since the biological efficiency in most personal work related to farming activities falls far below the 1/5th of total energy spent, the estimate of total human energy expenditure cannot be coupled to the mechanical power of machinery. Emphasis is on the small and marginal farmers, who are grossly deprived of taking advantage of powered machinery to support their means of production (Nkakini et al. 2006). In the case of small farm holdings, compulsions are very much pressing against the introduction of heavy machinery (Akdemir 2013) due to economies of scale (Mehta et al. 2014a, b; WAPCOS 2018), lack of resources, and fragmentation of land size. Because of the small size of the landholding and the involvement of a significant proportion of the workforce in farming in most regions of the developing countries, there is a need to introduce scale-appropriate technologies (Mottaleb et al. 2016).

Estimating the state of mechanization in smallholdings is a challenge since farm power utilization primarily comes from animate sources, which includes human and draught animals. Whereas there is a trend of the gradual influx of heavy machinery in the populous agriculture abundant economies through governmental and non-governmental support systems, in many cases, machines remain under-utilized due to the small and scattered nature of the farms (Singh 2005). Therefore, the assessment of the status of farm mechanization may also be judged against the extent of animate power utilization and drudgery proneness.

Level of Farm Mechanization

Mechanization has been viewed from various angles; it is crop-specific, region-specific, and energy-specific. The most popular approach to calculate the mechanization index is based on machinery power utilization in crop production practices. Mauzu et al. (2014a, b) used the classical mathematical equation, i.e. the proportion of mechanical energy used to the total energy expenditure (mechanical plus human) to evaluate the mechanization index of paddy cultivation in Malaysian agriculture. Chisango and Dzama (2013) described the farm mechanization Index as qualitative and quantitative criteria used to evaluate the impact of using the power sources, both mechanical and human, on production efficiencies. Again all energy inputs from mechanical sources were compared with the total energy demand of the farm. Rasouli and Ranjbar (2008) observed that the degree and level of mechanization are meagre in the Eastern Azarbayjan Province of Iran, and emphasized the need for more penetration of tractors in most farming areas. Hormozi et al. (2012) studied the effect on productivity and technical efficiency due to the application of farm machinery in rice production in Iran. Farm mechanization Index and technical efficiency showed a strong correlation in rice farming. In the similar context, Alikhani et al. (2013) showed that the share of diesel fuel energy as highest in the rice-producing farms, in comparison to chemical energy and others.

Economic Analysis

Singh (2006) made an effort to calculate the mechanization index by taking into account economic factors, i.e. the cost of using mechanical, animal, and human power. It is also a ratio of total mechanical power and the total power used in a particular crop but from an economic point of view. Baldi et al. (1989) conducted the study for three years in five grape-producing farms in the Tuscan region of Italy. The total energy equivalent of the product was used to analyze the productivity of human labour. Mechanization Index was derived using a conventional formula with a little modification, where it was expressed as the ratio of the energy cost of mechanical power and the total energy cost of the farm. Pishgar-Komleh et al. (2011) made an effort to calculate the energy requirement and energy input–output relationship in the rice-producing farms of Gulian province of Iran. The study concluded that the large farms with a farm size of more than one hectare are more successful in energy management with a better cost-benefit ratio. Troujeni et al. (2018) conducted the study in Behshahr city of Iran among pomegranate producers concerning energy use and economic cost involved in farming. All the inputs like fertilizer, pesticides, electricity, diesel, and human labour were taken into consideration for economic efficiency and cost-benefit analysis. Diesel fuel was identified as the highest contributor to input energy.

Gokdogan et al. (2017) studied the energy input–output ratio, and energy productivity in guar and lupin production, and further explored the energy balance of organic olive (Gökdoğan and Erdoğan 2018) in the Aydin-Karpuzlu Region in Turkey. Several studies (Mohammadi et al. 2008; Zangeneh et al. 2010; Hamedani et al. 2011) have been conducted in potato producing farms in Iran and performed the financial analysis to find out the input–output ratio for energy consumption and the cost-benefit ratio. Samavaetan et al. (2011) conducted a study among the garlic growers of Iran to find out the total input–output energy ratio per unit area of cultivation and also the economic status of the farm. Paman et al. (2014) examined the availability of mechanical power and estimated the labour requirements and costs for small-scale rice farming operations in the Riau Province of Indonesia. They reported that farming operations were predominantly performed by human labour because of the lack of farm machines and recommended farmers to adopt more machines to save labour, time, and cost.

Huang et al. (2002) described a cost comparison between self-ploughing and hired service. Fadavi et al. (2010) administered a study in an apple orchard to investigate the energy input–output ratio of the farm, and concluded that the economic accomplishment of the large farms was better. Masuda (2018) placed a strong argument of increasing the scale of rice farming to improve its energy efficiency in Japanese agriculture. The researcher used the data for six inputs, i.e. fossil fuels, electricity, chemical fertilizers, pesticides, agricultural services, and agricultural machinery, to arrive at his recommendation. Here, the reason for the improvement in energy efficiency was a reduction in the energy consumed per unit area by agricultural machinery and agricultural services with an increase in the scale of farming.

Human Energy Expenditure

Among a few studies, Yusof and Pebrian (2016) have attempted to see farm mechanization from a different angle. This study was based on sugar palm farms in Malaysia. During harvesting season, workload and productivity of the workers were analyzed. Nawi et al. (2012) surveyed the low land rice cultivating farms of Malaysia to evaluate the energy consumption of each task involved in rice farming. They identified the seed broadcasting task as one of the most strenuous jobs. Khosruzzaman et al. (2010) observed a steady increase in the mechanization index of existing agricultural practices in Bangladesh. Jatropha cultivation is known to be labour-intensive. Grimsby et al. (2012) conducted a study in Tanzania to investigate human energy expenditure in the process of producing jatropha oil.

Mechanization Indices

Several researchers (Nowacki 1974; Oritz and Salvador 1980) grouped the level of farm mechanization into three stages. The first level of mechanization (M1) included operations carried out by a human and is expressed as a ratio of the total working hours spent in cultivation per unit area. The use of man aided animal-drawn machinery was considered to be the second level of mechanization. The third level of mechanization corresponded to the use of machinery with mechanical energy sources under direct human control. Mondal (2019) aptly compiled different classical sets of equations (Table 7.1) that have been used to calculate the mechanization index. The review compilation of Mondal (2019) elucidates the mechanization index (Table 7.2) that referred to as the ratio of input and output of energy in large farms, where there is extensive use of machinery involving electrical and mechanical power.

Mauza et al. (2014) derived farm mechanization index as the ratio of total machinery energy expenditure and total energy expenditure, i.e. human and machine energy in each farm lot. Six primary paddy production operations were identified along with the measurement of the total land area. Canakci et al. (2005) gave the formula to calculate mechanical energy using weight, useful life, and field capacity of the machine. Again using the method provided by Gajaseni (1995), human energy expenditure was calculated for a particular operation. In this formula, the duration of the activity, along with the human energy conversion factor, was taken into consideration. The findings suggested that the farm size should be reasonably larger to bear the cost of buying and maintaining heavy machinery.

Table 7.1 Classical equations to estimate farm mechanization

Level of mechanization	Equation	Researcher
M1 (operations carried out mainly by a human)	$L_{MH} = \frac{0.1HT}{A}$ kW/ha; where L_{MH} = Work outlay of human effort; 0.1 = Human power in kW; H = Number of operators; T = time devoted to manual operations in hours; A = area of land (ha)	Nowacki 1974; Ortiz and Salvador 1980
M2 (Operations carried out by man aided animal-drawn machines)	$\mathrm{L_{MK}} = \frac{\mathrm{KT}}{\mathrm{A}}$ kWh/ha; where $\mathrm{L_{MK}}$ = work outlay of animal-drawn machinery; K = Number of working animals;T = time used (h); A = area of land (ha)	
M3 (the use of machinery with mechanical energy source under direct human control)	$L_{MM} = \frac{NT}{A}$ kWh/ha; where L_{MM} = work outlay of a motorized machine; N = horsepower of the tractor (kW); T = time used (h); A = area of land (ha)	

Table 7.2 Equations to calculate Mechanization Index (Mondal 2019)

	Variables Described	References
$W_{ME} = \frac{L_M}{(L_H+L_M)} * 100$	W_{ME} = Mechanization Index; L_M = Work done by machines obtained by summing up all the work outlay by human and machines; $L_M = (L_H + L_M)$; L_H = total sum of operation done by man; L_M = total sum of operation done by machines	Olaoye and Rotimi (2010)
$MI = \frac{ME}{ME+HE}$	MI = mechanization index; ME = mechanical energy; HE = human energy	Muazu et al. (2014)
$I_{mij} = \frac{C_{EMij}}{C_{EHij}+C_{EAij}+C_{EMij}}$	I_{mij} = mechanization index ith crop and jth state; C_{EMij} = cost of the use of machinery in the ith crop in the jth state; C_{EAij} = cost of the use of animal labour in the ith crop in the jth state	Singh (2005)
$MI = \frac{E_C}{E_I} \times 100$	MI = Mechanization Index; E_C = Commercial energy input $(MJha^{-1})$; E_I = Energy input $(MJha^{-1})$	Khosruzzaman et al. (2010)
$ML = \frac{A_m}{T_a} \times 100$	ML =Mechanization level Index; A_m = Mechanized area of the farm; T_a = Total area	Banaeian and Zangeneh (2011)
$MR = \frac{M_f}{T_f} \times 100$	MR = Mechanization Ratio Index; M_f = Mechanized Farmers; T_f = Total farms	
$P = \frac{1}{1+E^{-FM}}$ Where $FM = C(1) + C(2) * A_{\exp} + C(3) * E + C(4) * I + C(5) * L + C(6) * P_f + C(7) * L_c$	P = Probability that the house achieved mechanization $A_{\exp}$ = square of the age of the head of household (proxy for the experience); E = level of contact with the extension agencies; I = access to irrigation; L = land holding (farm size); P_f = proportion of HYV and fertilizer intensive crops to the gross cropped area; L_c = access to institutional loan; C(1) to C(7) are Logit regression analysis coefficients that affect the level of mechanization	Ghosh (2010)

Singh (2006) developed an index by taking into consideration the cost of machinery use, human labour, and animate energy. The study evaluated the mechanization index of a particular crop. According to the study, irrigation and farm power have a positive relationship with the production of crops and require the highest capital investment. Even the most mechanized crop, i.e., wheat in India, involves massive participation of human labour. The classical set of equations given by Nowacki (1974) was used by Olaoye and Rotimi (2010) to calculate the level of farm mechanization in some farm settlements of Nigeria. The traditional mathematical equation was used to derive the mechanization index, as the ratio of mechanical energy to the total energy expenditure, taking into account various sources of energy and human participation to establish the relationship.

Banaeian and Zangeneh (2011) introduced equations to estimate (a) mechanization level index and (b) mechanization ratio index. The first equation used to derive the percentage of non-mechanized land, while the other determined the percentage of farmers using machinery among the total number of farmers. Ghosh (2010) used the Logit regression model and determined the mechanization level of a farm, using an index based on ownership and machinery usage pattern of the farmer. Readers may refer to other explanatory variables, like the experience of the farmer, access to institutional credit, the proximity of the extension service that were also used in the equation. The study revealed that if a farm can achieve more than 50% of the maximum achievable score, it is considered to be a mechanized farm. Altogether, it is an entirely different approach to determine the level of farm mechanization, explained as the basis of the energy use pattern of a farm. Pebrian and Yusuf (2016) derived a model to predict the human energy expenditure in the labour-intensive task of harvesting sugar palm in Malaysia. Human energy expenditure per unit was calculated using the average heart rate, the mean increase in heart rate, workers' productivity, and palm height. Nawi et al. (2012) observed that total human energy expenditure per unit area is directly proportional to the heart rate of the person and accordingly derived an equation using average heart rate and field capacity in farm operations.

Several studies indicated the economic benefits of using machinery (Pishgar-Komleh et al. 2011; Troujeni et al. 2018). However, the countries of the developing world have a predominance of small and marginal farmers, who run on subsistence farming through the utilization of animate power sources. Many countries show a mixed pattern in their use of mechanical and electrical power in the farming sector, as elaborated above. By mere accounting of the utilization of mechanical power does not capture the entire scenario. The challenge remains regarding the utility of the stated approaches in the evaluation of farm mechanization, where there is no significant use of mechanical power.

Ergonomics and Safety Issues in Farm Mechanization

The farming scenario in the developing countries continues to be dominated by small farm holdings in the range of 1–4 ha. Labour intensiveness, family farming, dependence on age-old tools, methods, and practices make this sector resource-poor. Policymakers often view that farm mechanization is the avenue to liberate the resource-poor farming workforce from their backwardness. However, the small farm holdings are grossly constrained to adopting mechanization. Undoubtedly, the per unit output and efficiency of the small farms are insignificant in comparison to mechanized farms. The degree of mechanization is derived from the extent of change that has been taken place by the use of heavy machinery, including tractorization as well as other mechanical and electrical energy sources (Diao et al. 2014). In some countries, like India, the farms are mixed type, i.e. crop-specific machinery and traditional methods are used at the same time. In such mixed scenario, consideration of mechanization covers the incorporation of improved tools and implements, and the mechanism that assists in enhancing output and productivity, reducing drudgery, improving efficiency, and partial activity-specific substitution of human labour (Verma 2006) and safety at work (Hegazy et al. 2013).

Generally, the quality of farm machines manufactured by small-scale industries is not of the desired standard resulting in poor-quality work, more extended downtime, low output, high operational cost, and a lot of health and safety issues. Therefore, ergonomics and safety aspects require priority in the design and use of mechanical power sources and other farm machines. Also, due attention is required for the design of women-friendly tools and machines to reduce drudgery in various operations.

It can be seen that the farm mechanization has been adopted mostly to meet the shortage of labour, to increase production and productivity, and to reduce the cost of cultivation. During this process, the whole attention has been placed on developing and using low-cost farm machines to carry out the work in the shortest amount of time and to reduce the cost of production. The ergonomics and safety issues have generally been neglected during this phase of development. Thus, the problems of farm machinery injuries and hazards do not get much attention to date. No systematic data are available in most of the developing countries about farm injuries and other health issues of agricultural workers (Nag and Nag 2004). Data from India (Gite 2008) shows that the overall accident incidence rate in agriculture was 333 accidents per 100,000 workers per year, and the fatality rate of 18.3 per 100,000 workers per year. Thus, it was estimated that, in India, there might be about 0.76 million farm accidents, resulting in about 44,000 deaths every year, and also a large number of disabling injuries to others. On the reference date of 2008, the economic loss to the country might be to the tune of about Rupees 54 billion. The situation in other countries may also be quite dangerous like this. Therefore, while planning for farm mechanization in a country or region, due attention needs to be paid to incorporate ergonomics principles in the design and operation of tools and machines and to generate systematic data on farm injuries and health issues of workers. The readers may refer to Chaps. 11, 12, and 13 that elucidated the ergonomics criteria and its application in the design of farm tools and machinery.

References

AghaAlikhani, M., Kazemi-Poshtmasari, H., & Habibzadeh, F. (2013). Energy use pattern in rice production: A case study from Mazandaran province, Iran. *Energy Conversion and Management, 69,* 157–162.

Akdemir, B. (2013). Agricultural mechanization in Turkey. *IERI Procedia, 5,* 41–44.

Baldi, F., Spugnoli, P., & Zoli, M. (1989). Energy efficiency in the vineyard. *Journal of Agricultural Engineering Research, 44,* 301–310.

Banaeian, N., & Zangeneh, M. (2011). Mechanization and economic analysis of wheat production in Iran. *AMA-Agricultural Mechanization in Asia Africa and Latin America, 42*(4), 15.

Basu, D., & Nandi, A. K. (2014). Farm mechanisation and rationality of labour use in Indian agriculture: A frontier analysis of cost of cultivation data. *Indian Journal of Agricultural Economics, 69,* 336–346.

Biggs, S., Justice, S., & Lewis, D. (2011). Patterns of rural mechanisation, energy and employment in South Asia: reopening the debate. *Economic and Political Weekly*, pp. 78–82.

Canakci, M., Topakci, M., Akinci, I., & Ozmerzi, A. (2005). Energy use pattern of some field crops and vegetable production: Case study for Antalya Region, Turkey. *Energy Conversion and Management, 46*(4), 655–666.

Clarke, L. (2013). Agricultural mechanization in countries in transition in Eastern Europe and Central Asia. In J. Kienzle, J. E. Ashburner, & B. G. Sims (Eds.) *Mechanization for Rural Development- A review of patterns and progress from around the world, Integrated Crop Management*, (Vol. 20-2013, p. 336). Plant Production and Protection Division, Food and Agriculture Organization of the United Nations, Rome.

Diao, X., Cossar, F., Houssou, N., & Kolavalli, S. (2014). Mechanization in Ghana: Emerging demand, and the search for alternative supply models. *Food Policy, 48,* 168–181.

Fadavi, R., Keyhani, A., & Mohtasebi, S. S. (2010). Estimation of a mechanization index in apple orchard in Iran. *Journal of Agricultural Science, 2*(4), 180–185.

FAO. (2018). *World food and agriculture - Statistical pocketbook* (254 pp). Rome: Food and Agriculture Organization of the United Nations.

Fonteh, M. (2013). Agricultural mechanization in West and Central Africa. In J. Kienzle, J. E. Ashburner, & B. G. Sims (Eds.), *Mechanization for Rural Development- A review of patterns and progress from around the world, Integrated Crop Management* (Vol. 20-2013, p. 336) Plant Production and Protection Division, Food and Agriculture Organization of the United Nations, Rome.

Chisango, F. F. T., & Dzama, T. (2013). An assessment of agricultural mechanization index and evaluation of agricultural productivity of some fast track resettlement farms in Bindura district of Mashonaland Central Province: Zimbabwe. *International Journal of Social science and Interdisciplinary Research, 2*(7), 62–82.

Gajaseni, J. (1995). Energy analysis of wetland rice systems in Thailand. *Agriculture, Ecosystems & Environment, 52*(2–3), 173–178.

Ghosh, B. K. (2010). Determinants of farm mechanization in modern agriculture: A case study of Burdwan districts of West Bengal. *International Journal of Agricultural Research, 5*(12), 1107–1115.

Gite, L. P. (2008). Occupational Health and Safety Issues in Indian Agriculture – Present Scenario and Suggestions for Improvement. In *Proceedings National Workshop, Management of Occupational Health and Safety Delivery: Scopes and Challenges*, National Institute of Occupational Health, Ahmedabad, India, 30 Sept –1 Oct. 33–37.

GOI. (2019). Agricultural Statistics at a Glance. (2018). *Ministry of Agriculture & Farmers Welfare*. New Delhi: Government of India.

Gokdogan, O., & Erdogan, O. (2018). Evaluation of energy balance in organic olive (Olea Europaea L.) production in Turkey. *Erwerbs-Obstbau, 60*(1), 47–52.

Gokdogan, O., Seydosoglu, S., Kokten, K., Bengu, A. S., & Baran, M. F. (2017). Energy input-output analysis of guar (Cyamopsis tetragonoloba) and lupin (Lupinus albus L.) production in Turkey. *Legume Research-An International Journal, 40*(3), 526–531.

Grimsby, L. K., Aune, J. B., & Johnsen, F. H. (2012). Human energy requirements in Jatropha oil production for rural electrification in Tanzania. *Energy for Sustainable Development, 16*(3), 297–302.

Hatibu, N. (2013). Investing in agricultural mechanization for development in East Africa. In J. Kienzle, J. E. Ashburner, & B. G. Sims (Eds.), *Mechanization for Rural Development- A review of patterns and progress from around the world, Integrated Crop Management* (Vol. 20-2013, p 336) Plant Production and Protection Division, Food and Agriculture Organization of the United Nations, Rome.

Hamedani, S. R., Shabani, Z., & Rafiee, S. (2011). Energy inputs and crop yield relationship in potato production in Hamadan province of Iran. *Energy, 36*(5), 2367–2371.

Hegazy, R., Schmidley, A., Bautista, E., Sumunistrado, D., Gummert, M., & Elepaño, A. (2013). *Mechanization in Rice Farming—Lessons Learned from Other Countries*. Asia Rice Foundation's annual rice forum 2013.

Hormozi, M. A., Asoodar, M. A., & Abdeshahi, A. (2012). Impact of mechanization on technical efficiency: A case study of rice farmers in Iran. *Procedia Economics and Finance, 1*, 176–185.

Huang, M. Y., Huang, C. J., & Fu, T. T. (2002). Cultivation arrangements and the cost efficiency of rice farming in Taiwan. *Journal of Productivity Analysis, 18*(3), 223–239.

IBGE. (2006a). *Censo Agropecuário 2006* (p. 265). Rio de Janeiro: Instituto Brasileiro de Geografia e Estatística.

IBGE. (2006b). Censo agropecuário. Instituto Brasileiro de Geografia e Estatística.

Islam, M. S., & Shirazul, D. (2009). Farm mechanization for sustainable agriculture in Bangladesh: Problems and prospects. In *5th APCAEM Technical Committee Meeting and the Expert Group Meeting on Application of Agricultural Machinery for Sustainable Agriculture. United Nations Asian and Pacific Centre for Agricultural Engineering and Machinery, Manila, Philippines.* (pp 14–16).

Justice, S., and Biggs, S. (2013). Rural and agricultural mechanization in Bangladesh and Nepal: Status, processes and outcomes. In J. Kienzle, J. E. Ashburner, & G. G. Sims (Eds.), Mechanization for Rural Development- A review of patterns and progress from around the world, Integrated Crop Management (Vol. 20-2013, p. 336) Plant Production and Protection Division, Food and Agriculture Organization of the United Nations, Rome.

Kabir, W. (2004). *Country Report- Bangladesh.* Governing Board/Technical Advisory Committee Meeting. 3rd Asian Pacific Center for Agricultural Engineering and Machinery, 2004. Vietnam.

Khosruzzaman, S., Asgar M.A., Karim, N. and Akbar, S. (2010). Energy intensity and productivity in relation to agriculture– Bangladesh perspective. *Journal of Agricultural Technology, 6*(4),615–630.

Kienzle, J., Ashburner, J. E., & Sims, B. G. (2013). Mechanization for rural development: a review of patterns and progress from around the world. *Integrated Crop Management, 20.* Plant Production and Protection Division, Food and Agriculture Organization of the United Nations, Rome. p 336.

Kulkarni, S. D. (2009). *Mechanization of Agriculture: Indian Scenario.* Paper presented at the Fifth Session of the Technical Committee of the UNAPCAEM & Expert Group Meeting on Application of Agricultural Machinery for Sustainable Agriculture in the Asia-Pacific Region, Los Banos, the Philippines, October 14–16.

Masuda, K. (2018). Energy efficiency of intensive Rice production in Japan: An application of data envelopment analysis. *Sustainability, 10*(1), 120.

Mehta, C. R., Chandel, N. S., & Senthilkumar, T. (2014a). Status, challenges and strategies for farm mechanization in India. *Agricultural Mechanization in Asia Africa and Latin America, 45*(4), 43–50.

Mehta, C. R., Chandel, N. S., Senthilkumar, T., & Singh, K. K. (2014b). Trends of agricultural mechanization in India. *Economic and Social Commission for Asia and the Pacific (ESCAP) Policy Brief, 2.*

Mohammadi, A., Tabatabaeefar, A., Shahin, S., Rafiee, S., & Keyhani, A. (2008). Energy use and economical analysis of potato production in Iran a case study: Ardabil province. *Energy Conversion and Management, 49*(12), 3566–3570.

Mondal, T. (2019). Situational Analysis and Usability Assessment of Farm Mechanization in Small and Marginal Holdings, Ph.D thesis, Ramakrishna Mission Vivekananda Educational and Research Institute (RKMVERI), Belur, Howrah.

Mottaleb, K. A., Krupnik, T. J., & Erenstein, O. (2016). Factors associated with small-scale agricultural machinery adoption in Bangladesh: Census findings. *Journal of Rural Studies, 46,* 155–168.

Muazu, A., Yahya, A., Ishak, W. I. W., & Khairunniza-Bejo, S. (2014). Machinery utilization and production cost of wetland, direct seeding paddy cultivation in Malaysia. *Agriculture and Agricultural Science Procedia, 2,* 361–369.

Nag, P. K., & Nag, A. (2004). Drudgery, accidents and injuries in Indian agriculture. *Industrial Health, 42*(2), 149–162.

Nawi, N. M., Yahya, A., Chen, G., Bockari-Gevao, S. M., & Maraseni, T. N. (2012). Human energy expenditure in lowland rice cultivation in Malaysia. *Journal of Agricultural Safety and Health, 18*(1), 45–56.

NIOSH. (2019). https://www.cdc.gov/niosh/topics/aginjury/default.html. Page last reviewed: October 9, 2019.

Nkakini, S. O., Ayotamuno, M. J., Ogaji, S. O. T., & Probert, S. D. (2006). Farm mechanization leading to more effective energy-utilizations for cassava and yam cultivations in Rivers State. *Nigeria. Applied Energy, 83*(12), 1317–1325.

Nowacki, T. (1974). Examples of technical and economic analysis of mechanized processes in various agro-technical conditions. *AGRI/MECH Report (UN)(ECE). no. 32/Add. 1.*

Olaoye, J. O., & Rotimi, A. O. (2010). Measurement of agricultural mechanization Index and analysis of agricultural productivity of some farm settlements in South west, Nigeria. *Agricultural Engineering International: the CIGR E-journal, 12*(1), 125–134.

Ortiz-Canavate, J., & Salvador, I. (1980). Effects of different mechanization levels in Spanish dryland farms. *AMA, Agricultural Mechanization in Asia., 3*(5), 31–36.

Paman, U., Inaba, S., & Uchida, S. (2014). The mechanization of small-scale rice farming: Labor requirements and costs. *Engineering in Agriculture, Environment and Food, 7*(3), 122–126.

Pimentel, D., Doughty, R., Carothers, C., Lamberson, S., Bora, N., & Lee, K. (2002). Energy inputs in crop production in developing and developed countries. *Food Security and Environmental Quality in the Developing World*, 129–151.

Pinto, F. C., Queiroz, D., M. and Resende, R. C. (2013). The development of farm mechanization in Brazil. In J. Kienzle, J. E. Ashburner, & B. E. Sims (Eds.), *Mechanization for Rural Development-A review of patterns and progress from around the world, Integrated Crop Management* (Vol. 20-2013, p. 336) Plant Production and Protection Division, Food and Agriculture Organization of the United Nations, Rome.

Pishgar-Komleh, S. H., Sefeedpari, P., & Rafiee, S. (2011). Energy and economic analysis of rice production under different farm levels in Guilan province of Iran. *Energy, 36*(10), 5824–5831.

Renpu, B. (2014). Analysis of the Trends of Agricultural Mechanization Development in China (2000-2020). *ESCAP/CSAM Policy Brief, Issue, 1*(9).

Rasouli, S. V., & Ranjbar, I. (2008). Determination of the degree, level and capacity indices for agricultural mechanization in Sarab Region. *Journal of Agricultural Science and Technology, 10,* 215–223.

Roy, K. C., & Singh, G. (2008). Agricultural mechanization in Bangladesh. *AMA, Agricultural Mechanization in Asia, Africa & Latin America, 39*(2), 83.

Rukuni, M. (1994). *Getting agriculture moving in East and Southern Africa and a framework for action* (No. REP-5551. CIMMYT.).

Samavatean, N., Rafiee, S., Mobli, H., & Mohammadi, A. (2011). An analysis of energy use and relation between energy inputs and yield, costs and income of garlic production in Iran. *Renewable Energy, 36*(6), 1808–1813.

Simalenga, T. E. (2013). Agricultural mechanization in Southern African countries. In J. Kienzle, J. E. Ashburner, & B. G. Sims (Eds.), *Mechanization for Rural Development- A review of patterns and progress from around the world, Integrated Crop Management* (Vol. 20-2013, p. 336) Plant Production and Protection Division, Food and Agriculture Organization of the United Nations, Rome.

Singh, J. (2005). Scope, Progress and Constraints of Farm Mechanization in India. In K. Tyagi, H. Bathla, & S. Sharma (EDs.), *Status of Farm Mechanization in India* (pp. 48–56). New Delhi: Indian Agricultural Statistics Research Institute.

Singh, G. (2006). Estimation of a mechanisation index and its impact on production and economic factors—A case study in India. *Biosystems Engineering, 93*(1), 99–106.

Singh, G. (2013). Agricultural mechanization in India. In J. Kienzle, J. E. Ashburner & B. G. Sims (Eds.), *Mechanization for Rural Development- A review of patterns and progress from around the world, Integrated Crop Management* (Vol. 20-2013, p. 336) Plant Production and Protection Division, Food and Agriculture Organization of the United Nations, Rome.

Singh, G. (2015). Agricultural mechanisation development in India. *Indian Journal of Agricultural Economics, 70*(1), 64–82.

Singh, R. S., Singh, S., & Singh, S. P. (2015). Farm power and machinery availability on Indian farms. *Agricultural Engineering Today, 39*(1), 45–56.

Stavis, B. (1978). *The politics of agricultural mechanization in China.* London, U.K.: Cornell University Press.

Tewari, V. K., Kumar, A. A., Kumar, S. P., & Nare, B. (2012). Farm mechanization status of West Bengal in India. *Basic Research Journal of Agricultural Science and Review, 1*(6), 139–146.

Troujeni, M. E., Khojastehpour, M., Vahedi, A., & Emadi, B. (2018). Sensitivity analysis of energy inputs and economic evaluation of pomegranate production in Iran. *Information Processing in Agriculture, 5*(1), 114–123.

USDA. (2014). *U.S.Census of Agriculture 2012, United States. Summary and State Data.* Volume 1: Geographic Series, Part 51. Report AC-12-A-51. United states Department of Agriculture. National Agricultural Statistics Service, Washington.

Verma, S. R. (2006). Impact of agricultural mechanization on production, productivity, cropping intensity income generation and employment of labour. In K. Tyagi, H. Bathla, & S. Sharma, S. (Eds.), *Status of farm mechanization in India* (pp. 133–153). New Delhi: Indian Agricultural Statistics Research Institute.

Wang, M. (2013). China: Development of farm mechanization and the agricultural machinery industry. In J. Kienzle, J. E. Ashburner, & B. G. Sims, B. G. (Eds.), *Mechanization for Rural Development- A review of patterns and progress from around the world, Integrated Crop Management* (p. 336, Vol. 20-2013). Plant Production and Protection Division, Food and Agriculture Organization of the United Nations, Rome.

WAPCOS. (2018). Final report on Monitoring, concurrent evaluation and impact assessment of sub-mission on agricultural mechanization, Mechanization & Technology Division, Department of Agriculture, Cooperation & Farmers Welfare, Ministry of Agriculture & Farmers Welfare, Govt. of India, New Delhi.

Wikipedia.org. (2019). https://www.en.wikipedia.org/wiki/Agriculture in the United States. Last reviewed 19th Feb 2020.

Yusof, Z., & Pebrian, D. (2016). Analysis of human energy expenditure in harvesting sugar palm (Arenga Pinnata Merr) in Malaysia. *Agricultural Engineering International: CIGR e-Journal, 18*(1), 149–157.

Zangeneh, M., Omid, M., & Akram, A. (2010). A comparative study on energy use and cost analysis of potato production under different farming technologies in Hamadan province of Iran. *Energy, 35*(7), 2927–2933.

Part IV
Health and Safety

Chapter 8
Farm Accidents and Injuries

Introduction

Agriculture has been the single most important economic activity that employs about one billion people to produce food grains, to raise cattle, to provide milk and milk products, and to grow vegetables and fruits. A large number of people also work in agroforestry for raising fuelwood and timber. The farmworkers face hugely varying climatic conditions and adverse working conditions. The working environment may involve exposure of the workers to hazardous chemicals, dust, and other stress agents such as noise, vibration, and adverse thermal conditions. Often the farming activity carries the potential risk of accidents and injuries of the musculoskeletal or other bodily structures. As a result, the farmworkers meet with accidents either due to man–machine incompatibility, crop factors, or environmental factors. The accidents cost dearly throughout life, as one might get permanently or partially disabled. The losses due to accidents are twofold, that is, the loss of human beings or functional disability, and the other is the loss of human resources for farming, and thereby culminating in increased cost towards compensation and medical expenses, and also reduction in the incomes from farms.

Human sufferings and economic losses are constant reminders to set and implement work safety norms and standards and execute accident prevention programmes in agriculture. However, despite that getting employment is the foremost importance, enough attention is lacking in the safety and occupational health issues in agriculture. Prevention of accidents in agriculture can go a long way to save lives and to make the workplace safer and productive. The apparent call is on the social awareness and focus of the Government on the calamity that arises at the farms associated with risks of operating dangerous farm machinery and exposure of farmworkers to adverse working conditions and environment.

P. K. Nag and L. P. Gite, *Human-Centered Agriculture*, Design Science and Innovation,
https://doi.org/10.1007/978-981-15-7269-2_8

Accidents and Safety

An accident is an unplanned, unforeseen, or uncontrolled event that generally has unhappy consequences. During the operation of various agricultural machinery, accidents take place, resulting in either loss of life or making people injured and physically disabled. An accident can be expressed as a result of forces from three sources—the host, most often a person (man/woman), the agent, and the environment in which hosts and agents find themselves. Using the host—agent—environment model, the control of accidents involve reducing exposure to the host, reducing the hazard associated with the agent, and changing the environment. Haddon's Matrix is an approach that combines the epidemiological triangle (host, agent, and environment) and levels of prevention (Haddon 1968, 1970). It helps in planning for injury interventions and prevention strategies.

Accident Analysis

Accidents happen in agriculture as a culmination of various factors, and injury rates in agriculture are one of the highest among different occupations. A significant number of farm machinery is now being used for recognizable benefits on farms, i.e. to produce more and to reduce the drudgery of workers. These machines may be new to the workers, or many times the workers may not have adequate training to operate those machines. Many tractor incidents occur during field operation or when driving on the road during transport. Reapers, mowers, combine harvesters, balers, loaders, feed processing machines, and other types of equipment are the potential sources to cause injuries. In horticultural and forestry jobs, the workers are exposed to considerable safety risks of slips and fall, as well as the operational failure of power-operated tools. In livestock and dairy, injuries occur due to animal hits.

Accident statistics are essential in planning accident prevention activities and assessing their effectiveness. Before suitable precautions against accidents can be taken, it is necessary to know how exactly and why they occur. Different methodological approaches (e.g. statistical and epidemiological analysis, risk analysis, safety attitudes and behaviour, case studies, near accident studies, motion and work experiments, interviews) provide information about the work-related accidents (Myers et al. 2009; Nag and Nag 2004). The analysis adds to establishing the necessary actions to prevent accidents (Lundqvist and Gustafsson 1992).

Data on occupational injuries include the organization, the injured person, the working conditions, the nature and extent of the damage, and the circumstances of the injury. When the injured person has to remain away from work for a defined period following the accident, this gets included in the official statistics. In the United States, as per OSHA regulations, any work-related fatality, any work-related injury or illness requiring medical treatment beyond first aid or any work-related diagnosed case of cancer, chronic irreversible diseases, fractured or cracked bones or teeth or

Table 8.1 Indices of accident statistics[a]

Index	Accident statistic
Accident incidence rate	No. of accidents per 100,000 workers
Fatality rate	No. of fatalities per 100,000 workers
Accidents incidence rate	No of accidents per 1000 machines or 1000 workers
Accident frequency rate (F)	(No. of accidents × 10^6)/Total No. hours worked
Accident severity rate (S)	(No. of days lost × 10^6)/No. of hours worked
Frequency severity indicator	(F*S)/1000
Cost factor	(Cost incurred (in a given monetary unit) × 1000)/Total work hours

[a]In some countries/literature, instead of an accident, the *word* injury is used

punctured eardrums, are considered as recordable injury or illness (OSHA 2020). In the United Kingdom, for recording the accident, the absence due to the injury has to be three days or more (HSE 2013). There have been variations across countries in defining conditions to record an accident officially. In south Asian countries, such as India, Pakistan, and Bangladesh, as per the Factories Act, if there is a death or the absence from work due to injury exceeds two days (48 h) (from the time of occurrence of the accident), it is recorded as an accident (ILO 2020). Some indices used to represent accident statistics are listed in Table 8.1.

For reporting the severity of the injury, generally, the Abbreviated Injury Scale (AIS) (Gennarelli 1985) is used. The AIS is scaled from 0 to 6, referring to 0 = no injury, 1 = minor, 2 = moderate, 3 = serious, 4 = severe, 5 = critical, and 6 = maximum (untreatable). The injuries like cuts on the skin, scraping off of skin, superficial vein cuts, cuts on toes or fingers, and muscle stresses are considered as AIS 1 severity. The cuts on limbs, deep vein cuts, permanent loss of any part of the body, and infections at injured limbs are considered between AIS 2 and AIS 3 depending on the severity of the injury. AIS 4 and AIS 5 are the more severe injuries, whereas AIS 6 is the injury leading to fatality or a fatality.

World Scenario of Farm Accidents

Nearly 50% of the world workforce was engaged in agriculture in the nineties, and that has systematically gone down to 27% in recent years. In 2016, in Africa, the farm workforce was 63% of the total workforce, whereas, in industrialized nations, less than 5% of the workforce is employed in agriculture (FAO 2018). The nature of work in agriculture varies from the continuance of the traditional manually arduous tasks to highly mechanized form of agriculture, depending on the introduction of mechanization in the country. As mentioned elsewhere, agriculture covers many settings, such as family farms, including subsistence agriculture, large corporate farms, plantations, urban farms, including speciality enterprises, and migrant and seasonal work. Literature is evident that farming work is associated with a variety of health problems, including the risk of cancers, respiratory diseases, and injuries (refer to Chap. 9, for details). Sprains and strains are commonly observed problems among agricultural workers, and they are also exposed to mechanical and thermal stresses. The health and safety issues arise from man-machine incompatibility, handling heavy loads, repetitive motion, and postural stresses. Occupational accidents in agriculture are a problem all over the world (ILO 2011). Farmers, family members, and farmworkers face a higher risk than in most of the other occupations. Data available from different countries vary widely regarding the prevalence of work hazards and the economic consequences of accidents.

North America

United States

As per the US Bureau of Labour Statistics (BLS), agriculture consistently ranks among the most dangerous industries from year to year. In 2014, the rate of fatal occupational injuries in agriculture was 25.6 deaths per 100,000 workers, which was over sevenfold higher than the all-industry fatality rate of 3.4 deaths per 100,000 workers. Although fatal occupational injuries are well-documented, non-fatal injuries that occur on family farms are not generally included in the injury reporting systems, such as workers' compensation data, OSHA injury logs, or the survey of occupational injuries and illnesses (Leigh et al. 2014).

The US National Centre for Farm Workers Health (NCFH 2017) recorded that in 2015, there were 570 fatal injuries in the agricultural industry, estimating 22.8 fatal injuries for every 100,000 full-time farmworkers. The agricultural sector also has a high number of non-fatal injuries and illnesses that required either time off from work or job transfer. Over 18,000 injuries and illnesses in agriculture, forestry, and fishing industry yielded an incidence rate of 1890 injuries and illnesses per 100,000 full-time workers, as compared to 1040 injuries per 100,000 workers in all sectors. An estimated half of injuries among crop production workers were classified as strains or sprains. The National Agricultural Workers Survey (2013–2014) indicated about 4 and 17% of workers reported experiencing a workplace injury, and musculoskeletal pain or discomfort, respectively.

AgInjuryNews.org is an online tool that enables users to search the database of publicly available US agricultural injury and fatality reports and to get a near real-time snapshot of the distribution and nature of trauma incidents, both nationally and locally. The Census of Fatal Occupational Injuries reported national agricultural fatal injuries from 1992 to 1998 as 25.8 deaths per 100,000 workers (Hard et al. 2002). The Traumatic Injury Surveillance of Farmers (TISF) in 1993–1995 reported the occurrence of 7500 non-fatal injuries nationally per 100,000 workers. NIOSH (2019) data showed that 417 farmers and farm workers died from work-related injury in 2016, i.e. a fatality rate of 21.4 deaths per 100,000 workers. Transportation incidents (e.g. tractor overturns) were the leading cause of death among the farmworkers. The effective way to prevent tractor overturn penalties is the use of a Roll-Over Protective Structure (ROPS) with a seatbelt. In 2014, 62% of tractors on farms were equipped with ROPS.

Older farmers are at the highest risk for farm fatalities, and traumatic injuries continue as a primary concern for youth working on the US farms. Fatality data show that older farmers accounted for over half of all farming deaths between 1992 and 2004 (Myers et al. 2009). Myers and Hard (1995) reported the work-related fatalities rate from 1980 to 1989 in agricultural production as 22.9 deaths per 100,000 workers, in comparison to seven deaths per 100,000 for all workers. Farm machinery and tractor vehicles, frequently from tractor rollovers, were the leading causes of death in agricultural production. Other factors of farm fatalities include electrocutions, caught in, flying objects, environmental causes, and drowning. These data correspond to several earlier studies (Murphy 1981; Zhou Chun and Roseman 1994), indicating that accidents associated with farm machinery and tractor vehicles are the leading causes of fatality. Lehtola et al. (1994) analyzed tractor-related deaths published in newspapers from Iowa during five years and reported that overturning of tractors accounted for 56% of total fatalities. About 59% incidents of overturns occurred in a field or farmyard, while 37% on a public roadway with no other vehicle involved. Run-over caused 17% fatalities where victims were mostly the children. No casualties were reported from tractors equipped with ROPS. Lehtola et al. (1994) proposed YPLL (years of productive life lost) as a method to estimate the cost due to fatal injuries. The cost of one human life was calculated using the earning level of the victim, and the damage due to non-fatal injuries was worked out by taking into consideration the medical cost, downtime cost, and repair cost.

For injuries that require hospitalization, the financial burden can be catastrophic, since the majority of farmworkers do not receive health insurance, nor are they covered by workers' compensation benefits. Analyzing 841 deaths and 512,539 non-fatal injuries that occurred in the US in 1992, Leigh et al. (2001) estimated the costs of job-related injuries in agriculture. The non-fatal injuries include 281,896 that led to at least one full day of work loss. The average injury cost was estimated as $4.57 billion (ranging from the US $3.14 billion to $13.99 billion), in which the direct costs amounted to $1.66 billion and indirect costs, $2.93 billion. On a per-person basis, farming injury attributed to nearly 30% more cost than the national average to other occupational injuries. Despite a considerable expense, economic resources are limited to preventing and ameliorating farm injuries. From the analysis of 5,600 fatal and nearly 8,559,000, non-fatal injuries, Leigh (2011) arrived at the national costs of occupational injury and illness among civilians in the US for 2007. The cost estimated was $6 billion, and $186 billion, respectively. The fatal and non-fatal illnesses were estimated at 53,000 and nearly 427,000, respectively, with cost estimates of $46 billion and $12 billion. For injuries and diseases combined, the medical cost was estimated at $250 billion, the injuries being responsible for 3/4th of the total cost.

Under the US Occupational Safety and Health Act of 1970, the Occupational Safety and Health Administration (OSHA) established national workplace safety standards to ensure safe and healthful working conditions for men and women and to provide training, outreach, education, and assistance (OSHA 1970). Although agriculture was included in the original federal regulation, the amendment of 1976 exempted all farms that employ fewer than 11 workers. That is, family-run farms and farms with fewer than 11 employees are exempt from compliance with the OSHA standard that mandated safety features on machinery and precautionary training for agricultural equipment.

Federal guidelines have dramatically improved working conditions and decreased workplace hazards in other industries, but have had little impact on agri-business. The Migrant and Seasonal Agricultural Worker Protection Act (MSPA 1983) set forth stipulations regarding wages, working, and housing conditions. Leigh et al. (2014) mentioned that the federal agencies responsible for tracking workplace hazards failed to report over 3/4th of the injuries and illnesses of US agricultural farmers. The government records dramatically undercount farming injuries and illnesses, ranging from chemical exposures to musculoskeletal injuries. The primary reasons for the discrepancy are the government's focus on mid- to large-sized farming enterprises that represent less than 50% of employment in the agricultural industry. The analysis may be required on the appropriateness of exemptions from federal guidelines that are designed to promote workplace safety and protect workers.

In general, weak enforcement of safety regulations, inconsistent employee training regarding hazards and use of PPE, and the precarious nature of the agricultural industry combine to create a work environment that is fraught with potential for injury. Various workforce agencies have recognized the hazards of the farming industry. However, due to cultural and historical reasons there is a gross neglect of occupational health and safety regulation in agriculture. The sector has traditionally been resorted to voluntary injury prevention and safety education programmes.

Various farm-based safety improvement initiatives have been taken to address ergonomics risk factors in farming and to educate workers and children about farm safety. Rautiainen et al. (2008) reported that the intervention initiatives had increased participant knowledge about farm safety. However, there is no evidence to prove that the educational interventions had any injury reducing effect. Financial incentives reduce the injury level immediately following the intervention, but its significant impact on the injury trend over time is yet to be ascertained. There is an apparent awareness that the promotion for the use of PPE might protect an individual from injuries at the personal level. Further, improvization of engineering designs in agricultural machinery and devices may reduce mortality and morbidity associated with injuries.

Canada

The analysis of agricultural injury death rates in Alberta, Canada, over the period from 1990 to 2009 (ACICR 2010) indicated 10.9 deaths per 100,000 farmworkers in 1990 to 12.0 deaths per 100,000 farmworkers in 2009. The yearly rates were highest in 2007, with 16.8 deaths per 100,000 farm population. About 2/3rd of the agriculture-related deaths were machine related, such as machine rollovers, machine runovers, and being pinned or struck by a machine. The leading causes of non-machine related deaths include being struck by an animal, drowning, and asphyxiation in grain or soil mass. From analyzing agricultural injuries and interventions in Saskatchewan province of Canada, Hagel et al. (2008) observed that the educational interventions were not associated with any observable differences in farm safety practices, physical farm hazards, or farm-related injury outcomes. There has been a vivid realization to extend injury prevention initiatives to the full public health model of education, engineering, and regulations.

European Countries

Nearly half of the European territory accounts for the farmland, and these are vastly family farms (97% of the total of 12.2 million farms). In 2010, about 5% of the European workforce was engaged in agriculture. Farmworkers are exposed to potentially dangerous machinery, vehicles, livestock, inclement weather, hazardous substances, noise, and dust. During the year 2012, the fatalities in European agriculture were 519, with the fatality rate of 4.64 per 100,000 workers, and 150,317 non-fatal accidents, with the incidence rate of 1343 injuries per 100,000 workers (Schmitz-Felten 2016).

Tómasson et al. (2011) mentioned that among the Nordic countries, the work-related fatalities were 9.3 per 100,000 workers during the period 2003–2008. Lemola and Pehkonen (1987) analyzed 1500 fatal farm accidents between 1982 and 1985 in Finland. The most dangerous operations identified were tillage, harvest, and machinery repair during the intensive times of the growing season. Among the accidents that occurred in animal production, moving cattle was the most hazardous tasks. Suutarinen (1992) reported the statistical data on 403 tractor accidents and case studies of 50 tractor accidents in Finnish farms. The accidents took place mostly during tilling and sowing. About 2/5th of the accidents resulted from climbing into and out of the cab. Hitching and unhitching implements contributed to about 30% of the tractor accidents. These facts become especially important because the work operations such as mounting and dismounting from the tractor and coupling and uncoupling implements, take a proportionately small part of the time when a tractor is used (Gustafsson et al. 1991). Wernisch (1989) noted that about 10,000 falling accidents take place in Austrian farms every year.

Konstantinos et al. (2018) reported that about 38,000 agricultural accidents are recorded each year in about 4.15 million hectares of the utilized agricultural area in Greece. More and more non-professional farmers and civilians tend to undertake outdoors and farming duties, using mainly small or hand-operated machines, which cause severe traumas or fatality owing to careless or improper use. Pfortmueller et al. (2013) mentioned that only about 2% of Switzerland's population was involved in agriculture. Swiss accident statistics indicate agriculture as the most dangerous occupational sector, followed by the construction and wood industries. In 2010, nearly 16,570 per 100,000 people working in agriculture suffered the injury, and three deaths occurred per 100,000 people.

Several studies elucidated occupational injuries in agriculture and horticulture in Sweden (Lundqvist and Gustafson 1992). Hansson et al. (1989) observed a high prevalence of accidents in milk production (animal husbandry). Larsson (1990) reported severe hand injuries among Swedish farmers. The average cost of hand injuries sustained by farmers comes to around 90,000 SEK. The highest costs were involved in combine harvester accidents (245,000 SEK), followed by timber processing (148,000 SEK). Lundqvist (2008) reported that the total number of fatal injuries in agriculture in 2007 was 21 per 100,000 workers. The injury frequency was highest in the small farms in Sweden. About half of the injured were 55 years or older. The economic consequences of these accidents at a yearly level were 200 to 300 million Euro, and the comparable estimate from the Norwegian farms was also about 200 million Euro. In 63,000 active farms in Sweden, Pinzke et al. (2018) reported about 4400 injuries in 2013. The industry and government initiatives, and programmes to reduce occupational injuries in agriculture had only a marginal effect. Despite all efforts, the fatality rate in agriculture was 6.1 injuries per 100,000 employed persons, as compared to 0.8 per 100,000 employed persons for all industries in Sweden. The work environment in agriculture is often characterized by a variety of injury risks linked to conditions such as the use of machines, work tools, working at height, load lifting, handling of animals, handling of chemicals, tree felling, and other work stressors. Safety campaigns and more permanent safety initiatives in agriculture would raise awareness of workplace risk factors and encourage safe behaviours.

There has been a consistent alarm of gross under-reporting in the official statistics of occupational injuries in the agriculture sector (Pinzke and Lundqvist 2007). A study of 20,000 Swedish farmers and forestry workers in 1987, showed that the number of injuries was twice that reported in official statistics (Hansson et al. 1989). A follow-up study showed that approximately 5000 occupational injuries occurred in agriculture in 2004. In contrast, the injury statistics (SWEA 2006) indicated a significantly lower level of just over 400 injuries in agriculture, forestry, hunting, and fishing combined. This issue of under-reporting exists very much in several other countries (e.g. Tiwari et al. 2002; Jacinto and Aspinwall 2004; Rosenman et al. 2006; Solomon et al. 2007; Donnelly 2015; Donham and Thelin 2016; Patel et al. 2017). Chapman (2017) have shown that a lack of injury reporting was preventing the industry from making safety changes before injuries occur. By under-reporting of injuries, the real scope of the problem remains subdued and not visible to concerned authorities and stakeholders. ILO (2011) mentioned that in many countries, the reporting and compensation systems exclude the agriculture sector. Many countries group injury statistics of agriculture with other areas, such as hunting, forestry, and fishing, and thereby makes it difficult to evaluate the farm-related injury trends over time.

As per the UK Health and Safety Executive (HSE) data of farm accidents, the fatal injury rate during 2018–19 was 9.21 per 100,000 agricultural workers, in comparison to the five-year fatal injury rate from 2014–15 to 2018–19 as 8.29 per 100,000 agricultural workers (HSE 2019). The fatal injuries in the farming sector have fallen to nearly half since the early 1980s. British farms had a higher rate of fatal injuries in comparison to the main industrial sectors. Workers aged 60 and older accounted for nearly half of all the fatal injuries. From the analysis of 85,000 accidents, Monk et al. (1986) estimated the cost of these accidents to the British economy to be about 94 million Sterling Pound, with 2/3rd of the cost being borne by the farmers and the employers. Damage to the property was the predominant cost element, accounting for nearly half the overall cost.

The 2011 National Farm Survey in Ireland estimated that farm injuries increased by 35% to 2,459 injuries per 100,000 farms in 2010, compared to the previous survey estimate of 1,815 per 100,000 farms in 2006. Injuries from slips, trips, falls, and blows were most frequent, accounting for 42% of the non-fatal farm incidents. Livestock-related injuries attributed to 1/3rd of the total injuries. Machinery injuries have declined significantly over time. About 95% of reported injuries required medical treatment, half of those requiring hospital inpatient treatment, and as a result, 41% of the injured lost more than 20 workdays (Donnelly 2015).

Across the European continent, the incidence of fatal accidents in agriculture and forestry is comparable only with the construction industry. However, the numbers of the latter appear to continuously decline, while there is an increase in agriculture and forestry. The severe injuries occur among the people due to loss of control, slipping, and falling of the operator, especially with hand-operated machines. Many machinery-related accidents occur as a result of the operator's negligence, including taking shortcuts during work, touching sharp and rotating objects, ignorance on the safety instructions and rules, improper clothing worn, and improper machinery maintenance.

Australia and New Zealand

Arthur (2019) reported that in Australia, agriculture has only 2.6% of the total workforce but accounts for an alarming 21% of the fatalities. Earlier, Franklin et al. (2001) reported farming fatalities in Australia as 20.6 per 100,000 workers. Langley et al. (1997) reported that 18.6% of all fatal injuries occurring on farms in New Zealand were related to tractors, and over half of these fatalities were due to the overturning of tractors.

African Countries

Small-scale and self-employed farmers predominate in production agriculture in most African countries. The farmers frequently encounter occupational diseases caused by exposure to high levels of workplace toxins. Parasitic infections, acute health effects, and deaths from high-level pesticide exposures are common occurrences. Many avoidable injuries ensuing from working with unprotected machinery and lack of adequate personal protective equipment at the workplace still prevail. A small sample survey of 20 farmers, 20 nurses, and 20 agricultural extension workers in the Gambia (Kuye et al. 2006) revealed that 2/3rd of the farms reported one injury during the past year. Predisposing factors to the injuries were climatic conditions, working in static positions, bending and twisting, and carrying heavy objects. Cuts and lacerations were identified as the most frequent injury types, and the most common sources were hand tools (hand hoe, cutlass, axe, and knife) and animal-powered carts. Farm work in the Gambia is performed mainly with draught animals and hand tools. Farmers rarely use PPE when engaged in farm work. The farmers generally lack proper training on the techniques of agrochemical handling and pesticides application.

Yiha and Kumie (2010) assessed occupational injuries in the Tendaho region of Ethiopia. The overall injury prevalence rate was reported as 7,830 per 100,000 exposed workers per year. A total of 6,153 workdays were lost, at an annual average of 11.4 days per injured worker. Working more than 48 h per week, absence of health and safety training, sleeping disorder, alcohol consumption, job dissatisfaction, and lack of protective devices were some significant factors attributing to the common occupational injuries.

Asian Countries

South Korea

The agriculture sector in South Korea recorded a sharp decrease in employment from about 14 million in 1970 to about 3 million in 2010. Ageing among the farming population is in increasing trend. As of 2009, nearly half of the agricultural community was $\geq$60 years of age. The farmers encounter several uncontrollable factors of the working conditions, potentially carrying risks of accidents and injuries. The increased age of farmers is also a strong risk factor for agricultural injury. Chae et al. (2014) reported that the average accident rate in the farming sector was higher than all other industries taken together. Of the 8,064 responses in 2009, 269 farmers suffered an injury that required more than four days of absence during 2008, and the estimated agricultural injury rate was 3.2%. The agricultural injury rates among orchard farmers were the highest when the types of farming classified into rice, dry fields, orchards, greenhouses, and livestock. Males had a significantly higher rate of occupational injury than females because males tended to use farm equipment more often or conduct more high-risk operations than the females do.

Lee et al. (2012) analyzed the entry and compensation data compiled in 2005 by the Safety Aid System of Farm Workers, the government insurance for agricultural injury in South Korea. There were 11,931 compensated events, including 219 compensated deaths. Farm injuries occurred most frequently in October, and most injuries were unintentional. The incidence and fatality rates were 1667 per 100,000 person-years and 30.6 per 100,000 person-years, respectively. The most common cause of deaths was accidents caused by machinery use (35.6%) for males and other transport accidents (23.5%) for females, while the most common cause of injuries for females was falling (45.4%). The incidence and mortality rates of agricultural injuries were higher in elders. Kim et al. (2016) reported data from 68 injury cases, indicating that 3/4th of all injuries were attributable to falls, strangulation, amputation, and collision from flying and falling objects. The primary root causes were lacunae in training/procedure (25%), design faults (20%), gaps in mitigation procedures (12%), human factors (9%), maintenance/inspection issues (6%), and lack of warning/notification (1%).

Bangladesh

About 45% of the population in Bangladesh is associated with agricultural work, and the majority of those are women. The farm labourers often face injuries during work, like cuts on the limbs, scraping off of the skin, blisters, superficial and deep vein cuts, cuts on toes or fingers, permanent loss of body part, and also musculoskeletal disorders. Parvez and Shahriar (2018) analyzed 434 agricultural injuries from the Panchagarh district of Bangladesh. About 2/3rd of injuries were due to hand tools, and the remaining were due to machinery or other sources. The severity of about 1/4th of injuries was more significant than or equal to the AIS 2.

India

The agriculture-related injuries in India constitute about 14% of all episodes of ill health in rural areas (Mohan and Patel 1992). Data on accidents in the state of Madhya Pradesh (Tiwari et al. 2002) show that the overall annual accident incidence rate in agriculture was 125 per 100,000 farmworkers. About 9.2% of the incidents were fatal, and the remaining non-fatal. About 78% of the accidents were due to farm machinery, 12% due to hand tools, and the remaining 10% due to other sources like snakebites and falls in wells. The tractor-related incidents were highest (46%) followed by threshers (14.5%), sickles (10.5%), electric motors and pump sets (7.9%), snakebites (5.3%), and chaff cutters and wells (4.0% each). Tiwari et al. (2004) analyzed the tractor-related injuries and reported that the annual incidence rates for fatal and non-fatal injuries were 2.2 and 23.6 per 1000 tractors. They suggested the use of ROPS on tractors, development of safe tractor-trailers, and proper training to operators to minimize these accidents.

Nag and Nag (2004) mentioned that the work drudgery, the traumatic accidents, and injuries were the primary concerns to explore options for ergonomics intervention and betterment of work in crop production activities in Indian agriculture. Most of the fatal accidents resulted from the powered machinery, with the annual fatality rate estimated as 22 per 100,000 farmers. Despite the enactment of the legislation, the shortcomings in production and monitoring of the machinery in-field use may be responsible for the high rate of accidents (e.g. 42 thresher accidents per 1,000 mechanical threshers per year in southern India). Due to the lack of technical capability of the local artisans, adhering to safety and design standards in the fabrication of implements in the rural areas is impractical. The analysis emphasizes that active safety and health management may be possible through legislative enabling of the local infrastructure, such as block development authority and primary health services, and thereby to permeate occupational health and safe work practices in the farming sector.

Patel et al. (2010) reported that in the Etawah district of Uttar Pradesh state, the annual farm injury incidence rate was 8,000 per 100,000 workers, and 9% of the injuries were fatal. The injury victims in the age group of 30–44 years were the highest. Tewari et al. (2010) reported that in the state of West Bengal, farm machinery accidents were 15%, hand tools 55%, and snake bites and others were 30%. The annual accident incident rate was 128 per 100,000 workers.

Gite (2008) reported a more precise nature of accidents in Indian agriculture, indicating 30.5% were due to farm machines, 34.2% due to hand tools, and 35.3% were due to other sources like snake bites, animal bites, falls in well, lightening, and heatstroke. Of the total accidents, 5.6% were fatal, whereas 94.4% were non-fatal (Gite et al. 2009). The overall accident incidence rate per year was 333 per 100,000 workers, whereas the fatality rate per year was 18.3 per 100,000 workers. The annual estimated number of accidents in Indian agriculture was about 0.76 million and fatalities about 44,000. The estimated economic burden to the country was about Rs. 54 billion. Gite et al. (2009) detailed the farm machinery accidents, showing that the highest number of accidents were due to tractor and tractor operated implements (31%) followed by animal-drawn implements (22%), threshers (14%), electric motors/pump sets (12%), chaff cutters (9%), power tillers (6%), sprayers (4%), and other machines (2%). Tractors and tractor operated implements were responsible for 44% of the total fatalities caused due to farm machinery. The annual fatal and non-fatal injury rates were 5.7 and 96 per 100,000 farm workers, respectively. The farm machines which needed immediate ergonomics intervention for better work safety were tractors and tractor operated implements, electric motors/pump sets, sprayers, power tillers, and threshers.

Injury Hazards

Injuries in Tractor Operation

The problem of tractor-related injuries is a real concern in the agricultural sector. Some examples of tractor-trailer accidents are depicted in Fig. 8.1. Most injuries and deaths are due to tractor accidents, such as overturns, runovers, power take-off (PTO) systems, and unintended contact with tractor attachments or implements. The tractor overturns incidents primarily take place when the vehicle overturns on the side or tip over backwards. The overturning comes about in seconds, as the operator has little chance to manoeuvre the vehicle to avoid tips or rolls. The use of ROPS and the seat belt system in the tractor have proved as vital safety intervention. A ROPS is a roll bar or cage frame that forms a zone of protection around the operator when a tractor tends to overturn. Old tractors often lack safety devices. Slowing down of the tractor speed when turning or crossing slopes, and on rough, slick, or muddy surfaces can avoid overturns. Nichol et al. (2005) suggested the use of a tractor instability monitoring device to alert the operator about overturning and to prevent accidents.

The tractor that is designed as farm machinery is also used as a vehicle of highway transport, construction work, or a stationary power source. Often in rural areas, tractors are typically introduced with minimal operator training and are used as a means of rural passenger transport. The run overs of the riders who have fallen

Fig. 8.1 **a–c** Tractor overturns and trailer accidents, and **d** tractor with ROPS

from the moving vehicles have become the second leading cause of tractor-related fatalities. Older tractor models can also be retrofitted with ROPS to minimize hazards. Other design features that can improve the safety of tractor operators include wider wheelbases, design measures to lower the centre of gravity of the vehicle, and also ergonomically designed seating system and controls to improve stability. All-weather operator enclosures have been introduced to reduce exposure to noise and dust during tractor operation. All tractors, as well as tractor-trailers, must have a slow-moving vehicle (SMV) emblem affixed to the rear for safety reasons (BIS 1999, 2000). The distinctive sign of a red triangle with a yellow-orange centre designates a class of vehicles as SMV, i.e. the vehicle travels at a speed less than 40 km/h on the road (Fig. 8.2).

A power take-off (PTO) shaft allows farmers to use power from the tractor's engine to drive other machines and implements. PTOs are useful, but the unprotected PTOs can pose a severe safety hazard. All the guards should be kept in place when the tractor or any tractor-operated machine is working. The tractor driver must ensure that everyone nearby is away from the device before starting the tractor's engine and engaging the PTO shaft. The PTO-driven machinery should not be operated alone. A rotating PTO shaft can catch long hair or loose clothing, and in the event of such entanglement, a co-worker should be trained to stop the PTO shaft and prevent serious injury. Also, having children on a tractor is unsafe, and therefore, it should be avoided as far as possible.

Fig. 8.2 An emblem of a red triangle with a yellow-orange centre designates a class of vehicles as a slow-moving vehicle (SMV)

Injuries in Harvesting and Threshing

In the more labour-intensive harvesting and threshing practices in the agriculture-dependent countries, like India, the primary hazards have changed only a little with time. There are ample arduous, uncomfortable, and exhausting tasks during harvesting grain or sugar cane with a sickle, scythe or machete, picking fruit or vegetables by hand, and manually removing peanuts from the vine, and threshing crops to separate the grains from plants. One of the most substantial motivating factors for the incorporation of modern machinery in harvesting and threshing practices has been to increase productivity and eliminate the physical drudgery associated with the operations. However, the intensive mechanization in the harvesting and threshing process has resulted in more injuries in crop production processes. Injuries related to harvesting and threshing operations typically relate to tractors, harvesters, threshers, combines, grain-handling equipment, and grain-storage structures. The injuries take place during activities as well as servicing, repairing, or adjusting the machines when the power to machine components is engaged. The alarming scenario of accidents, injuries, and large-scale loss of crops from farm fire are severe concerns in the regions of Northern India. Shutske et al. (1994) reported a loss of over $10 million from grain combine and tractor fires in the U.S. Automated fire detection and dry chemical extinguishing system prevents such losses. Some design concepts have been taken into account for passive and active operator warnings and interlocks, such as safety switches placed near the operator to prevent machine operation and to reduce the maintenance points to minimize operator's exposure to running machinery. Many design interventions suggested for harvesting machinery remain voluntary. The manufacturers of the machinery often ignore them, or the operators bypass the essential safety precautions due to ignorance or negligence.

Hay and forage harvesting equipment expose workers to hazards since the device contains components for cutting, crushing, grinding, chopping, and blowing crop material at high speed. Hay balers are operated under harsh field conditions, such as high temperatures, rough terrain, and dusty conditions. The handling and compression of large bales of hay into sizeable rolls have been identified as a frequent source of injuries, such as hand and arm amputations. Injuries are common during manual work in potato cultivation as well as in mechanical potato harvesting. The tractor's PTO shaft that connects the tractor and the harvester is the source of kinetic energy, bearing the risk of injuries in the case of unsafe manoeuvring of the system.

Thresher Injuries

Threshing machines have been the most successful examples of farm mechanization, due to a variety of benefits, like reduction in time of threshing, reduced drudgery to the workers, and lower cost of threshing. The machines separate grains from earheads/plant stalks and put cleaned grains and straw at the respective output spouts.

A typical mechanical thresher consists of a threshing cylinder, concave, flywheel, blower/aspirator assembly, straw rack/sieve assembly, and a feeding chute/hopper or a conveyor mounted on a sturdy frame. These machines are generally operated by an electric motor or diesel engine mounted on the frame or from a tractor through the belt–pulley transmission. Power to various moving components is transmitted with the help of belts and pulleys.

During the 70s, in India, a large number of thresher accidents occurred, and as a result, many workers lost their lives/upper limbs during grain threshing using power threshers (Jasra and Kothiyal 1981; Mittal 1996; Singh et al. 2005). Mohan and Patel (1992) noted that the incidence rate of moderate to severe thresher injuries was 13.1 per thousand threshers. Lakhtakia (2000) reviewed 110 thresher injuries to hand (crushing) and found that about 1/3rd of victims were unable to use their injured hands. Factors responsible for mechanical thresher accidents included lack of safety awareness and skill, feeding chute deficiencies, lack of safety guards on moving parts, use of faulty and protective wears, wearing of loose garments, and other crop factors (Verma et al. 1978; Gite et al. 2009). Major causes for such accidents were improper methods of crop feeding, putting a cloth or gunny bag around the hand to protect it from abrasion due to crop, unstable platform, and open electric wire (Fig. 8.3).

A comprehensive standard on safe feeding devices on threshers, as per Indian Standard IS 9020 Power Threshers—Safety Requirements (BIS 2002) helped to minimize these accidents in India. Also, the Government passed the *Dangerous Machines (Regulation) Act* (DMRA 1983) and made safe feeding chutes/feeding systems compulsory on power threshers (Gite et al. 2006). Feeding chutes are used on low horsepower threshers, whereas the automatic feed conveyor systems are used for crop feeding in high horsepower threshers. These safe feeding systems ensure that the operator's hands do not come in contact with the threshing mechanism to cause an accident. The automatic conveyor system consists of an endless chain or belt rotating on a pair of pulleys or sprockets. All the rotating parts are to be guarded with safety covers. Also, the workers are advised not to stand on unstable platforms (grain

Fig. 8.3 Accident risks in power thresher operation

bags, bullock carts, tractor-trailer, tyres, or on crop bundles). Special care is needed when the power transmission to thresher pulley is from tractor PTO. Adequately guarded belt from PTO to pulley minimizes thresher accidents.

Fodder Cutter Injuries

Silage making is a necessary operation performed by the farmers and dairy owners for better assimilation of green/dry fodder. Manual, as well as power operated chaff cutters, are used for this purpose. Studies have shown that the chaff cutter (fodder cutter) injuries accounted for about 9 to 11% of the total agricultural injuries. Mohan and Patel (1992) analyzed a total of 576 farm injuries from northern India and recorded the incidence of moderate to severe injuries at 5.1 per 1,000 manual cutters and 8.6 per 1,000 powered cutters.

The operators sustain injuries while feeding the fodder into the rollers of the cutter. Most of the accidents with chaff cutters occur at the *push-in point* or feeding point (Fig. 8.4). Injury types include amputations of the fingers and hands, deep cuts, lacerations, or crushing of fingers (Mohan et al. 2004). The faulty design of feeding chute leads to entanglement or chopping off hands or fingers. The other potential source of the accident is an entanglement of loose garments with rotating parts, such as a flywheel, driving belts/pulleys, and gears.

The safety requirements of manually operated chaff cutter, described in IS 7898 (BIS 2001) include flywheel locking device, front safety guard, gear cover, warning roller, about 90 cm long feeding chute with 45 cm cover (if no warning roller is provided), and covers for flywheel and blades (Fig. 8.4c). The safety requirements for the power-operated chaff cutters, mentioned in IS 15542 (BIS 2005) include 90 cm long feeding chute or conveyor type feeding system, front safety guard, warning roller, feeding chute cover of minimum 55 cm length (if no warning roller is used), and guards for blade and transmission system. The eleven commandments for safe chaff cutter operation are

- To avoid forceful overfeeding;
- To use appropriate guard and other safety devices;

Fig. 8.4 Unsafe practices in **a** powered and **b** manually operated chaff cutting, and **c** the manual machine with the safety features

- To keep the hands at a safe distance while feeding the fodder;
- To avoid crossing over the belt during operation;
- To use a pushing device for feeding small fodder;
- To lock the flywheel with a pin after work;
- To keep hands away from feed roll and open blade;
- Not to allow the children and elderly persons to feed the machine;
- Not to make any adjustments to chaff cutter or alignment to belt during operation;
- To ensure that workers do not wear loose clothing, bangles, watches while working with chaff cutter; and
- To restrict persons intoxicated with liquor or opium to operate the chaff cutters.

Hand Tool Injuries

The farmworkers sustain injuries (e.g. cuts on the hands, feet, and shins) in using a variety of hand tools, such as spades, sickles, digging forks and crowbars, rakes, axes, shovels and hoes. As reported, about 30 to 67% of the total injuries on the farm were due to hand tools (Gite 2008; Xiang et al. 2000; Parvez and Shahriar 2018). Gite et al. (2009) reported that out of the 783 hand tool-related injuries, only one injury was fatal. Parvez and Shahriar (2018) observed that about 78% of the hand tool injuries were minor injuries AIS 1, 17.5% were AIS 2, and less than 1% were AIS 6. About 40% of the injuries occurred in the 16–30 year age range of workers and 26% of injuries occurred in the 31–45 year age group. Other studies (Demers and Rosenstock 1991; Tiwari et al. 2002) also provided evidence that about 3/4th of all the injury victims were in the age range between 15 and 45 years.

Accidents occur during the use of hand tools due to the unpredictable nature of the soil, standing water, or blade hitting hard surfaces or stones, and awkward work posture. The improper dimensions of the handle, weak gripping position, and excessive handle weight result in wrist deviations and cause musculoskeletal disorders. Sometimes the homemade hand tools with inadequate gripping facilities cause injuries. The tool handles may be of bamboo, wood, steel, or PVC material. A handle without a smooth surface might cause blisters on the palm. Sickles and hoes used in removing weeds or cutting crops can cause skin abrasion, superficial vein cuts, and cuts on toes, fingers, or limbs. Prolonged work and continuous sweating of palm might cause tool slippage from hand during operation.

Designing hand tools using hand anthropometry is one of the effective ways for the reduction in hand tool injuries (Nag et al. 2003; Mandahawi et al. 2008; García-Cáceres et al. 2012). Safety shoes and gloves available are too expensive and not suitable for farmworkers in the tropical climate. There is a need for developing comfortable shoes (soft, light in weight, and porous for exchange of air), guards, and gloves for farmworkers. The design changes in the hand tools should be such that the village level artisans and technicians can incorporate the changes quickly in the fabrication of tools/devices.

Child Accidents

In small landholdings, particularly in developing countries, farming is a family activity, and all members of the family, including children, participate in the business. This makes the life of the countryside children different. Letts (1989) described farm machine accidents involving children. Mohan and Patel (1992) reported that children sustained 17% of all farm injuries, and about 1/3rd of the child's injuries were severe. About 6% of the total injuries were fodder cutter injuries, and the reason being the children were playing with the machines. Due care needs to be taken to prevent these injuries.

Accidents in MMH Tasks

Often agricultural activities involve manual materials handling (MMH) tasks, such as lifting, lowering, turning, pulling, pushing, and carrying of heavy loads (Fig. 8.5). High prevalence of musculoskeletal strain, sprains, falls, spinal injuries, and abdominal hernia, due to strenuous MMH tasks has been evident. Physical fatigue results in a lack of coordination and an increase in the frequency of accidents. Enforcing any legislation and safety guidelines in agricultural activities has its challenges since the sector primarily falls under the unorganized or informal sector. However, making the information known among farmers can help to mitigate workplace hazards.

Studies on different categories of farming injuries in MMH tasks show that there is a dramatic increase in the severity of fall injury when the fall height is more than 2 m. The impact forces reduce much if the victim falls on soft earth, hay or sand (Mohan et al. 1979; Snyder et al. 1977). Workers, when subjected to work at greater heights, should have safety harnesses and provisions of soft surfaces where there is a likelihood of falling injuries. Handrails and guard rails are necessary when the work is performed on platforms that are high off the ground. Loads heavier than 15 to 20 kg should not be lifted alone, and assistance of another person should be taken for the same. Containers should have two handles when the weight is more than 10 kg. Guidelines for ergonomics design of container handles are also available (Deeb et al. 1985, 1986). Attempts have been made to improve shoulder load-carrying techniques, designs of wheelbarrows, and even enhanced models of bicycle trailers. The success of these is dependent on the technical skills, equipment, and materials locally available. Sen and Nag (1975) provided a detailed account of the work organization of different modes of load carrying of grains, and other farm produces in rural settings in India. The handling of loads ranged 30 to 125 kg, and accordingly suggested 1200 kcal (5020 kJ) as the net energy output in an 8-h workday for the load handlers engaged in the extremely heavy type of work. Balogun (1986) indicated that load transportation methods using a transverse yoke and head pack were more efficient than the frontal yoke.

Fig. 8.5 Some load-carrying modes in rural farming areas

Injuries in Storage

The farm products such as grains, hays, fruit, and vegetables are often stored in structures, including pits, bins, silos, barns, refrigerated units, carts, trucks, and wagons. Foodgrains, such as rice, wheat, or corn, are usually stored in loose containers or bags. Hay, straw, or plant fibres are bundled and baled to store, whereas vegetables, grains, and feeds are crated, bagged, or stored loose. Fruit and vegetables are often stored in refrigerated units, such as cold storages. Stacked foodstuffs and piles of farm products often cause injuries due to the collapse of the storage and fall of materials. In bins, silos, barns, and other storage structures, accidents and injuries might occur due to falls from heights and falling from ladders. Falls from the same level occur from slippery and uneven surfaces, tripping over objects or stairways.

Adequate care and precautions are necessary in the closed and confined storage facilities, such as silos, tanks, and refrigeration units. People working in such facilities might encounter breathing difficulties due to a possible lack of fresh air supply. Depending on the moisture content in the product, the storage areas can accumulate life-threatening hazardous gases, such as carbon monoxide (CO), carbon dioxide (CO_2), and oxides of nitrogen (NOx). Regular monitoring of such storage areas would ascertain the exposure levels of the gases. Foods and feeds may also be sprayed or treated with safe pesticides to reduce spoilage or mould, spore, or insect damage. Prolonged exposure to cold may be a potential hazard, including incidences of frostbite, cold-induced injury, or breakdown of body temperature regulation. Personal protective clothing should be worn by all concerned in the cold-storage units.

Grain bins are the potential risk areas when a worker enters such facility for loading or unloading of grains. One may accidentally be engulfed into the loose grains. Fall protection measures, such as a safety belt and rope, are essential for the workers to enter into such places. Sometimes, children come and play in the bins and get sucked into the grains. Such situations may suffocate victims and become fatal. Open storages, such as barns, open carts, and wagons, carry risks of a collapse of the structure.

Many storage facilities require machinery for product handling. Hazards and accident risks arise from conveyor belts, pulleys and gears, flywheels, crank and lever mechanisms, contact with moving components, electrical transmissions, inadvertent starting of machinery, and excessive noise and vibration. Besides, workers engaged in the MMH tasks are at risk of musculoskeletal strains. Workers suffer from respiratory disorders and dysfunctions due to accumulated contaminants in the storage areas. Lung disorders (such as occupational asthma, farmer's lung, green tobacco sickness, brown lung, organic dust toxic syndrome, silo filler's or unloader's disease, bronchitis, and airway obstruction) have been reported among workers handling grain, cotton, flax, hemp, hay, and tobacco. Prevention of lung dysfunction demands periodic assessment of the work environment and taking health promotion programmes targeting primary prevention.

Injuries During Transportation

The transportation in the farming sector ranges from goods being carried/transported by humans, animals, manual devices such as cycle rickshaw, animal-drawn carts, mechanical equipment such as trailers pulled by tractors, and with the use of small and large trucks. Each of the modes of transportation has its hazards. For example, the movement of slow-moving vehicles, (SMVs), i.e. tractors and similar equipment, cause a significant number of collisions with the faster moving vehicles (FMVs), particularly during darkness, early morning or evening. The US National Safety Council (NSC 1993) reported 8,000 farm tractors and other agricultural vehicles met with highway accidents in 1992. Accidents during transportation with tractor-trailers in India are widespread, resulting in a large number of fatalities (Tiwari et al. 2004;

Kumar et al. 1998). Over 1/3rd of the total accidents reported through newspapers were due to collision with other vehicles on roads, and nearly the same number of accidents occurred due to the overturning of tractor and tractor-trailers. One of the reasons for these overturning accidents was reported to be the improper design of tractor-trailers. Now, some research organizations, as well as industries, have come out with trailers having safety features.

During transportation operation, operators need to follow safety principles and obey local traffic rules. No riders or passengers other than those necessary to accomplish the transport and unloading duties should be permitted on the tractor. Vehicles should stay as close to the shoulder of the road as road conditions allow. All marking and lighting on moving machinery should be appropriately maintained. The lighting and marking requirements vary from state to state, and country to country. The slow-moving agricultural vehicles need to have two headlights, two tail-lights, two turn signals, two brake lights, and an SMV symbol. Also, placing of reflectors, flags and retroreflective fluorescent markings may be used. The highway movement of FMVs combined with inadequate lighting or no marking of SMVs is often a deadly combination for accidents. Lights are critical for dusk, dawn, and night-time movement/operation of farm equipment.

The farmworkers who are involved in the transportation of farm products may be at risk for respiratory disorders, due to their prolonged exposure to organic dust, chemicals, pesticides, fungal spores, and endotoxins. The exposure scenario depends on whether the transport vehicle has an enclosed cab, where the contaminants may be trapped inside the cab. The workers may also be exposed to pollutants during the loading and unloading processes. Steps for prevention of lung diseases include a periodic assessment of the workers' health and promotion of PPE during work.

The tractor-trailers are not meant for the transport of people. In remote rural areas, the trailer is a purposive vehicle for passenger transport (Fig. 8.6), and also in construction activities. The combination of poor road conditions and live loads makes the tractor-trailers prone to overturning, and these accidents result in considerable loss of lives. Such accidents can be prevented by the use of safe trailers as well as educating the people to avoid travelling in trailers.

Risk of Injuries in Other Farm-Related Activities

Agricultural activities encompass widely different areas, such as nursery raising, horticulture, and greenhouse operations. In nurseries, generally, plants are grown to replant at other locations, and greenhouses are used to grow crops, such as flowers, vegetables, and herbs, to avoid harsh weather conditions. AgSafe (1992) data indicate that nearly 2/5th of all reported injuries in horticulture and nursery activities were sprains and strains, and cumulative overexertion injuries. For planting, pruning, and cutting operations, repetitive hand actions (flexion, extension, and ulnar deviation) are executed in awkward postures, like bending and kneeling on the floor during the prolonged working cycle. Accidents occur in the greenhouse passages due to slippery

Fig. 8.6 Tractor-trailer is also a passenger transport vehicle in remote settlements

and uneven surfaces, obstacles, and sharp objects. Many times, hand tools and power machines are also involved in accidents. Lundqvist and Gustafson (1992) reported the occurrence of accidents and near accidents in Swedish greenhouse cultivation. The back, fingers, hip joints, legs, knees, and feet were the affected body parts; and the average number of compensation (sick leave) days per injury was about 20.

In livestock rearing, animal handling has been responsible for quite a good number of accidents. It demands a better understanding of the behaviour of farm animals to minimize accidents. In Swedish animal husbandry activities (e.g. milking, feeding, dung collecting, moving the animals), milking was responsible for most of the near accidents (Gustafson et al. 1991). Gustafsson (1990) elucidated some guidance about the choice of accident prevention measures in milking system, parlour or stall milking, and recommended to use a system having an overhead rail with moveable hanging bars attached to wheeled wagons carrying the milking equipment.

Accident Prevention

Agriculture is as much a way of life as it is an occupation, particularly in the rural-based family farming. Farmers and farmworkers typically perform a wide range of activities in irregular work hours throughout the farming season. Often the farm-workers serve as part-time workers on the farm on seasonal demands, and then they move away to other non-agricultural jobs during off-seasons. The unorganized workforce is generally poorly trained in safe procedures for performing tasks that are intrinsically extremely hazardous. This workforce encounters accident risks in different farm practices. While the operation of farm machinery is a potential source of severe injuries and fatalities, the conventional manual methods of farm practices using age-old farm tools also bring a multitude of stressors, workplace hazards, and risk of injuries.

Improvements in farming techniques and methods and collective awareness among the farmworkers about safe work practices are the apparent approaches in preventing accidents. Organizational measures that are common in industrial action programmes are often out of the ambit to apply in agriculture to avoid accidents. The prevention of agricultural accidents requires due consideration on technical, human, and organizational aspects. As early as 1969, the International Labour Organization provided guidelines for safety in agriculture (ILO 1969). Again, in 2011, ILO published a code of practice on Safety and Health in Agriculture (ILO 2011). This code of practice aims to improve occupational health and safety in agriculture and complements the Safety and Health in Agriculture Convention 184 (ILO 2001), and guides their application in practice.

Various safety standards have been prepared for agricultural equipment, symbols for operator controls, alert systems, protective guards, vibration of hand tools and vehicles, working postures, brake testing system, and rollover protection (e.g. ASAE standards, Indian Standards). Letts (1989) suggested encouraging manufacturers of farm equipment to improve the safety design of the machinery. Governments of various countries need to develop effective medical assistance programmes, and realistic rehabilitation programmes for the injured farmers, and safety education for farmers as well as their family members. With these points in view, accident prevention in farming may be addressed in a three-pronged manner, i.e. through engineering, enforcement, and educational routes.

Engineering Aspects

Technical measures for accident prevention involve the following aspects.

- Redesigning of machinery, methods of work, and installing safety guards on machinery.
- Designing tractor-trailers keeping in consideration the stability aspects and hitching arrangement.
- Designing the tractor operator's workplace to match with the anthropometric and strength data of the user population.
- Providing proper safety guards for rotating parts of prime movers and farm equipment. The machine guards shall prevent access to the danger zone during operation. The guards should not cause any discomfort to the operator, and

hindrance to work and in maintenance. These should be durable, and fire and corrosion-resistant, and should not constitute a hazard by itself.

- Encouraging manufacturers to design and manufacture machines, such as trailers, threshers, chaff cutters having safe design features, to make those less prone to accidents (Letts 1989).
- Installing electric pump sets and motors correctly to avoid electrocution accidents happening in agriculture.

Enforcement Aspects

Proper training and mandatory driving license for driving a tractor or other self-propelled machinery are necessary. Wherever necessary, motor vehicle rules may be modified to make the safety features such as turning indicators, rear lights, SMV emblem on tractor-trailers mandatory. Fuel meter has to be an integral part of the display console of a tractor. Provision of ROPS on tractors can help to reduce over-turning of the vehicle and minimize fatal accidents. Therefore, the ROPS needs to be made mandatory on tractors all over the world. Machine safety standards can be set for designing and constructing safe machinery. Understandably, the monitoring of the condition of machinery in use has its limitation, and also, adhering to safety standards in the equipment fabricated locally at small workshops, is difficult, especially in the unorganized sector.

As described elsewhere, the built-in safety devices for manually and power-operated chaff cutters need to be made mandatory. The power-operated threshers must have integrated safe feeding chutes or other secure feeding devices. Threshers' designs and relevant standards need to be revised periodically. Suitable PPE like aprons, gloves, gumboots, goggles, facemask need to be developed, evaluated, and popularized among the farmers to prevent accidents due to pesticides, hand tools, and snake bites. Provision may be made to make available PPE along with the sale/supply of sprayers. The procedure to gather and document the agricultural accidents needs to be introduced at the local level and integrated into the local government management system for compensation to accident victims.

Educational Aspects

The farm safety education at all levels is essential, beginning with primary schooling in rural areas. Acknowledging that most farm accidents are preventable, success lies in one's understanding of the probabilities of undesirable outcomes. However, one requires to recognize the dangers of a situation and accordingly to apply skills to deal with such a situation, and prevent accidents. Often accidents happen while children play in the farmyard, and quite often get injured by farm machinery and by livestock. Many of these accidents have severe or permanent consequences. The children must grow up learning of the danger of the farm environment, arising from the tools and machinery, livestock, pesticides, and chemicals, and electricity.

The farmers have the onus on themselves about safety promotion, and therefore, educating themselves on farm safety, use of PPE and safety guards are an obvious necessity. Whether or not the farmworkers are aware of the risk factors associated with the machinery and other equipment, the machinery manufacturers, the farm managers, and local agriculture support agencies must ensure that the safety guidelines of operations of the machinery (Leier 1990) are given due consideration. Encouraging the farmers to undergo periodic training and orientation courses on safe operation and maintenance of tractors, sprayers and dusters, and threshers can significantly reduce farm accidents and injuries. Farm accidents need to be dramatized and broadcasted through suitable public media to create safety awareness among farmers and workers.

References

ACICR. (2010). *Agricultural related Injuries in Alberta.* The Alberta Centre for Injury Control & Research (ACICR), Edmonton Alberta, Canada, p 66.

AgSafe. (1992). Occupational injuries in California agriculture 1981–1990. *Coalition for Health and Safety in Agriculture.* Berkeley, USA: University of California.

Arthur, P. (2019, July 18). *Australian Farm online News.* New safety strategy targets farm deaths and injuries.

Balogun, J. A. (1986). Ergonomic comparison of three modes of load carriage. *International Archives of Occupational and Environmental Health, 58*(1), 35–46.

BIS (2000). IS: 8213. Agricultural trailer – specification. Bureau of Indian Standards, New Delhi.

BIS (2001). IS:7898. Manually operated chaff cutter – Specification, Bureau of Indian Standards, New Delhi.

BIS. (2002). *IS 9020: 2002, Power Threshers – Safety Requirements.* New Delhi: Bureau of Indian Standards.

BIS (2005). IS:15542. Power-operated chaff cutter- Safety requirements. Bureau of Indian Standards, New Delhi.

BIS. (1999). IS:12239(Part 2). Guide for safety and comfort of agricultural tractors and power tiller, Part 2: Requirements relating to agricultural tractors. Bureau of Indian Standards, New Delhi.

Chae, H., Min, K., Park, J., Kim, K., Kim, H., & Lee, K. (2014). Estimated rate of agricultural injury: The Korean farmers' occupational disease and injury survey. *Annals of Occupational and Environmental Medicine, 26*(1), 1–8.

Chapman, J. (2017). How farm safety can be improved. Report No. 04/11/2017. Nuffield Farming Scholarships Trust, Taunton, UK.

Deeb, J. M., Drury, C. G., & Begbie, K. L. (1985). Handle positions in a holding task as a function of task height. *Ergonomics, 28*(5), 747–763.

Deeb, J. M., Drury, C. G., & McDonnell, B. (1986). Evaluation of a curved handle and handle positions for manual materials handling. *Ergonomics, 29*(12), 1609–1622.

Demers, P., & Rosenstock, L. (1991). Occupational injuries and illnesses among Washington State agricultural workers. *American Journal of Public Health, 81*(12), 1656–1658.

DMRA. (1983). The Dangerous Machines (Regulation) Act, 1983, Act No. 35, Govt of India.

Donham, K. J., & Thelin, A. (2016). *Agricultural medicine: rural occupational and environmental health, safety, and prevention.* John Wiley & Sons.

Donnelly, M. (2015). There is gross under-reporting of farm accidents and injuries. Agriland -Ireland Farming News Portal, May 26, 2015.

FAO. (2018). *World Food and Agriculture- Statistical Pocketbook 2018* (p. 2018). Rome: Food and Agriculture Organization of United Nations.

Franklin, R. C., Mitchell, R. J., Driscoll, T. R., & Fragar, L. J. (2001). Agricultural work-related fatalities in Australia, 1989–1992. *Journal of Agricultural Safety and Health, 7*(4), 213–227.

García-Cáceres, R. G., Felknor, S., Córdoba, J. E., Caballero, J. P., & Barrero, L. H. (2012). Hand anthropometry of the Colombian floriculture workers of the Bogota plateau. *International Journal of Industrial Ergonomics, 42*(2), 183–198.

Gennarelli, T. A. (1985). American Association for Automotive Medicine. *Abbreviated Injury Scale. American Association for Automotive Medicine.*IL, USA: Arlington Heights.

Gite, L. P. (2008). Occupational Health and Safety Issues in Indian Agriculture – Present Scenario and Suggestions for Improvement. In *Proceedings of National Workshop, Management of Occupational Health and Safety Delivery: Scopes and Challenges*, National Institute of Occupational Health, Ahmedabad, India, 30 Sept –1 Oct 33–37.

Gite, L. P., Pharade, S. C., & Majumder, J. (2006). Revision of Dangerous Machine (Regulation) Act-1983. *Agricultural Engineering Today, 30*(3, 4), 44–55.

Gite, L. P., Tiwari, P. S., Khadatkar, A., Tyagi, K. K., Bathla, H. V. L. & Kher, K. K. (2009). Farm machinery accidents in Indian agriculture. In *Proceedings of the International Ergonomics Conference HWWE 2009*, Kolkata, India, Dec 17–19.

Gustafsson, B. (1990). The milking of cows by people over the age of 60. In *Proceedings of the XXIV CIOSTA-CIGR V Congress, August 27–30, Helsinki, Finland,* pp. 167–174.

Gustafsson, B., Lindgren, G., & Lundqvist, P. (1991). Near-accidents in agriculture. A survey of Swedish studies. *Swedish Journal of Agricultural Research, 21,* 85–93.

Haddon, W., Jr. (1968). The changing approach to the epidemiology, prevention, and amelioration of trauma: the transition to approaches etiologically rather than descriptively based. *American Journal of Public Health and the Nations Health, 58*(8), 1431–1438.

Haddon, W., Jr. (1970). On the escape of tigers: An ecologic note. *American Journal of Public Health and the Nations Health, 60*(12), 2229–2234.

Hagel, L. M., Pickett, W., Pahwa, P., Day, L., Brison, R. J., Marlenga, B., et al. (2008). Prevention of agricultural injuries: an evaluation of an education-based intervention. *Injury Prevention, 14*(5), 290–295.

Hansson, R., Broberg, E., Johansson, A., Jorner, U., Selander, R. & Karlberg Nilsson, B. (1989). Undersokning av olycksfallen i jord- och skogsbruket 1987. Lantbrukshalsan. Rapport 28. Stockholm. (Cited by Lundqvist, P. and Gustafsson, B.(1992). Accidents and accident prevention in agriculture a review of selected studies. *International Journal of Industrial Ergonomics, 10*(4), 311–319).

Hard, D. L., Myers, J. R., & Gerberich, S. G. (2002). Traumatic injuries in agriculture. *Journal of Agricultural Safety and Health, 8*(1), 51.

HSE. (2013). *Reporting accidents and incidents at work. A brief guide to the Reporting of Injuries, Diseases and Dangerous Occurrences Regulations 2013 (RIDDOR).* Health and Safety Executive, Govt. of UK.

HSE. (2019). Fatal injuries in agriculture, forestry and fishing in Great Britain 2018/19. Health and Safety Executive, Govt. of UK. http://www.hse.gov.uk/agriculture.

ILO. (1969). *Guide to Safety in Agriculture.* Geneva: International Labour Office.

ILO. (2001). Convention No. 184 - Safety and Health in Agriculture. International Labour Office, Geneva.

ILO. (2011). *Safety and Health in Agriculture - ILO code of practice.* Geneva: International Labour Office.

ILO. (2020). India-The Factories Act, 1948 (Act No. 63 of 1948), as amended by the Factories (Amendment) Act, 1987 (Act 20 of 1987). NATLEX Database. https://www.ilo.org/dyn/natlex/docs/WEBTEXT. International Labour Office, Geneva.

Jacinto, C., & Aspinwall, E. (2004). A survey on occupational accidents' reporting and registration systems in the European Union. *Safety Science, 42*(10), 933–960.

Jasra, R. V. & Kothiyal, K. P. (1981). *Threshers are not for maiming. Science Today*, November, 9–11.

Kim, H., Lee, K., & Räsänen, K. (2016). Agricultural injuries in Korea and errors in systems of safety. *Annals of Agricultural and Environmental Medicine, 23*(3), 432–436.

Konstantinos, A. P., Konstantinos, B., Ilias, B. & Anestis, C. (2018). Perineal and lower extremity severe injuries in accident with agricultural machinery-A case report. *ARC Journal of Clinical Case Reports, 4*(4): 12–15.

Kumar, A., Mohan, D., & Mahajan, P. (1998). Studies on tractor related injuries in Northern India. *Accident Analysis and Prevention, 30*(1), 53–60.

Kuye, R., Donham, K., Marquez, S., Sanderson, W., Fuortes, L., Rautiainen, R., et al. (2006). Agricultural health in the Gambia II: A systematic survey of safety and injuries in production agriculture. *Annals of Agricultural and Environmental Medicine, 13*(1), 119–128.

Lakhtakia, P. K. (2000). Evaluation of thresher hand injuries in rural areas of Rewa region in Madhya Pradesh (India). In *Proceedings of the 5th World Conference on Injury Prevention and control*, IIT, New Delhi, March 5–8.

Langley, J. D., Clarke, J., Marshall, S. W., Cryer, P. C., & Alsop, J. (1997). Tractor Fatalities and Injury on New Zealand Farms. *Journal of Agricultural Safety and Health, 3*(4), 237–251.

Larsson, T. J. (1990). Severe hand injuries among Swedish farmers. *Journal of Occupational Accidents, 12*(4), 295–306.

Lee, S. J., Kim, I., Ryou, H., Lee, K. S., & Kwon, Y. J. (2012). Work-related injuries and fatalities among farmers in South Korea. *American Journal of Industrial Medicine, 55*(1), 76–83.

Lehtola, C. J., Marley, S. J., & Melvin, S. W. (1994). A study of five years of tractor-related fatalities in Iowa. *Applied Engineering in Agriculture, 10*(5), 627–632.

Leier, S. M. (1990). Health and safety in agriculture—The farmers role in safety on the farm. In J. A. Dosman & D. W. Cockcroft (Eds.), *Principles of health and safety in agriculture* (pp. 342–344). Boca Raton: CRC Press.

Leigh, J. P. (2011). Economic burden of occupational injury and illness in the United States. *The Milbank Quarterly, 89*(4), 728–772.

Leigh, J. P., Du, J., & McCurdy, S. A. (2014). An estimate of the US government's undercount of nonfatal occupational injuries and illnesses in agriculture. *Annals of Epidemiology, 24*(4), 254–259.

Leigh, J. P., McCurdy, S. A., & Schenker, M. B. (2001). Costs of occupational injuries in agriculture. *Public Health Reports, 116*(3), 235.

Lemola, J. & Pehkonen, A. (1987). The most dangerous farm works. In *Proceedings of the IXth Joint International Ergonomics Symposium "Working postures in agriculture and forestry"*. June 9–12, Kuopio.

Letts, R. M. (1989). Farm machinery accidents in children. In J. A. Dosman, D. W. Cockcroft (Eds.) *Principles of health and safety in agriculture* (pp. 357–361). Inc Boca Raton, FL, United States: CRC Press.

Lundqvist, P., & Gustafsson, B. (1992). Accidents and accident prevention in agriculture: A review of selected studies. *International Journal of Industrial Ergonomics, 10*(4), 311–319.

Lundqvist, P. (2008). Zero fatal injuries in agriculture - A commitment for Sweden. https://www.slideserve.com/tallis/zero-fatal-injuries-in-agriculture-a-commitment-for-sweden-powerpoint-ppt-presentation.

Mandahawi, N., Imrhan, S., Al-Shobaki, S., & Sarder, B. (2008). Hand anthropometry survey for the Jordanian population. *International Journal of Industrial Ergonomics, 38*(11–12), 966–976.

Mittal, V. K., B. S. Bhatia, & S. S. Ahuja. (1996). A study of the magnitude, causes and profile of victims of injuries with selected farm machines in Punjab. Final report of ICAR adhoc Research Project. Department of Farm Machinery and Power Engineering, Punjab Agricultural University, Ludhiana, India.

Mohan, D., & Patel, R. (1992). Design of safer agricultural equipment: Application of ergonomics and epidemiology. *International Journal of Industrial Ergonomics, 10*(4), 301–309.
Mohan, D., Bowman, B. M., Snyder, R. G., & Foust, D. R. (1979). A biomechanical analysis of head impact injuries to children. *Journal of Biomechanical Engineering, 101*(4), 250–260.
Mohan, D., Kumar, A., Patel, R., & Varghese, M. (2004). Development of safer fodder-cutter machines: A case study from north India. *Safety Science, 42*(1), 43–55.
Monk, A. S., Morgan, D. D. V., Morris, J., & Radley, R. W. (1986). The cost of accidents in agriculture. *Journal of Agricultural Engineering Research, 35*(4), 245–257.
MSPA. (1983). Migrant and Seasonal Agricultural Worker Protection Act of 1983. Pub.L.No.87-692,97-470,96 Stat. 2584, USA.
Murphy, D. J. (1981). Farm safety attitudes and accident involvement. *Accident Analysis and Prevention, 13*(4), 331–337.
Myers, J. R., & Hard, D. L. (1995). Work-related fatalities in the agricultural production and services sectors, 1980–1989. *American Journal of Industrial Medicine, 27*(1), 51–63.
Myers, J. R., Layne, L. A., & Marsh, S. M. (2009). Injuries and fatalities to US farmers and farm workers 55 years and older. *American Journal of Industrial Medicine, 52*(3), 185–194.
Nag, A., Nag, P. K., & Desai, H. (2003). Hand anthropometry of Indian women. *Indian Journal of Medical Research, 117,* 260–269.
Nag, P. K., & Nag, A. (2004). Drudgery, accidents and injuries in Indian agriculture. *Industrial Health, 42*(2), 149–162.
NCFH. (2017). Agricultural worker occupational health & safety. The occupational health & safety factsheet of November 2017. National Centre for Farm Worker Health (NCFH), Texas, USA. p. 7.
Nichol, C. I., Sommer, H. J., III, & Murphy, D. J. (2005). Simplified overturn stability monitoring of agricultural tractors. *Journal of Agricultural Safety and Health, 11*(1), 99–108.
NIOSH. (2019). *Agricultural Safety*. National Institute for Occupational Safety and Health, United States. https://www.cdc.gov/niosh/topics/aginjury/). Oct 9, 2019.
NSC. (1993). *Accident facts*. USA: National Safety Council.
OSHA. (1970). OSH Act of 1970. Occupational Safety and Health Administration, Washington DC, USA.
OSHA. (2020). OSHA injury and illness recordkeeping and reporting requirements. Occupational Safety and Health Administration (osha.gov/recordkeeping/). *US Bureau of Labor Statistics*.
Parvez, M. S., & Shahriar, M. M. (2018). Agricultural farm-related injuries in Bangladesh and convenient design of working hand tools. *Journal of Healthcare Engineering, Article ID, 3543610,* 10.
Patel, K., Watanabe-Galloway, S., Gofin, R., Haynatzki, G., & Rautiainen, R. (2017). Non-fatal agricultural injury surveillance in the United States: A review of national-level survey-based systems. *American Journal of Industrial Medicine, 60*(7), 599–620.
Patel, S. K., Varma, M. R., & Kumar, A. (2010). Agricultural injuries in Etawah district of Uttar Pradesh in India. *Safety Science, 48*(2), 222–229.
Pfortmüller, C., Kradolfer, D., Kunz, M., Lehmann, B., Lindner, G., & Exadaktylos, A. (2013). Injuries in agriculture-injury severity and mortality. *Swiss Medical Weekly, 143,* w13846.
Pinzke, S., & Lundqvist, P. (2007). Occupational accidents in Swedish agriculture. *Agricultural Engineering Research, 13*(5), 159–165.
Pinzke, S., Svennefelt, C. A., & Lundqvist, P. (2018). Occupational injuries in Swedish agriculture: Development and preventive actions. *Journal of Agricultural Safety and Health, 24*(4), 193.
Rautiainen, R., Lehtola, M. M., Day, L. M., Schonstein, E., Suutarinen, J., Salminen, S., & Verbeek, J. H. (2008). Interventions for preventing injuries in the agricultural industry. *Cochrane Database of Systematic Reviews*, (1).
Rosenman, K. D., Kalush, A., Reilly, M. J., Gardiner, J. C., Reeves, M., & Luo, Z. (2006). How much work-related injury and illness is missed by the current national surveillance system? *Journal of Occupational and Environmental Medicine, 48*(4), 357–365.

Schmitz-Felten, E. (2016). Farm Safety-OSH issues. OSHWIKI. https://oshwiki.eu/wiki/Farm_Safety_-_OSH_issues. Last edited March 1, 2016.
Sen, R. N., & Nag, P. K. (1975). Work organization of heavy load handling in India. *Journal of Human Ergology, 4,* 99–109.
Shutske, J. M., Field, W. E., & Chaplin, J. (1994). Grain combine fires: A loss reduction approach. *Applied Engineering in Agriculture, 10*(2), 175–182.
Singh, R., Sharma, A. K., Jain, S., Sharma, S. C., & Magu, N. K. (2005). Wheat thresher agricultural injuries: A by-product of mechanised farming. *Asia Pacific Journal of Public Health, 17*(1), 36–39.
Snyder, R. G., Foust, D. R., & Bowman, B. M. (1977). *Study of impact tolerance through free-fall investigation* (No. UM-HSRI-77–8 Final Rpt.). University of Michigan, Highway Safety Research Institute, Ann Arbor, USA.
Solomon, C., Poole, J., Palmer, K. T., & Coggon, D. (2007). Non-fatal occupational injuries in British agriculture. *Occupational and Environmental Medicine, 64*(3), 150–154.
Suutarinen, J. (1992). Tractor accidents and their prevention. *International Journal of Industrial Ergonomics, 10*(4), 321–329.
SWEA. (2006). Arbetsskador 2004 [Occupational injuries 2004]. Arbetsmiljöstatistik Rapport 2006:1. Stockholm, Sweden: Swedish Work Environment Authority.
Tewari V.K., Vidhu, K.P., & Shrivastava A.K. (2010). Agricultural accidents scenario of West Bengal. 2010. Proc. *44th Annual Convention- Symposium of ISAE, 28–30 Jan.*
Tiwari, P. S., Gite, L. P., Dubey, A. K., & Kot, L. S. (2002). Agricultural injuries in Central India: nature, magnitude, and economic impact. *Journal of Agricultural Safety and Health, 8*(1), 95–111.
Tiwari, P. S., Gite, L. P., Kot, L. S., & Dubey, A. K. (2004). Overall assessment of tractor related injuries in Central India. *International Agricultural Engineering Journal, 13*(4), 135–144.
Tómasson, K., Gustafssan, L., Christensen, A., Solberg Rov, A., Gravesenth, H. M., Bloom, K., et al. (2011). *Fatal occupational accidents in the Nordic countries 2003–2008.* Kopenhagan: Nordic Council of Ministers.
Verma, S. R., Rawal, G. S., & Bhatia, B. S. (1978). A study on human accidents in wheat threshers. *Journal of Agricultural Engineering, 15*(1), 19–23.
Wernisch, A. (1989). Arbeitssicherheit am Bauernhof -Gefahrenanalyse Sturz und Fall. Bayerisches Landwirtschaftliches Jahrbuch. Heft 3. BLV Verlagsgesellschaft MBH Miinchen, pp. 359–362. (cited by Lundqvist, P. and Gustafsson, B. (1992). Accidents and accident prevention in agriculture a review of selected studies. *International Journal of Industrial Ergonomics, 10*******(4), 311–319).
Xiang, H., Wang, Z., Stallones, L., Keefe, T. J., Huang, X., & Fu, X. (2000). Agricultural work-related injuries among farmers in Hubei, People's Republic of China. *American Journal of Public Health, 90*(8), 1269–1276.
Yiha, O., & Kumie, A. (2010). Assessment of occupational injuries in Tendaho agricultural development SC, afar regional state. *Ethiopian Journal of Health Development, 24*(3), 167–174.
Zhou, C., & Roseman, J. M. (1994). Agricultural injuries among a population-based sample of farm operators in Alabama. *American Journal of Industrial Medicine, 25*(3), 385–402.

Chapter 9
Health Hazards in Farming

Introduction

Agriculture, a primitive form of industry, continues as the world's largest employer of human resources to produce food grains, to raise cattle, to get milk and milk products, and to grow vegetables and fruits. People also work in agroforestry to gather fuelwood and timber. The farm work ranges from predominantly arduous manual jobs (family farms, subsistence agriculture, migrant, and seasonal work) to high dependence on mechanized farming (corporate farms and plantations). There is a multitude of strenuous activities in the farming sector, including those in plantation work, like in banana, vineyard, apple orchard, and other variety of crops. *(Refer to Chap. 2—Manpower utilization: working methods and practices in small landholdings, and Chap. 4—Energy cost of human labour in farming).* Several health hazards including injury and mechanical stress, respiratory illness, dermatological infections, toxic and neoplastic, and other dangers associated with work environmental stresses manifest among the farmworkers.

Evidence is indicative that agricultural workers encounter adverse work conditions and health risks due to their harmful exposure to physical agents, like noise and vibration originating from farm machinery, and climatic extremes. The farmworkers face health risks of respiratory diseases due to their exposures to organic or other dust, chemicals, pesticides, fungal spores, and endotoxins (Linaker and Smedley 2002; Ye et al. 2013). During sowing and cultivation, workers get exposed to manure, fertilizers, and pesticides. Health risks prevail among the workers during the transportation of agricultural products and their engagement in the loading and unloading processes. Generally, this sector of employment is deprived of essential health and safety services (ILO 2011a) , due to the overwhelming fact of its informal character. In the populous countries like India, where the farming activities spread over to remote locations, a daunting challenge remains to extend health and safety services to more extensive areas. Prevention of lung dysfunction would require a periodic assessment of the working environment, health check-up, targeting to primary

P. K. Nag and L. P. Gite, *Human-Centered Agriculture*, Design Science and Innovation,
https://doi.org/10.1007/978-981-15-7269-2_9

prevention, and the use of personal protective devices to mitigate the risks of lung dysfunction among the farmworkers substantially.

Heavy manual materials handling, awkward working postures as well as the continued use of poorly designed hand tools and devices are the everyday stressors that farmers encounter in their workplace rituals. The overexertion to musculoskeletal structures and functions carries a high risk of accidents and injuries (Biswas et al. 2016). With the mechanization and complexity of farm machinery, the safety of the operator is emerging as the mainstream concern. Machinery and powered implements are the sources of excessive noise, raising the risk of hearing loss.

A healthy farm or livestock operation requires appropriate hygienic and health measures to protect plants, animals, and farmworkers. The concept of agro-medicine, as a subspecialty of occupational/environmental medicine, emphasizes the core areas, like traumatic injury, pulmonary exposures, and agrochemical injury. The interdisciplinary teamwork involves an ecological approach to pest management, agricultural ergonomics, and engineering interventions in designing farm machinery and mitigating trauma and respiratory illnesses, and preventing and controlling tuberculosis among livestock, and workers involved. Other areas of concern include zoonoses, rural health services, food safety, health education and environmental protection, population growth, biotechnology, and sustainable agriculture. This speciality branch of study explores the interventions to mitigate the health hazards of the vast farming workforce. This contribution brings a brief account of the farming hazards for awareness among professionals of other branches of agricultural sciences.

Occupational Health Issues

Vibrations in Farm Machinery

With the increasing trend in mechanization, farm machinery operators are exposed to high levels of vibration (Fig. 9.1). Vibrations get transmitted to the human body as

Fig. 9.1 Vibration exposure to operators during the field operation of tractor-drawn implements

whole-body vibration and local vibration (mainly hand-arm vibration). The vibration characteristics of concern are the direction of vibration, the vibratory surfaces that come in contact with the body, the vibration frequency, its magnitude, and duration (Griffin 1986). The low-frequency vibration produced by an agricultural tractor can be extremely severe, based on the speed and terrain at which the tractor traverses. The operator gets exposed to whole-body vibration via the seat or footrest, and local vibration felt in the hands and arms, via the steering wheel and other controls. The vibration level depends on factors such as driving surface, travel and engine speed, type of seat, kind of tyres, and construction of the machine.

When the vibration exceeds a certain level, the health and the working efficiency of the person, get adversely affected. The physiological damages, including damage to the central nervous system (CNS), joints, and muscles, or the wrist and elbow, may result from the impacts of vibration. The ill-effects of vibration on the health of tractor drivers manifest as a vibration-induced white finger (VWF), low back pain, and spinal disorders. In an early study, based on observations drawn from 1155 tractor drivers, Bovenzi and Betta (1994) emphasized the association of the occurrence of low back pain and disorders among the tractor drivers with their exposure to vibration dose and postural stress. The back, hip and bottom, forehead and chest were the maximum pain areas in most of the tractor operators.

Prolonged exposure to vibration might cause advanced degenerative disc disease (DDD) of the tractor operators. The MRI analysis indicated bilateral features of early S_1 radiculopathy among the tractor operators. The "H" reflex in most of the cases had average latencies and of severely low amplitude. Clinically correlating to radiculopathy, low back pain among the respondents might be attributed to their occupation (CIAE 2012). The cardiovascular load, as heart rates of the tractor operators during transportation, first till, second till, and puddling operations ranged between 100 and 120 beats/min. Frequent looking back to see the operation of the implement, coupled with managing lever controls, might attribute to the increased cardiovascular load during the puddling operation. The body part discomfort rating indicated that transportation was less strenuous as compared to other operations studied. The higher gripping force exertion on the steering wheel along with frequent usage of lever controls in manoeuvring the vehicle caused an increased electrical activity of arm muscles (e.g. flexor carpi radialis). A decrease of the median power frequency in the muscles with time was indicative of fatigue during tractor operation. In comparison to first till, second till, and transportation, the fatigue index was highest in puddling operation. Real-time video analysis of farm operations revealed that the neck rotation >90° and flexion >10° were the dominant human postures that amounted to 2/3rd of the total work cycle with an approximate frequency of 4.41 events/min. Such severity of operation with the combined effect of vibration and postural strain is a likely risk factor to cause spinal injury over time. Therefore, human vibration needs measurement and analysis to determine sources of transmission and levels of safe exposure. The apparent importance is to isolate the driver from the origins of vibration (Fairley

1995) to reduce the risks of health and discomfort, and improve the performance of the driver.

ISO Standards on vibration in agricultural machinery
The evaluation of vibration on tractors and other farm machines is specified in the ISO standards, as listed below:

ISO 2631/1 (ISO 1997/2017). Evaluation of human exposure to whole-body vibration—Part 1: General requirements. International Organization for Standardization, Geneva.
ISO 5007 (ISO 2003). Agricultural wheeled tractors—Operator's seat—Laboratory measurement of transmitted vibration. International Organization for Standardization, Geneva.
ISO 5008 (ISO 2002). Agricultural wheeled tractors and field machinery—Measurement of whole-body vibration of the operator. International Organization for Standardization, Geneva.
ISO 5349 (Part 1). (ISO 2001a). Mechanical vibration—Measurement and evaluation of human exposure to hand-transmitted vibration—Part 1: General requirements. International Organization for Standardization, Geneva.
ISO 5349 (Part 2). (ISO 2001b). Mechanical vibration—Measurement and evaluation of human exposure to hand-transmitted vibration—Part 2: Practical guidance for measurement at the workplace. International Organization for Standardization, Geneva.

Country standards
Various countries have their national standards for measurement and evaluation of vibration. As an example, some Indian standards are mentioned below.

IS 13276 (Part 1), (BIS 2000c). Mechanical vibration and shock—Evaluation of human exposure to whole-body vibration—Part 1: General requirements. Bureau of Indian Standards, New Delhi.
IS 14917 (Part 1), (BIS 2001). Mechanical vibration—Laboratory method for evaluating vehicle seat vibration—Part 1: Basic requirements. Bureau of Indian Standards, New Delhi.

ISO 2631/1 (ISO 1997/2017) covers the whole body vibration. The criteria included in the standard to assess vibration in different situations, are the preservation of (a) working efficiency (human fatigue decreased the proficiency boundary), (b) health and safety (exposure limits), and (c) comfort (reduced comfort boundary). The human body is most sensitive to the frequency ranges of 4 to 8 Hz for vertical (a_z) vibration, and below 2 Hz for transverse (a_x, a_y) vibration. ISO 2631/1 proposed the fatigue decreased proficiency limits for whole-body vibration (Table 9.1) for different exposure times. The reduced comfort boundary limits may be obtained by dividing these values by 3.15 and the exposure limits by multiplying it with a factor of 2. The assessment of risk for hand-arm vibration injury is described in ISO 5349 (ISO

Table 9.1 Fatigue decreased proficiency limits for whole-body vibration

Exposure time, normal workday, h	Acceleration, m/s^2 rms	
	Vertical (a_z)	Horizontal (a_x, a_y)
8	0.315	0.224
6	0.400	0.285
4	0.530	0.355

2001a, b), i.e. the levels of vibration accelerations in the contact surface between the hands and the vibrating object that might cause injury of VWF (vibration white finger) in 10% of those exposed after ten years of regular exposure (Table 9.2). The BS 6841-1987 classified the likely human reactions to rms weighted accelerations into different categories (Table 9.3).

Total weighted rms acceleration value since the seat position of the tractor operator is directly above the rear wheels, the operator subjects to short- and long-term discomfort associated with vibration. The short-term effects reflect on blood circulation, difficulty in speech, head-nodding movement, loss of accuracy in reading, and consistency in foot pressure. Gastrointestinal and spinal disorders arise from the long term effects of vibration. Installation of an anti-vibration seat suspension system or with suspended cabs can substantially reduce seat vibrations and increase the comfort of the tractor operators. Verma (1970) compared four different types of seats, having

Table 9.2 Frequency-weighted hand-arm vibration acceleration limits[a]

Exposure time, normal workday, h	Acceleration, m/s^2 rms (a_x, a_y, a_z)
8	2.1
4	2.9
2	4.1
1	5.8

[a]*Note* frequency-weighted vibration accelerations in the range of 6.3–1250 Hz

Table 9.3 Likely human reactions to weighted vibration acceleration levels, as per BS 6841, (BSI 1987)

rms values	Human reactions
$a_w < 0.315$ m/s^2	Not uncomfortable
$0.35 < a_w < 0.63$ m/s^2	A little uncomfortable
$0.5 < a_w < 1$ m/s^2	Fairly uncomfortable
$0.8 < a_w < 1.6$ m/s^2	Uncomfortable
$1.25 < a_w < 2.5$ m/s^2	Very uncomfortable
$a_w > 2.5$ m/s^2	Extremely uncomfortable

Note Total weighted rms acceleration value —$a_w = \surd\,(1.4\,a_x)^2 + (1.4\,a_y)^2 + (a_z)^2$

suspensions of solid rubber, double leaf spring, single leaf spring, and parallelogram linkage supported on helical coil spring and shock absorber. The observation indicated that the seat with helical coil spring and shock absorber had minimum transmissibility. The maximum transmissibility was 1.21, and that occurred at the frequency of 1.9 Hz, which was beyond the natural frequency range. Thapar (1995) studied the dynamic performance of tractor seats on a simulation rig and found that improvisation of seat suspension reduced vibration transmissibility by nearly half, in one of the tractor seats. Prasad (1995) developed a tractor seat having better vibration attenuation than those available on existing tractors. TNAU (2002) showed the whole body vibration ranging from 1.36 to 3.77 m/s^2 in the tractor operation, whereas the hand-transmitted vibration varied from 0.54 to 3.48 m/s^2.

Earlier, measurement of tractor vibration has primarily been carried out on standard test tracks. Some information on the analysis of tractor vibration under actual operating conditions is available. Mehta et al. (1997a, b) tested four makes of in-use tractors under laboratory and field operations, with the vibration measurements taken in vertical and horizontal directions at no load and during field operations. The observations indicated that vibration levels exceeded the maximum limit of 100 μm (micrometre) on many of the components. The magnitude of vibration at no load was higher for all the operator's comfort-related parts and most of the other functional components in case of the tested tractors, as compared to identical new tractors. The observations corresponded to the findings of Yadava et al. (1990) and Sharma and Shyam (1993) that vibration levels on different components of tractors in use in India exceeded the prescribed limit of 100 μm. Also, the study covered quantifying ride vibration of low horsepower (hp) tractor-implement system, measuring vibration at human-seat interface along the longitudinal, lateral, and vertical axes. The acceleration levels were higher with the increase in the forward speed of travel; however, acceleration levels on tar-macadam road and farm road during transport mode showed no conclusive difference. From the evaluation of the ride vibration, as per ISO 2631/1, and BS 6841 (BSI 1987), the study suggested that the exposure time of the tractor operator during ploughing and harrowing tasks should not exceed 2.5 h.

The vibration dose received by the tractor drivers during ploughing operation may exceed the limits specified by the ISO standards for an 8-hour work period in a day (Mehta 1993). Fairley (1995) conducted 704 field tests to predict vibration-related discomfort from the measurement of vibration in the tractor cabin. Based on the correlation between subjective judgments and predicted values, the study recommended using the frequency-weighted rms values of vibration (0.5–20 Hz) measured on the seat surface in three orthogonal directions (ISO 2631). Lines et al. (1995) also noted that whole-body vibration on tractors and other agricultural vehicles exceeded the level of 8-h exposure limit in different field operations. They suggested that the assessment of tractor vibration may be based on BS 6841 standard, using the root mean quad method. Park et al. (2013) mentioned about the high probability of adverse health effects on the operators of small tractors, and suggested to schedule work, including the vibration-free time and thereby to reduce overall vibration exposure time.

Vibration in Power Tillers

Power tiller is a multipurpose hand tractor, primarily used for rotary tilling and other farm operations in small- and medium-sized farms (Fig. 9.2). The machine in operation generates excessive mechanical vibration due to imbalance in the inertial force of the engine. Vibration transmitted through the hand-arm interface causes severe drudgery in power tiller operation (Dibbits et al. 1978). When the seat was provided for the worker to ride and operate the power tiller, the person was subjected to whole-body vibration in addition to hand-arm vibration components (Mehta et al. 1997a, b). By measuring acceleration transmitted from the handle during tillage operations, Ragni et al. (1999) observed that VWF, a vascular disorder of the hand could appear in 10% of the exposed population after three years of continuous use of small tilling machines.

The vibration in power tiller is maximum at the top of the engine, followed by the chassis, the root of the handlebar, and the gearbox of power tiller. Design interventions, including the insertion of vibration-isolating rubber on the handles, modification in the mass and stiffness of handles, installation of isolators in between the engine and the foundations, and balancing of engine flywheel have a different degree of utility to reduce power tiller vibrations.

During field operations, the generation of vibrations may vary in the frequency range of 20–250 Hz (Pawar 1979) with the dominant centre frequency at 125 Hz, and the levels of acceleration between 2.4 to 3.5 G rms and 1.1 to 1.4 G rms in horizontal and vertical directions, respectively (G = 9.81 m/s^2). TNAU (2002) study recorded the hand-transmitted vibration in the range of 3.66–5.26 m/s^2, during power tiller operation. At the seat of the power tillers, the whole-body vibration varied within 0.63 to 1.21 m/s^2. The rms values of vibration at the chassis ranged from 16 to 30 m/s^2 and at the grip, it was 10 to 24 m/s^2 (Tewari 2003). However, at the seat, the vibrations ranged from 2.2 to 10 m/s^2, with an average of 4 m/s^2.

Kathirwel et al. (2004) developed vibration isolators, which were fitted at the engine, the handlebar, and the handle of the power tillers. The provision of the three-stage vibration isolators reduced handle vibration by 45 to 60%, and the exposure

Fig. 9.2 Hand-arm vibration stress in power tiller operation

limit increased to 8 h during rotatilling operations. As per the guidelines of ISO 5349 (ISO 1986), studying rototilling in an untilled and tilled field, Sam and Kathirvel (2006) recorded the hand-transmitted vibration 20% more in an untilled than in the tilled field. The study indicated that the use of power tiller for four hours a day at the forward speed of 2·4 km/h could result in vascular disorders of the hands in 4 years, for about 10% of the operators.

At a power tiller speed of 4.0 to 8.3 km/h, the peak vibration levels were 8.07 and 5.27 m/s^2 during rotatilling and rotapuddling, respectively (Dewangan and Tewari 2009). Handle grips made of foam rubber and styrene-butadiene rubber could reduce vibrations by about 11% and 5%, as compared to the vibrations, as per ISO 5349 (ISO 2001a, b) at the existing handle grip during rotatilling and rotapuddling operations (Dewangan and Tewari 2010). Chaturvedi et al. (2012) observed power tiller vibrations at 8.00 m/s^2 during rotatilling with rotavator, as compared 5.96 and 6.81 m/s^2 during tilling with cultivator and transportation on farm roads. The handle grip with polyurethane (PU), rubber, and a combination of PU and rubber could reduce vibration, and about 30% reduction was achieved with the rubber gadgets in all operational conditions.

Power-Operated Knapsack Sprayers

Both manual- and power-operated sprayers are used extensively for spraying in-field crops for controlling insects and pests allover the world. India alone manufactures about 7,00,000 such sprayers per annum. Carrying of power sprayers on the shoulder results in the transmission of vibration to body parts of the operator and also causes noise pollution. The operators experience backache and neck stiffness due to the weight of sprayer and awkward working posture. Studies indicated that the vibration transmission to the operator could be reduced by better cushioning of the sprayer.

In the case of a power-operated knapsack sprayer, the vibrations might reach to peak between 50 and 130 Hz at engine speeds between 3000 and 8000 rpm. Singh and Kaul (1972) examined the physical strains associated with the vibration of power sprayers when the engine was operating at 6500 rpm. The vibration arising out of power sprayer affected pulmonary ventilation and comfort of the operators (Bawa and Kaul 1974). The frequency of back vibration and padding thickness also had effects on the heart rate response of the operators (Gupta and Pathak 1981). There were different levels of the amplitude of vibrations at different points on the backrest. The suitable provision of cushion between back and the sprayer was found to reduce the vibrations to a large extent. With the use of the resilient engine and pump mountings and better padding between the machine frame and the users' back, the transmission of vibrations can be attenuated.

Control Measures for Vibration

Comprehensive management of whole-body and hand-arm vibration in farm machinery involves engineering and administrative measures. The *engineering control* measures involve attenuation of vibration through proper design and fabrication of machines, and isolation of vibration from the worker's body. The magnitude of vibration, its frequency, direction, and duration, and also the body posture are the main factors to determine the transmission of vibration to the human operator. The control of vibration exposure would mean addressing one or more of these factors. One approach to reducing vibration at the source is to analyze the machine components of the likely cause of vibration and take control measures in reducing the transmission. Another approach is to ensure that (a) machinery moving parts are well lubricated and balanced, (b) the vibration frequency at the operating speed of the machine should not affect the human body, and (c) additional mass is used to shift the natural frequency of the machine.

Isolation: It involves separating the source of the vibration from the worker. In other words, the operators are moved away from the vibrating machines. Spring mounting the vibrating machines or placing the device on separate slabs can isolate the source from the surroundings. The hand-transmitted vibration in the machinery can be reduced by coating the handles or control levers with rubber isolators and other damping materials.

Vibration isolators for tractors and power tillers (Manian et al. 2004; Kathirvel et al. 2004) were found useful to reduce vibration substantially. For the tractor, the vibration isolator consisted of 12 isomer blocks of 25 mm diameter and 20 mm height sandwiched between two mild steel plates. Placing of the isolator underneath the tractor seat resulted in the reduction of vibration about 12 to 83% during different operations (Fig. 9.3). The vibration isolator for power tiller had three components (the engine, handlebar, and the handle) and reduced the vibration by about 50 to 60%. Suitable placement of the vibration isolators on the handlebar of power-operated hand tools, such as tea pruning brush cutter can substantially reduce the transmission of vibration to the hands and minimize the risk of VWF (CIAE 2010).

Pneumatic tyres in tractors and other self-propelled machines help to attenuate the vertical vibration transmission at low frequencies. The vehicle suspension is another option in reducing vibration transmission. However, most of the off-road machines, tractors, and forklift trucks are built without suspension between the wheel and the chassis. Only cars, trucks, and some recently made dumpers are fitted with parabolic or pneumatic suspensions that give better control of vibration transmission. Besides, isolation of the cab from the chassis can control the vibration transmission, and which can be achieved by providing suspended axles or suspended cabs in tractors.

The other ways of reducing vibration transmission are through the provision of seat suspension and seat cushioning. The suspensions consist of upper and lower frames guided by a combined scissor and rolling action, springs, and a damper or a shock absorber. Typically, mechanical or pneumatic types of seat suspension systems are used. For an ordinary foam and spring type of seat, natural resonance

Fig. 9.3 Vibration isolator fitted between the body of the tractor and the seat

frequency occurs at around 4 Hz (Griffin 1990). In contrast, a suspension seat with additional spring and damper mechanism may have a resonance frequency at about 2 Hz. Suspended seats should suitably be adjusted in the middle of the suspension range, and thereby to avoid any amplification and driver discomfort. When a tractor is driven fast, the suspensions may hit the end stops, causing jolts, and discomfort to the operator. Mehta and Tewari (2002) studied nine commercially available seat cushion materials of different density, thickness, and composition for transmissibility of vibration by measuring input vibration acceleration on the base plate and output acceleration at mass at different frequencies from 1 to 7 Hz. Accordingly, synthetic rubber foam cushions having a density of 69 kg/m^3 was recommended for tractor seat pan and backrest, with the cushion thickness of 101 and 54 mm, respectively.

Administrative controls—Another way of reducing vibration exposure is by reducing the duration of exposure, that is, through adjusting work schedule and rotating operators, and thereby each one would spend less time exposed to vibration. The use of PPE, such as gloves/shoes, as an interface medium, can partially reduce vibration transmission to the body regions. Training of operators should cover the proper way to fit seats, maintenance of the position, and correct body posture in sitting (preferably to avoid a too erect posture), to minimize transmission of vibration to the body. Awkward body posture is partially responsible for back stress and strain associated with exposure to vibration.

Noise in Agricultural Machinery

The human ear responds to noise in terms of its loudness and frequency. The usual range of hearing for a healthy young person is within 20 to 20,000 Hz. The most sensitive frequency range for the human ear is 2–5 kHz. Noise is known to distract and make mental concentration difficult, and the main effect being the hearing loss. Noise-induced hearing loss happens gradually over several years. Exposure to loud noise that can cause permanent hearing loss often produces a temporary threshold shift (McBride et al. 2003). Hearing loss is usually characterized by a decrease in hearing acuity at 4 kHz, followed by a spread into the conversational (lower) frequencies that can severely affect one's ability to communicate.

Effect of tractor noise on operators' hearing has received attention since long. Lierle and Reger (1958) reported that tractor noise negatively effects on the auditory sensitivity of the operators. The issue is getting more attention nowadays as farm mechanization has taken a great leap forward, and the number of tractors and self-propelled farm machines of various complexity has increased tremendously. Research emphasizes on the assessment of the hearing loss to the workers caused due to farm machinery noise. Evaluation of 200 farm machinery operators (Dogra 1996) observed a 4 kHz dip in the hearing threshold at all audiometric frequencies. Higher hearing loss was observed among the operators having increased duration of noise exposure. The noise levels of most Indian tractors exceeded the OSHA recommended safe limits (Kumar et al. 2005) and the farmers associated with tractor driving had higher high-frequency hearing loss than those not driving the tractors. Khadatkar et al. (2017) analyzed hearing impairment of 30 tractor drivers having more than ten years of driving experience. The audiogram indicated that the average hearing threshold of the right ear of tractor drivers was higher at 3 to 4 kHz, compared to the left ear. The lateral differences in the hearing threshold were partly attributed to the positioning of the exhaust system of the tractors, which was generally on the right side.

ISO standards related to noise in agricultural machinery

Some of the essential international standards related to noise in tractors and other farm machines are as follows:

ISO 5131 (ISO 1996). Acoustics—Tractors and machinery for agriculture and forestry—Measurement of noise at the operator's position—Survey method. International Organization for Standardization, Geneva.

ISO 7216 (ISO 1992). Acoustics—Agricultural and forestry wheeled tractors and self-propelled machines—Measurement of noise emitted when in motion. International Organization for Standardization, Geneva.

ISO 1999 (ISO 1990/2013). Acoustics—Determination of occupational noise exposure and estimation of noise-induced hearing impairment. International Organization for Standardization, Geneva.

Country standards

Various countries have their national standards on noise in agricultural machinery operation. As an example, some relevant Indian standards are mentioned below:

IS 12180 (Part 1), (BIS 2000a). Tractors and machinery for agriculture and forestry—noise measurement—method of test. Part 1: Noise at the operator's position—survey method. Bureau of Indian Standards, New Delhi.
IS 12180 (Part 2), (BIS 2000b). Tractors and machinery for agriculture and forestry—noise measurement—method of test. Part-2: Noise emitted when in motion. Bureau of Indian Standards, New Delhi.
IS 12207, (BIS 1999). Recommendations on selected performance characteristics of agricultural tractors. Bureau of Indian Standards, New Delhi.

The noise level in the tractors can be attributed mainly to combustion and cooling fan noise, and also other sources, such as bearings, gears, air suction, fuel injection, and valve operating mechanisms. The peaks in noise levels generally correspond to the fundamental or the harmonics of the combustion frequencies and the cooling fan blade frequencies. The test reports indicated that the noise levels in different makes of tractors in India ranged from 87 to 100 dB(A) (Gupta 1978; Bansal et al. 1982). Also, the noise levels increase with engine speed and the power generated (Mehta et al. 1997a, b). Pessina and Guerretti (2000) examined noise levels of 60 used tractors in Northern Italy, and recorded around 88 dB(A), with maximum peaks of up to 101 dB(A). Using hearing protection devices, namely earmuffs, earplugs, and ear canal caps, the noise at ear level could be reduced to 78 dB(A). Behroozi et al. (2012) studied the noise at the operator's ear level for tractors without and with cabs during different farm operations in Iran. The values ranged from 91 to 93 dB(A), and 76 to 77 dB(A), respectively, and accordingly recommended that tractors should be provided with cabins to protect the operators from hearing loss. Egela and Hamed (2017) mentioned that about 76% of the total number of tractors used in Egypt were without cabs. The noise levels of these tractors at the operator's ear level during field operation ranged from 96 to 107 dB(A).

High noise levels are also a problem in the operation of other agricultural machines, like power tillers and threshers. Studies (Pawar 1978; Tiwari 2003) reported the noise levels in power tiller ranging from 85 to 95 dB(A) at an engine speed of 1,600–2,000 rpm. At such noise levels, continued exposure for more than 5 h would cause hearing damage. The noise levels of threshers are quite high, mainly when they run on diesel engines or tractors (Bansal 1981). The multi-function-brush cutters generate noise in the wide range between 70 and 93 dB(A) when the machine is operated at speeds from 2,600 to 8,200 rpm (Tandon 1991). At speeds exceeding 7,000 rpm, the noise levels exceeded the recommended levels for safe exposure of 8 h. Use of ear defenders could help to reduce the noise level by 2 dB(A) to bring it within a safe limit.

Table 9.4 Occupational noise exposure limits

Source of recommendation	Noise limit—dB(A) 8-h/day exposure	The exchange rate for doubling or halving exposure period, dB(A)
ISO 1999 (ISO 1990/2013)	90	3
OSHA (2013) (Permitted exposure level—PEL)	90	5
NIOSH (1998, 2018) (Recommended exposure level-REL)	85	3
ACGIH (2006)	85	3
WHO (Neitzel and Fligor 2017)	83	3
Govt. of Singapore (2020)	85	3
India: IS 12207 (BIS 1999)	90	3

Safe Exposure Limits of Occupational Noise

Occupational noise exposure is the most frequent hazard present in the workplace. Although regulations and recommended standards (NIOSH 1998; ACGIH 2006) have been in place for decades, the prevalence of noise overexposures and noise-induced hearing loss remains high (Nelson et al. 2005). Nearly 22 million workers in the US are exposed to potentially hazardous levels of noise each day (Tak et al. 2009). There are some variations in the safe exposure limits, as given in Table 9.4. The NIOSH recommendations have been arrived at based on the studies related to hearing loss, and therefore are more protective of hearing. The Threshold Limit Value (TLV) set for noise by American Conference of Governmental Industrial Hygienists (ACGIH) as 85 dB(A) takes into account 8-h daily exposure and 3 dB(A) exchange rate. The TLV is supposedly taken as a reference to protecting the median of the population against a noise-induced hearing loss exceeding 2 dB after 40 years of occupational noise exposure.

Control Measures for Noise

The exposure to noise from farm machinery can be reduced by following a variety of measures, as described herewith.

Engineering control measures—Engineering controls when practical and economically feasible are useful to reduce the sound level at the source. Examples are

- Well-lubricated machine parts cut down friction and accordingly reduce noise;
- Replacement of worn out and unbalanced machine parts cuts down the generation of vibration;

- Use of adequately designed silencer on tractors and other power sources;
- Isolation of the noise source by providing an acoustically designed cab; and
- Appropriate design of farm machines considering noise reduction aspects.

Where the engineering controls are not adequate to bring down the noise level, the administrative actions are called for in organizing work schedules and job rotation as the alternative options to keep the farmworkers within the allowable exposure limit.

Personal protective equipment (PPE)—Intervention at the individual level, i.e. making farmworkers use PPE (hearing protection) can reduce noise exposure. Earmuffs vary in size, shape, and design, and can attenuate noise by as much as 40 dB. Earplugs may be pre-formed inserts type made of rubber or foam and hand-formed inserts of disposable materials such as wax or wool.

Dust in Agricultural Operations

From an occupational health point of view, the dust is classified by size into three primary categories. That is

(a) Total dust: It includes all airborne particles;
(b) Inhalable dust: It is that size of a fraction of airborne dust that can enter through the nose, but to a great extent trapped in the nose, throat, and upper respiratory tract. The median aerodynamic diameter of the dust is about 100 μm. Sometimes, this fraction is again bifurcated as inhalable dust and thoracic dust, and the median aerodynamic diameter of the later is taken as 10 μm; and
(c) Respirable dust: It refers to the dust particles that are small enough to penetrate through the upper respiratory system and deep into the lungs. Dust particles of size lesser than 5 μm are categorized as respirable dust.

During farming activities, such as ploughing, harrowing, moving of earth, loosening of soil, threshing, rice milling, agro-processing, and poultry farming, workplace dust emanates, causing health problems to workers (Mołocznik 2002; Faria et al. 2006; CIAE 2019; Gandhi et al. 2012). The nature of the airborne dust during the cultivating season varies with the type of soil, kind of farm work, and also the prevailing meteorological conditions. Significant dust hazards mainly come during harvesting and threshing activities. Dust formation may be most intensive during the operation of combines. Even the enclosures like tractor cab may contain dust concentration exceeding permissible levels for total as well as a respirable fraction (less than 5 μm size). Dust from rice, wheat, corn harvesting contains plant scraps, pollen, and spores, in which particles may be larger than 10 μm in size and non-respirable. The dust containing a significant fraction of free crystalline silica is potentially harmful to breathing disorders.

The upper permissible limits for total and respirable dust are 10 and 5 mg/m^3 of air, respectively (Hansson 1991; OSHA 2019). The dust concentration in the work environment of rice mills in India varied from 50 to 89 mg/m^3 of air (CIAE 2012), whereas, the respirable fraction of dust ranged from 3 to 10 mg/m^3 of air. Pranav and Biswas (2016) reported that ergonomics intervention through design modification of the feeding cum sieving section could reduce the total dust content in rice mill by about 56%. Also, the respirable fraction reduced from 8.9 to 4.3 mg/m^3, thus bringing it within the safe exposure limit. Providing suitable dust masks to mill workers can help to protect the workers from dust exposure further.

Respiratory Hazards

In multifactorial causal situations like in farming (open field as well as closed processing enclosures), several components, including farm crops, contribute to creating complex environmental conditions. During the handling of the farm products (grain, cotton, flax, hemp, hay, and tobacco), the farmworkers face the hazards of inhaling toxic gases and dust. Exposures to dust, gases, chemicals, fungal spores, and endotoxins can cause respiratory disorders. Depending on the moisture content in grains and fibres, the crop products placed in storages and silos may produce gases, such as carbon monoxide (CO), carbon dioxide (CO_2), and oxides of nitrogen (NO_x). Agricultural lung disorders have some familiar names, including occupational asthma, bronchitis, farmer's lung, green tobacco sickness, organic dust toxic syndrome, silo filler's disease, and the like. Besides, the crops might have been treated with pesticides to reduce spoilage or mould, spore, or insect damage. Also, oxygen deficiency in confined spaces on farms is a continuing problem. A periodic assessment of the work environment, health promotion through primary prevention, and adopting PPE and other protective devices are the proactive measures to prevent lung dysfunction of farmworkers. For understanding by the professionals of agricultural sciences, the respiratory illnesses (Nag 2019) that the agricultural workers might encounter are briefly summarized herewith (http://www.mayoclinic.org/diseases-conditions/copd/symptoms-causes/dxc-20204886).

Asthma and asthma-like disorders—The exposure to outdoor and indoor air pollutants, airborne dust, and allergens have long been known to cause adult-onset, or occupational asthma (Eder et al. 2006). Farm antigen exposures through pollen, storage mites, and grain dust can trigger asthma. Inflammation of the mucous membrane is a common reaction to airborne dust among persons with allergic rhinitis or a history of atopy. From a clinical point of view, bronchial asthma is a kind of chronic inflammatory disease. Contraction and tightening of the bronchial muscles, along with increased mucus production or decreased clearance of mucus, cause variable airflow obstruction. The signs and symptoms of asthma include frequent coughing, wheezing, chest tightness, shortness, and shallow breathing and tachycardia. The inflammation causes airway hyperresponsiveness with the appearance of variable and reversible airflow obstruction. Occupational asthma has been linked to several natural products, such as grain dust, tobacco dust, soybean, coffee bean, vegetable gums. Exposure to the substances of low molecular mass at high concentrations may lead to irritant-induced occupational asthma. Byssinosis is caused by cotton, flax, and hemp dust. Exposure to a high concentration of CO generated from the combustion sources is also potentially harmful to agricultural workers, particularly in a closed environment.

Irritant-induced acute occupational asthma, also named as reactive airway dysfunction syndrome (RADS) manifests from a single high-level irritant exposure (e.g. NO_x), or multiple exposures to an environment of irritant capacity. *Sensitizer-induced occupational asthma* may occur from exposure to high molecular mass sensitizing agents, e.g. glycoproteins of biological origin, grain mites/insects, grain dust, natural rubber/latex gloves. The latency period of the *sensitizer-induced occupational asthma* condition can last from weeks to months or years, between the first exposure to the sensitizing agent and the development of immunologically mediated symptoms.

The conditions of occupational asthma may be diagnosed by analyzing the history of workplace exposures, and measurement of pulmonary function tests, serial peak expiratory flow (PEF), histamine or methacholine challenge, and immunological tests, such as serum specific immunoglobulin E (IgE), skin prick, and epicutaneous tests (Fishwick et al. 2008; Tarlo et al. 2008). The diagnosis is typical of the evidence of reversible variable airway limitations, along with asthmatic trends between the workdays and relief on days off. A serial PEF (about four recordings a day over three weeks) has high specificity and sensitivity in the diagnosis of occupational asthma. An elevated diurnal variation, and differences in PEF between the workdays and days off, would indicate the likely presence of asthma. Serum specific IgE may assist in making a diagnosis, due to its possible presence in persons exposed to allergens with high molecular weight. Skin prick tests may also be positive for workplace allergen. Increased bronchial reactivity to sequential histamine, methacholine challenge is the evidence of sensitizer-induced occupational asthma. The prevention of occupational asthma requires environmental intervention and medical management.

All asthma-like symptoms may not be associated with asthma (e.g. chronic obstructive pulmonary disease, chronic bronchitis, pulmonary embolism, gastro-oesophageal reflux, idiopathic environmental illness). For example, the environmental tobacco smoke (ETS) is an aetiological factor of *chronic obstructive pulmonary disease, which* manifests a progressive reduction of pulmonary ventilation due to a combined effect of emphysema and obstruction of the small airways. *Chronic bronchitis* is common among farmers, due to their prolonged exposure to grain dust or work in swine confinement buildings.

Extrinsic Allergic Alveolitis (EAA)

The EAA is a condition also named as hypersensitivity pneumonitis. Repeated exposure to environmental allergens from organic dust and substances, and micro-organisms from spoiled hay, grain, and silage, may be related to the occurrence of hypersensitivity reaction of the alveolar and bronchiolar tissue, and interstitium of the lungs of farmworkers (Simon-Nobbe et al. 2008). The hypersensitivity reactions among farmers may manifest as an acute, subacute, and chronic form. The acute responses, like a cough, chest tightness, feverish chills, and flu-like illness, appear within a few hours after exposure and gradually fade away and reappear on subsequent exposures. The subacute reactions show progressive shortness of breath, dry cough, and weight loss. A chest x-ray may show fine interstitial infiltrates in acute and subacute forms of the disease. In the case of a chronic form of the disease, irregular scarring in lungs indicates the slow development of interstitial fibrosis (ILO 2011b). The patients show restrictive impairment, with the FEV_1/FVC ratio reduced in comparison to the normal range. An inhalation provocation test using a specific antigen can confirm the diagnosis of the illness. The EAA is a diffuse parenchymal disease, and thus affect the diffusion capacity of the lungs. Bronchoscopy with bronchoalveolar lavage shows increased total cell count, with the lymphocytes, increased more than 40%. The transbronchial biopsy may be carried out in suspected cases of EAA. The T-Lymphocyte Helper/T-Suppressor ratio is usually reduced to less than 1.

Some well-studied conditions are (a) *farmer's lung* caused from mouldy forage (mushroom, paprika, coffee dust), (b) *organic dust toxic syndrome (silage unloader's syndrome)* associated with exposure to mouldy fodder, and (c) *grain fever* caused due to exposure to stored grain dust. *Mushroom worker's lung* is associated with exposure to airborne spores (e.g. thermophilic actinomycetes, Excellospora flexuosa, Thermomonospora alba, T. curvata, T. fusca) that grow during the conditioning of compost in mushroom cultivation (Van den Bogart et al. 1993). Rylander (1986) observed that for causation of EAA symptoms, about 108 spores/m^3 of air was sufficient. The use of respirators equipped with a fine dust filter may prevent mushroom workers from the condition that otherwise irreparably damages lungs due to inflammation and reactive fibrosis. Management of EAA primarily requires removing the affected persons from the source of exposure. Corticosteroids intervention is a chosen approach at the acute episodes.

Thermal Stress

Primarily the agriculture is an outdoor activity, under varied weather conditions. Farmworkers are exposed to climatic extremes during the summer months, in the tropical and subtropical zones. The heat stress of farm workers is a recognized occupational hazard due to outdoor exposure to strong sunlight, elevated environmental temperatures, and other inclement weather conditions (Nag et al. 2007, 2009; Baptiste 2018). The health risks reflect as an increase in core body temperature and excessive sweat loss and dehydration, as consequences of exposure to heat (Parsons 2014). When the physiological regulation of body temperature is inadequate to achieve thermal equilibrium, significant health disturbances occur. At dehydration exceeding 4 to 6% of the body weight, the exposed workers might manifest signs and symptoms, such as excessive sweating, tachycardia, weakness, and pain in the limbs, abdominal pain, disorientation, headaches, dizziness. In extreme cases, loss of consciousness, fatigue of sweat glands and blockage of the sweat ducts (hydromeiosis) might appear among the exposed persons. Dehydration and subsequent thermoregulatory problems adversely affect cellular as well as systemic functions with increased probability of premature fatigue and developing heat injury. A variety of heat stress and disorders have been reported among the farmers. Exposure to solar radiation during the open-field occupational work and the cumulative human-environment heat exchange (Nag et al. 2007; Fiala et al. 2012) lead to severe sunburn and heat-related stress and strain, culminating in conditions like heat exhaustion and heatstroke.

Cultivation and harvesting of crops are performed when the outdoor temperature is high, and solar radiation is intense. The tractor cabs without air conditioners can be very hot and uncomfortable for the tractor drivers to operate the machine during the warmer months in the tropical regions. The flux level of direct solar radiation is 500 W/m^2 and more when the ordinary glass is used for cab windows. Tinted glass can lower the air temperature. Forced air ventilation of about 350 m^3/h can create a temperature gradient of about 5 to 7 °C between the inside and outside of the tractor cabin. Prolonged exposure to solar radiation can result in premature ageing of the skin and the likelihood of developing skin cancers. Conversely, the farmworkers in the temperate zones and particularly during extreme cold encounter consistent hazards of frostbite or fatality from hypothermia.

Exposure to hot and humid environments put the farmworkers at high risk for heat-related illness or death. Morabito et al. (2006) reported an association between hot climate (daytime heat index 25–28 °C) and increased accidents related hospital admissions from June to September 1998 and 2003, in Central Italy. In the United States, CDC (2008) report indicated that during the 15 years (1992–2006), 423 workers in agricultural and non-agricultural industries died from exposure to high heat stress. The heat-related average annual death rate in crop workers was 0.39 per 100,000 workers, which is 20 times higher than all US civilian workers combined. In India, as per the National Crime Records Bureau, a total of 7,686 people lost their lives between 2010 and 2015 because of heatstroke, i.e. each year about 1537 people lost their lives (Dubbudu 2018). Xiang et al. (2014) reported a 0.2% increase in daily injury claims with an increase of 1 °C daily maximum temperature in the range between 14.2 °C and 37.7 °C among workers in agriculture, forestry, and fishing in Adelaide, Australia. Spector et al. (2016) undertook a case-crossover study among 12,213 outdoor agricultural workers who claimed for traumatic injury compensation in the state of Washington between 2000 and 2012. The study indicated a strong association of traumatic injuries among the agricultural workers labouring during tree fruit harvest duties in the June and July period.

Loss of Productivity

Thermal stress has implications on health and productivity of people engaged in diverse farming and allied activities (Watts et al. 2018; Glaser et al. 2016). Several studies demonstrated that worker's productivity declines as heat exposure increases (Sahu et al. 2013; Zander et al. 2015; Krishnamurthy et al. 2017). Dally et al.

(2018) evaluated the relationship between temperature exposure, kidney function, and productivity of 4,095 sugarcane cutters over a 6-month harvest in Guatemala. The study assessed associations between WBGT and the productivity of farmworkers. The cumulative effect of exposure to WBGT of 34 °C was 1.16 tonnes less sugarcane cutting output in five days, compared to exposure at 29 °C WBGT.

In the macro analysis, rising climatic extremes or prolonged exposure to hot and humid tropical climate tend to reduce labour capacity (Dunne et al. 2013; Kjellstrom et al. 2009). The decline in the working capacity of farmers reflects in health implications, tropical diseases, and job attrition of people (Lundgren et al. 2013; Alexandratos and Bruinsma 2012). The farmworkers in tropical India have effective heat load due to external or environmental, and internal or metabolic conditions of high severity of work. However, the information on the effects of heat on health, performance and safety for people engaged in tropical farming is scarce (Nag et al. 1980; Nag 1996). The tobacco harvesters and tea pluckers have a high prevalence of malaria, iron deficiency, and anaemia, with endemic consequences on the working capacity and heat disorders. Also, in the central and southern European countries, the recurrent heatwaves in warm-season represent a significant hazard of farmworkers (Adam-Poupart et al. 2013).

Heat-Related Chronic Kidney Disease

Evidence is overwhelming of the pandemic chronic kidney disease of unknown origin (CKDu), often recognized as Mesoamerican nephropathy (MeN), among young male farmers and conspicuously sugarcane cutters, along the Pacific coastal lowlands of Central America—El Salvador, Costa Rica, Nicaragua (Correa-Rotter et al. 2014; Kupferman et al. 2018; Aguilar-Ramirez and Madero 2018;). The devastating occurrences of the CKD costed life of several thousand people in recent years.

The sugarcane is primarily produced in Brazil and other Central American countries, followed by India, China, and Thailand. In India alone, sugar and associated industries support about 50 million farmers' families. Data on incidences of CKD and other kidney ailments in the Asian regions are scanty. There are CKD hotspots in clusters of tropical regions of Sri Lanka (Wijetunge et al. 2013), the Uddanam region of Andhra Pradesh (India) (Dyer 2014), Rajasthan (India), and Egypt (Weaver et al. 2015) where the prevailing weather conditions are similar to the regions of Central America. Thus, Glaser et al. (2016) termed CKDu as heat stress nephropathy. Acute kidney injury (AKI) has been recorded in Hispanic agricultural workers in Florida (Mix et al. 2018). Luo et al. (2014) indicated that the outdoor workers of a shipbuilding company in Guangzhou, China, with more prolonged cumulative heat exposure had higher incidences of urolithiasis (formation of stony concretions in the bladder or urinary tract), as compared to the indoor employees.

Apart from heat stress and resulting recurrent dehydration among workers, researchers have associated the epidemic of chronic kidney ailments with exposure to other agents. These are agrochemicals, heavy metals through contaminated pesticide formulations, and fertilizers (García-Trabanino et al. 2015; Raines et al. 2014), contaminated drinking water (Laws et al. 2015), burning of the cane (Santos et al. 2015), and high fructose intake (Jimenez et al. 2014). Dehydration and hyperosmolarity may induce tubular injury via activation of the polyol-fructokinase pathway in the kidney. Hyperuricaemia and cyclical uricosuria associated with volume loss and dehydration have also been proposed as the likely contributors to the CKD epidemic (Johnson et al. 2013; Roncal-Jimenez et al. 2016). Further research would elucidate the mechanism of the occurrence of kidney ailments interacting with heat exposures in the farmworkers. Physically demanding jobs in hot climates would require adequate water intake, preferably with nutritional supplements (Bodin et al. 2016). The controlled climatic chamber investigation (Nag and Ashtekar 1999) indicated that hyperhydration by ingestion of a mix of glycerol plus water supplementation (1.0 g glycerol plus 21.2 g water/kg of body weight) delayed the development of dehydration in extremely hot-dry, and hot-humid exposures. Thereby, the heat exposed individuals could preserve the ability to dissipate heat and reduce cardiovascular strain. The proactive strategy currently in place to deal with heat stress is hyperhydration with a suitable mix of water with an electrolyte solution, and the introduction of frequent rest breaks in shades.

Occupational Skin Hazards

Human skin represents nearly 1/6th of the body weight. It is vital for specific physiological functions, such as the external protective barrier, the thermoregulatory function, touch and sensitivity, and the vitamin D synthesis (Ngatu 2018). The large surface area of the skin is also an active pathway for the absorption of hazardous chemical substances that threaten skin sensitization, injury, and other systemic disorders. Undoubtedly, the farmers frequently come in contact with plant excretions, and agrochemicals, resulting in skin irritation and other varieties of health hazards. The most common form of skin disease among the farmers is irritant contact dermatitis. Besides, allergic contact dermatosis is a reaction of exposures to sensitizers, including plants and pesticides. Readers may refer to Chap. 10 for details on pesticides and chemical toxicity in farming.

Skin injuries occur among people engaged in different farm-related activities, such as green tobacco pluckers, tea garden workers, coconut tree climbers, and others. Matsushita et al. (1989) reported the outbreak of skin hazards among workers in okra cultivation in Japan. Positive patch test reactions substantiated that the okra components (leaves, trichomes, and pods) cause irritant contact dermatitis and allergic contact dermatitis. The high prevalence of skin irritation and allergies, poisoning, injuries, breathing problems, sunstroke, and eye irritation have been reported among farm women in Jhansi district of Bundelkhand (India) during the activities of harvesting, weeding, and digging of groundnut (Pandey et al. 2010). Egharevba and Iweze (2004) reported that farming women from six rural communities of Edo State, Nigeria experienced muscular fatigue, dermatitis, respiratory diseases, impaired vision and hearing due to exposure to climatic stress, use of pesticides and fertilizers, dust, and insect bites. The outdoor workers require safety interventions, such as the use of PPE, protective clothing, in addition to the primary prevention of washing the body thoroughly, immediately after contacts with the irritation causing plants.

Other skin diseases include photo-contact, sun and heat-induced, and arthropod-induced dermatoses. Skin burn can take place from dry fertilizer, which is hygroscopic. Liquid substances, such as anhydrous ammonia that are applied to the soil, may come into contact with the eyes, skin, and respiratory tract. These can cause burns and permanent cell destruction. Farmers and farmworkers face increased risks of skin cancer from exposure to solar ultraviolet radiation when working outdoors. From a review of studies, Kearney et al. (2014) revealed that outdoor farmworkers often wear some type of locally designed headwear, as a primary method of protection from the sun. Some female farmers use sunscreen for protection.

Behavioural Hazards

Farmers are subjected to a multitude of stressors, either in the traditional or mechanized farming. The workers in small-scale family farming in regions of Asia often encounter stresses due to financial constraints, soil degradation, inclement weather conditions causing impacts on the farm productivity, excessive dependence on the application of pesticides, and traumatic injuries from heavy machinery. Under these circumstances, the workers face problems of dysfunctional relationships, local land conflicts, home violence, and suicide. There are overwhelming incidences of farmer suicides in India (Behere and Bhise 2009), apparently due to crop damages, market pressure of diminishing returns, and their inability to pay bank loans. Parvathamma (2016) reported that, in India, from 1995 to 2013, the average suicide rate of farmers was 13,754 per year. Policy intervention calls for initiatives towards adequate crop insurance, improved health care, and quality of living. Far contrast is the context of suicides among farmers in North America, who mostly use firearms to commit suicide due to depression (Boxer et al. 1995).

The labour force in farming is often migratory, and they move in different regions, depending on the demand for labour in peak farming seasons. This continues to be a dominant feature in agriculture in many regions of the world. Their employment is often seasonal and on a contractual basis. Predominantly on economic considerations, the labourers move a long distance away from home and engage in farming activity, ignoring the danger associated with it. In general, both male and female workers manifest low occupational risk perceptions. Various ILO reports elucidate a range of behavioural issues among migrant workers, including mental stress and strain, socio-domestic disturbances, child abuse, substance abuse, injuries at the workplace, and the like. These workforces are at high risk of tuberculosis and sexually transmitted diseases, and the females face problems of perinatal outcome and high infant mortality.

Fire Hazards

The farming sector often faces unwarranted fire situations, mainly in many types of harvesting operations. Crops that are required to be adequately dried to about 15% moisture content for storage make it an excellent fuel for ignition (Fig. 9.4). PAU (2019) reported 180 fire accidents in April and May 2019 in Punjab (India), causing loss of crops of about 1700 ha, with the monetary loss amounting to about US$ 3 million. Primary reasons for the crop fires included sparks from combine harvesters, overhead electrical wires, and uncontrolled burning of straw by farmers in the neighbouring land plots. The crops lying in the field get ignited spreading the fire on large areas. Such a situation results in dust and organic emissions, causing excruciating problems of air pollution to the regions (Shutske et al. 1994). Large machinery (combines and harvesters) that leave remains in the field are especially vulnerable to fires, due to the use of diesel engines in operations. Well-maintained equipment and electrical systems, and easy operator access to fire extinguishers can reduce the risk of fire-related damage or injury.

Fig. 9.4 Fire hazards in the farm fields

Workplace Hazards in Crop Cultivation

Tobacco Cultivation

Only four countries, namely China, the United States, Brazil, and India, produce over 60% of total world production of tobacco. The manual labour required to produce tobacco and tobacco products varies greatly, depending on the level of mechanization used for transplanting, harvesting, and processing for products. Large area of tobacco cultivation in India is labour intensive. Working in tobacco cultivation is drudgery prone, having risks of musculoskeletal discomfort, and cutting injuries due to transplanting , harvesting, and other activities. Manual tobacco harvesting (Fig. 9.5a) is an apparent hazard of skin absorption of leaf extracts.

A small percentage of field tobacco labourers often complain of green tobacco sickness (GTS), and that has been prevalent in Asian and South American tobacco farms. Water from rain or dew on the tobacco leaves probably dissolves nicotine and other substances that get absorbed through the skin (Achalli et al. 2012). The condition named as *green tobacco sickness* manifests among sensitive individuals. The symptoms include headache, pallor, nausea, vomiting, muscle weakness, dizziness, headaches, increased perspiration, abdominal pain, skin rash, diarrhoea, and increased salivation, and further making extreme conditions like prostration, shortness of breath, and fluctuations in blood pressure or heart rate (Khan et al. 2010). Symptoms are usually temporary but cause discomfort for several hours after exposure. The duration of the illness is usually between one and three days.

Besides, tobacco farmers are more prone to adverse health effects due to exposure to pesticides, as observed in Malayasian tobacco farms (Reeves and Schafer 2003). The questionnaire survey among Greek tobacco farmers (Damalas et al. 2006) recognized skin contact as the most common route of exposure to the pesticide. However, a significant proportion of the farmers reported of not adopting PPE when handling and spraying pesticides. From the analysis of the clinical and KAP information, Khan et al. (2010) revealed that the tobacco farmers in Swabi, Pakistan had mild to moderate pesticide poisoning that correlated significantly with the depression in plasma cholinesterase (PChE) levels.

Fig. 9.5 Workers engaged in **a** tobacco harvesting and **b** float platform in makhana harvesting

Sensitive workers may minimize contact with wet green tobacco by wearing lightweight, protective, water-resistant clothing, chemical-resistant gloves, plastic aprons, and rain suits with boots and socks. One may wear long trousers and long-sleeve shirts when working in dry tobacco. As a precaution, sensitive workers may involve in the work when the leaves are dry. Complete washing of the body immediately after leaving the field is a good practice for the sensitive workers.

Makhana Harvesting

Makhana is a seed produced from an aquatic crop that usually grows in ponds. India is one of the major countries that produce these seeds. The harvesting of makhana is generally done manually from August to October. For harvesting of makhana, a worker goes deep into the bottom surface of the pond, holds his breath, and drags the mud at one place which is then sieved with the locally made bamboo screen. Collecting mud at the bottom surface of the pond is drudgery prone, and a health hazard. Mud enters into the ears, eyes, nose, and mouth of the person when going deep into the pond. Khadatkar et al. (2015) reported an ergonomics intervention of using a floating platform (Fig. 9.5b) supported by a 10-litre cylinder having compressed breathing air with the regulator, 10 m hose pipe, and a mini diving kit (having a cap, mask, and content gauge). A comparative analysis showed that the intervention reduced the worker's overall discomfort rating significantly. The average output in the traditional method was 3.8 kg/h, and with the ergonomics intervention, the improved method had higher output, i.e. 11.3 kg/h.

Compost Preparation

The preparation of compost involves many processes. Gully pits used for collecting materials to prepare compost are usually devoid of oxygen, and the accumulated water contains a high concentration of hydrogen sulphide and ammonia. Piling of manure and other animal excreta and tissues in a closed area become a lethal environment due to high concentrations of carbon dioxide, hydrogen sulphide, and ammonia. All compost processing areas should be provided with local exhaust ventilation. It is also necessary to have a defined standard operating procedure for the personnel involved in the handling of compost.

Cashew Nut Processing

Cashew nut is a high valued cash crop. Shelling of cashew nuts by hand-operated or hand cum foot-operated cutters is an essential operation in cashew processing

Fig. 9.6 **a** Cashew nut green pods, **b** nuts with shells, **c** a worker cuts the shells, **d** manually scooping out kernels from the shells, **e** a mechanical device for the shelling of cashew nuts, and **f** skin injury from manual shelling of the cashew nutshell

industries (Fig. 9.6). The cutter is fixed on the table, and the work is performed in standing posture, leading to bodily musculoskeletal strains. Since the sizes of the nuts vary, careful manipulation is required to avoid hand injury. Besides, scooping out of intact kernels using needles, separating the kernels and shells, and grading of cashews are other operations that require attention for workplace improvement and reduction of drudgery. Some amount of oily extract comes out of the cashew nutshell during shelling operation, and that causes a reaction to human skin (Fig. 9.6f). Wearing gloves can avoid this problem. A mechanical sheller for shelling of cashew nuts (CIAE 2014) reduces drudgery and increases work output in shelling operation (Fig. 9.6c and e).

Herbs and Spices Cultivation

The workforce engaged in herb cultivation is tiny; however, herbs cultivation has a diversity of methods and geographical locations. Depending on the type of herb to be cultivated, seeds, cuttings, seedlings, or rhizome portions are planted. Labour-intensive cultivation methods, such as soil preparation, planting, construction of support structures, periodic weeding, harvesting, and other operations, place high demands on health and safety, including stressors of the musculoskeletal system.

The improvement in the method of cultivation and the introduction of mechanization are the possible directions that might mitigate workplace stressors.

The herb and spice crop processing industry includes processing of a variety of farm products, such as ginseng, parsley, black pepper, cinnamon, chicory, cloves, coriander, garlic, ginger, paprika and red chillies (capsaicin), mint, rosemary and thyme, nutmeg and vanilla. Generally, the industry represents the scenario of the high magnitude of exposure to specific hazards during grinding, crushing, and mixing of leaves, seeds, and other plant materials. The workplace is noisy and extremely dusty. Health conditions include traumatic injuries from moving machinery parts, skin diseases (occupational dermatitis), irritation of the eyes, mouth and gastrointestinal tract, and respiratory and immunological problems, associating to dust, fungi, and other air contaminants.

For extraction of volatile oils from herbs, the steam is forced into the sealed container of the cut and chopped plant materials, and the released oil is extracted from the resulting vapour. Hazards associated with the oil extraction include burning from steam, boiler spillover, and explosions. Adequate local exhaust ventilation, improvization in dust collection, mopping and cleaning of work areas, and provision of PPE to workers can reduce risks from dust explosions and contaminants in breathing air.

References

ACGIH. (2006). 2005–2006 Threshold limit values for chemical substances and physical agents and biological exposure indices (B.E.I.s): Noise. American Conference of Governmental Industrial Hygienists. 2006.

Achalli, S., Shetty, S. R., & Babu, S. G. (2012). The green hazards: a meta-analysis of green tobacco sickness. *International Journal of Occupational Safety and Health, 2*(1), 11–14.

Adam-Poupart, A., Labrèche, F., Smargiassi, A., Duguay, P., Busque, M. A., Gagné, C., et al. (2013). Climate change and occupational health and safety in a temperate climate: Potential impacts and research priorities in Quebec, Canada. *Industrial Health, 51*(1), 68–78.

Aguilar-Ramirez, D., & Madero, M. (2018). Untangling Mesoamerican Nephropathy. *American Journal of Kidney Diseases, 72*(4), 469–471.

Alexandratos, N., & Bruinsma, J. (2012). World agriculture towards 2030/2050: the 2012 revision. Rome: F.A.O., 2012 Contract No.: E.S.A. Working paper No. 12–03.

Bansal, A. S. (1981). Survey of noise levels of wheat threshers. In *Proceedings XVIII Annual Convention of Indian Society of Agricultural Engineers*, Karnal, Feb 26 to 28.

Bansal, A. S., Behniwal, N. S. & Kumar, A. (1982). Survey of noise levels of tractors. In *Proceedings of the XIX Annual Convention of Indian Society of Agricultural Engineers*, Udaipur, India, Feb 15–17.

Baptiste, N. (2018). Farm workers are dying from extreme heat. https://www.motherjones.com/food/2018/08.

Bawa, H. S., & Kaul, R. N. (1974). Some studies on vibration of a Knapsack power sprayer effecting operator comfort. *Journal of Agricultural Engineering,* 11(1).

Behere, P. B., & Bhise, M. C. (2009). Farmers' suicide: Across culture. *Indian Journal of Psychiatry, 51*(4), 242–243.

Behroozi, M. L., Payandeh, M., Bagheri, J., & Pour, Z. K. (2012). Comparison of noise level of tractors with cab and without in different gears on driver ear and bystander. *African Agricultural Research, 7*(7), 1150–1155.

BIS. (1999). IS 12207, 1999. Recommendations on selected performance characteristics of agricultural tractors. Bureau of Indian Standards, New Delhi.

BIS. (2000a). IS 12180 (Part 1), 2000. Tractors and machinery for agriculture and forestry – noise measurement – method of test. Part 1: Noise at the operator's position – survey method. Bureau of Indian Standards, New Delhi.

BIS. (2000b). IS 12180 (Part 2), 2000. Tractors and machinery for agriculture and forestry – noise measurement – method of test. Part-2: Noise emitted when in motion. Bureau of Indian Standards, New Delhi.

BIS. (2000c). IS 13276 (Part 1), 2000. Mechanical vibration and shock – Evaluation of human exposure to whole-body vibration – Part 1: General requirements. Bureau of Indian Standards, New Delhi.

BIS. (2001). IS 14917 (Part 1). Mechanical vibration – Laboratory method for evaluating vehicle seat vibration - Part 1: Basic requirements. Bureau of Indian Standards, New Delhi.

Biswas, G., Bhattacharya, A., & Bhattacharya, R. (2016). A review on the occupational health of sugar cane workers. *International Journal of Biomedical Research, 7*(8), 2016.

Bodin, T., García-Trabanino, R., Weiss, I., Jarquín, E., Glaser, J., Jakobsson, K., et al. (2016). Intervention to reduce heat stress and improve efficiency among sugarcane workers in El Salvador: Phase 1. *Occupational and Environmental Medicine, 73*(6), 409–416.

Bovenzi, M., & Betta, A. (1994). Low-back disorders in agricultural tractor drivers exposed to whole-body vibration and postural stress. *Applied Ergonomics, 25*(4), 231–241.

Boxer, P. A., Burnett, C., & Swanson, N. (1995). Suicide and occupation: a review of the literature. *Journal of Occupational and Environmental Medicine, 37*(4), 442–452.

BSI. (1987). BS 6841, Guide to Measurement and Evaluation of Human Exposure to Whole-Body Mechanical Vibration and Repeated Shock. British Standards Institution, London.

CDC. (2008). Heat-related deaths among crop workers—United States, 1992—2006. MMWR Morb. Mortal. Wkly. Rep. Centers for Disease Control and Prevention, Atlanta, U.S.A., 57(24), 649–653.

Chaturvedi, V., Kumar, A., & Singh, J. K. (2012). Power tiller: Vibration magnitudes and intervention development for vibration reduction. *Applied Ergonomics, 43*(5), 891–901.

CIAE. (2010). Research Highlights 2007–09. All India Coordinated Research Project on Ergonomics and Safety in Agriculture. Technical Bulletin No. 2009/148. Central Institute of Agricultural Engineering, Bhopal, India.

CIAE. (2012). Research Highlights 2010–12. All India Coordinated Research Project on Ergonomics and Safety in Agriculture. Technical Bulletin No. 2012/152. Central Institute of Agricultural Engineering, Bhopal, India.

CIAE. (2014). Research Highlights 2012–14. All India coordinated research project on ergonomics and safety in agriculture. Technical Bulletin No. CIAE/2014/176. Central Institute of Agricultural Engineering, Bhopal, India.

CIAE. (2019). Research Highlights 2017–19. All India coordinated research project on ergonomics and safety in agriculture. Technical Bulletin No. CIAE/ESA/2019/284. Central Institute of Agricultural Engineering, Bhopal, India.

Correa-Rotter, R., Wesseling, C., & Johnson, R. J. (2014). CKD of unknown origin in Central America: the case for a Mesoamerican nephropathy. *American Journal of Kidney Diseases, 63*(3), 506–520.

Dally, M., Butler-Dawson, J., Krisher, L., Monaghan, A., Weitzenkamp, D., Sorensen, C., & Newman, L. S. (2018). The impact of heat and impaired kidney function on productivity of Guatemalan sugarcane workers. *PloS one, 13*(10).

Damalas, C. A., Georgiou, E. B., & Theodorou, M. G. (2006). Pesticide use and safety practices among Greek tobacco farmers: A survey. *International Journal of Environmental Health Research, 16*(5), 339–348.

Dewangan, K. N., & Tewari, V. K. (2009). Characteristics of hand-transmitted vibration of a hand tractor used in three operational modes. *International Journal of Industrial Ergonomics, 39*(1), 239–245.

Dewangan, K. N., & Tewari, V. K. (2010). Handle grips for reducing hand-transmitted vibration in hand tractor. *International Agricultural Engineering Journal, 19*(2), 48–57.

Dibbits, H. J., Van Loon, J. H., & Curfs, H. P. F. (1978). The physical strain of operating a two-wheeled tractor. *World Crops and Livestocks, 30,* 250–252.

Dogra, B. (1996). Studies on effect of noise on hearing among farm machinery operators. M.Tech. Thesis. Punjab Agricultural University, Ludhiana, India.

Dubbudu, R. (2018). 1 out of every 5 Heat Stroke deaths in India occurs in Andhra Pradesh. FACTLY, March, 7. https://factly.in/1-every-5-heat-stroke-deaths-india-occurs-andhra-pradesh/.

Dunne, J. P., Stouffer, R. J., & John, J. G. (2013). Reductions in labour capacity from heat stress under climate warming. *Nature Climate Change, 3*(6), 563–566.

Dyer, O. (2014). C.D.C. will explore kidney failure epidemic among agricultural workers. *The British Medical Journal, 348*:g3385.

Eder, W., Ege, M. J., & von Mutius, E. (2006). The asthma epidemic. *New England Journal of Medicine, 355*(21), 2226–2235.

Egela, M., & Hamed, A. (2017). Tractor Noise Levels Impact on Operator Safety. *Journal of Soil Sciences and Agricultural Engineering, 8*(8), 355–362.

Egharevba, R. K., & Iweze, F. A. (2004). Sustainable agriculture and rural women: Crop production and accompanied health hazards on women farmers in six rural communities in Edo State Nigeria. *Journal of Sustainable Agriculture, 24*(1), 39–51.

Fairley, T. E. (1995). Predicting the discomfort caused by tractor vibration. *Ergonomics, 38*(10), 2091–2106.

Faria, N. M. X., Facchini, L. A., Fassa, A. G., & Tomasi, E. (2006). Farm work, dust exposure and respiratory symptoms among farmers. *Revista de Saúde Pública, 40,* 827–836.

Fiala, D., Havenith, G., Bröde, P., Kampmann, B., & Jendritzky, G. (2012). UTCI-Fiala multi-node model of human heat transfer and temperature regulation. *International Journal of Biometeorology, 56*(3), 429–441.

Fishwick, D., Barber, C. M., Bradshaw, L. M., Harris-Roberts, J., Francis, M., Naylor, S., et al. (2008). Standards of care for occupational asthma. *Thorax, 63*(3), 240–250.

Gandhi, S., Dilbaghi, M., Mehta, M., & Pruthi, N. (2012). Occupational Health hazards & efficacy of protective masks in threshing operation. *Work, 41*(Supplement 1), 4983–4986.

García-Trabanino, R., Jarquín, E., Wesseling, C., Johnson, R. J., González-Quiroz, M., Weiss, I., et al. (2015). Heat stress, dehydration, and kidney function in sugarcane cutters in El Salvador–a cross-shift study of workers at risk of Mesoamerican nephropathy. *Environmental Research, 142,* 746–755.

Glaser, J., Lemery, J., Rajagopalan, B., Diaz, H. F., García-Trabanino, R., Taduri, G., et al. (2016). Climate change and the emergent epidemic of CKD from heat stress in rural communities: the case for heat stress nephropathy. *Clinical Journal of the American Society of Nephrology, 11*(8), 1472–1483.

Govt. of Singapore. (2020). Factories Act (Chapter 104, Section 102(1)) Factories (Noise) Regulations-1997.Singapore Statues Online. https://sso.agc.gov.sg/SL/FA1973-RG14.

Griffin, M. J. (1986). Evaluation of vibration with respect to human response. *S.A.E. Transactions,* 323–346.

Griffin, M. J. (1990). *Handbook of human vibration.* Academic press.

Gupta, O. S. (1978). Agricultural equipment testing at Budni. *Agricultural Engineering Today, 2*(5), 5–11.

Gupta, P. K., & Pathak, B. S. (1981). Studies on transmission of vibration of back mounted power equipment. *Journal of Agricultural Engineering,* 18(3, 4), 9–16.

Hansson, J. E. (1991). *An ergonomic checklist for tractors and agricultural machinery.* National Swedish Institute of Occupational Health (Arbetsmiljöinstitutet).

ILO. (2011a). *Safety and Health in Agriculture - ILO code of practice*. International Labour Office, Geneva.

ILO. (2011b). *Guidelines for the use of the ILO international classification of radiographs of pneumoconioses*. Geneva: International Labour Office.

ISO. (1990/2013). ISO 1999. Acoustics—Determination of occupational noise exposure and estimation of noise-induced hearing impairment. International Organization for Standardization, Geneva.

ISO. (1992). ISO 7216. Acoustics—Agricultural and forestry wheeled tractors and self-propelled machines—Measurement of noise emitted when in motion. International Organization for Standardization, Geneva.

ISO. (1996). ISO 5131. Acoustics—Tractors and machinery for agriculture and forestry—Measurement of noise at the operator's position—Survey method. International Organization for Standardization, Geneva.

ISO. (1997/2017). ISO 2631/1. Evaluation of human exposure to whole-body vibration—Part 1: General requirements. International Organization for Standardization, Geneva.

ISO. (2001a). ISO 5349 (Part 1). Mechanical vibration—Measurement and evaluation of human exposure to hand-transmitted vibration—Part 1: General requirements. International Organization for Standardization, Geneva.

ISO. (2001b). ISO 5349 (Part 2). Mechanical vibration—Measurement and evaluation of human exposure to hand-transmitted vibration—Part 2: Practical guidance for measurement at the workplace. International Organization for Standardization, Geneva.

ISO. (2002). ISO 5008. Agricultural wheeled tractors and field machinery—Measurement of whole-body vibration of the operator. International Organization for Standardization, Geneva.

ISO. (2003). ISO 5007. Agricultural wheeled tractors—Operator's seat—Laboratory measurement of transmitted vibration. International Organization for Standardization, Geneva.

Jimenez, C. A. R., Ishimoto, T., Lanaspa, M. A., Rivard, C. J., Nakagawa, T., Ejaz, A. A., et al. (2014). Fructokinase activity mediates dehydration-induced renal injury. *Kidney International, 86*(2), 294–302.

Johnson, R. J., Nakagawa, T., Jalal, D., Sánchez-Lozada, L. G., Kang, D. H., & Ritz, E. (2013). Uric acid and chronic kidney disease: Which is chasing which? *Nephrology Dialysis Transplantation, 28*(9), 2221–2228.

Kathirvel, K., Manian, R., sam, B., & Gite, L. P. (2004). Anti-vibration devices for increased comfort of power tiller operators. Technical bulletin No. A.E.C. &RI Pub. 28/2004. AICRP on Human Engineering and Safety in Agriculture, TNAU, Coimbatore. India.

Kearney, G. D., Xu, X., Balanay, J. A. G., & Becker, A. J. (2014). Sun safety among farmers and farmworkers: a review. *Journal of Agromedicine, 19*(1), 53–65.

Khadatkar, A., Gite, L. P., & Gupta, V. K. (2015). Interventions to reduce drudgery of workers in the traditional method of harvesting Makhana (Euryale ferox salisb.) seeds from ponds. *Current Science*, 1332–1337.

Khadatkar, A., Mehta, C. R., Gite, L. P., Narwariya, B. S., & Kumar, A. (2017). Hearing impairment of Indian agricultural tractor drivers. *Current Science, 113*(5), 969–974.

Khan, D. A., Shabbir, S., Majid, M., Naqvi, T. A., & Khan, F. A. (2010). Risk assessment of pesticide exposure on health of Pakistani tobacco farmers. *Journal of Exposure Science & Environmental Epidemiology, 20*(2), 196–204.

Kjellstrom, T., Holmer, I., & Lemke, B. (2009). Workplace heat stress, health and productivity–an increasing challenge for low and middle-income countries during climate change. *Global Health Action, 2*(1), 2047.

Krishnamurthy, M., Ramalingam, P., Perumal, K., Kamalakannan, L. P., Chinnadurai, J., Shanmugam, R., et al. (2017). Occupational heat stress impacts on health and productivity in a steel industry in southern India. *Safety and Health at Work, 8*(1), 99–104.

Kumar, A., Mathur, N. N., Varghese, M., Mohan, D., Singh, J. K., & Mahajan, P. (2005). Effect of tractor driving on hearing loss in farmers in India. *American Journal of Industrial Medicine, 47*(4), 341–348.

Kupferman, J., Ramírez-Rubio, O., Amador, J. J., López-Pilarte, D., Wilker, E. H., Laws, R. L., et al. (2018). Acute kidney injury in sugarcane workers at risk for Mesoamerican nephropathy. *American Journal of Kidney Diseases, 72*(4), 475–482.

Laws, R. L., Brooks, D. R., Amador, J. J., Weiner, D. E., Kaufman, J. S., Ramírez-Rubio, O., et al. (2015). Changes in kidney function among Nicaraguan sugarcane workers. *International Journal of Occupational and Environmental Health, 21*(3), 241–250.

Lierle, D. M., & Reger, S. N. (1958). The effect of tractor noise on the auditory sensitivity of tractor operators. *Annals of Otology, Rhinology & Laryngology, 67*(2), 372–388.

Linaker, C., & Smedley, J. (2002). Respiratory illness in agricultural workers. *Occupational Medicine, 52*(8), 451–459.

Lines, J., Stiles, M., & Whyte, R. (1995). Whole body vibration during tractor driving. *Journal of Low Frequency Noise, Vibration and Active Control, 14*(2), 87–104.

Lundgren, K., Kuklane, K., Gao, C., & Holmer, I. (2013). Effects of heat stress on working populations when facing climate change. *Industrial Health, 51*(1), 3–15.

Luo, H., Turner, L. R., Hurst, C., Mai, H., Zhang, Y., & Tong, S. (2014). Exposure to ambient heat and urolithiasis among outdoor workers in Guangzhou, China. *Science of the Total Environment, 472,* 1130–1136.

Manian, R., Kathirvel, K., Balasankari, P. K. & Mehta, C. R. (2004). Anti-vibration device for increased comfort of tractor operators. Terchnical bulletin No. AEC&RI Pub: 21/2004. AICRP on Human Engineering and Safety in Agriculture, TNAU, Coimbatore, India.

Matsushita, T., Manda, F., Aoyama, K., Ueda, A., Ueda, T., Qing, L. I., et al. (1989). Skin hazards in workers cultivating okra (Hibiscus esculentus Linn). *Journal of the Japanese Association of Rural Medicine, 38*(1), 24–29.

Mayoclinic.org. http://www.mayoclinic.org/diseases-conditions/copd/symptoms-causes/dxc-20204886.

McBride, D., Gilmore, T., & Waite, D. (2003). The immediate retest in temporary threshold shift audiometry. *Journal of Occupational and Environmental Medicine, 45,* 1211–1212.

Mehta, C. R. (1993). Effect of farm implements on tractor ride vibrations. M.E. (Ag.) Thesis. College of Technology and Agricultural Engineering, Udaipur, India.

Mehta, C. R., Shyam, M., & Sharma, P. K. (1997a). *Studies on noise and vibration levels of different tractors.* CIAE, Bhopal, India: Final Report of Research Project.

Mehta, C. R., & Tewari, V. K. (2002). Vibrational characteristics of tractor seat cushion materials and ride comfort. *Journal of Low Frequency Noise, Vibration and Active Control, 21*(2), 77–85.

Mehta, C. R., Tiwari, P. S., & Varshney, A. C. (1997b). Ride vibrations on a 7· 5 kW rotary power tiller. *Journal of Agricultural Engineering Research, 66*(3), 169–176.

Mix, J., Elon, L., Vi Thien Mac, V., Flocks, J., Economos, E., Tovar-Aguilar, A. J., & McCauley, L. A. (2018). Hydration status, kidney function, and kidney injury in Florida agricultural workers. *Journal of Occupational and Environmental Medicine, 60*(5), e253-e260.

Molocznik, A. (2002). Qualitative and quantitative analysis of agricultural dust in working environment. *Annals of Agricultural and Environmental Medicine, 9*(1), 71–78.

Morabito, M., Cecchi, L., Crisci, A., Modesti, P. A., & Orlandini, S. (2006). Relationship between work-related accidents and hot weather conditions in Tuscany (central Italy). *Industrial Health, 44*(3), 458–464.

Nag P. K. (1996) Criteria for recommended standards for human exposure to environmental heat. A Report: Ministry of Environment and Forests. Government of India (No. 19/45/91-RE), National Institute of Occupational Health, (ICMR), Ahmedabad, India.

Nag P. K. & Ashtekar S. P. (1999). Hyperhydration, International Journal of Industrial Engineering, Industrial and Occupational Ergonomics: Users' Encyclopaedia (CD-ROM), ISBN 0-9654506-0-0), p. 5, 1999.

Nag, P. K., Sebastian, N. C., & Malvankar, M. G. (1980). Effective heatload on agricultural workers during summer season. *Indian Journal of Medical Research, 72,* 408–415.

Nag, P. K., Nag, A., & Ashtekar, S. P. (2007). Thermal limits of men in moderate to heavy work in tropical farming. *Industrial Health, 45*(1), 107–117.

Nag, P. K. (2019). *Office building – Health, safety and environment* (pp. 533). Singapore: Springer. ISBN 978-981-13-2576-2.

NIOSH. (1998). Criteria for a recommended standard: Occupational noise exposure, *DHHS (NIOSH) Publication No. 98-126, National Institute for Occupational Safety and Health, U.S.A.*

Neitzel, R., & Fligor, B. (2017). *Determination of risk of noise-induced hearing loss due to recreational sound: review*. Geneva: World Health Organization.

Nelson, D. I., Nelson, R. Y., Concha-Barrientos, M., & Fingerhut, M. (2005). The global burden of occupational noise-induced hearing loss. *American Journal of Industrial Medicine, 48*(6), 446–458.

Ngatu, N. R. (2018). Skin Function and Occupational Skin Hazards. In *Occupational and Environmental Skin Disorders* (pp. 7–19). Singapore: Springer.

NIOSH. (2018). Noise and hearing loss prevention - Guidance and Regulations. https://www.cdc.gov/niosh/topics/noise/reducenoiseexposure/regsguidance.html. National Institute for Occupational Safety and Health, U.S.A.

OSHA. (2013). OSHA Technical Manual-Section III, Chapter 5, Noise. Occupational safety and health administration. United States Deptt. Of Labour, Washington DC., U.S.A.

OSHA. (2019). Permissible exposure limits – Annotated tables. https://www.osha.gov/dsg/annotated-pels/index.html. Occupational Safety and Health Administration. United States Deptt. Of Labour, Washington DC., U.S.A.

Pandey, S., Meena, B. S., Sharma, P., & Dwivedi, R. N. (2010). Health hazards among farmwomen in different on-farm operations. *Journal of Community Mobilization and Sustainable Development, 5*(1), 38–40.

Park, M. S., Fukuda, T., Kim, T. G., & Maeda, S. (2013). Health risk evaluation of whole-body vibration by ISO 2631-5 and ISO 2631-1 for operators of agricultural tractors and recreational vehicles. *Industrial Health*, 2012-0045.

Parsons, K. (2014). *Human thermal environments: the effects of hot, moderate, and cold environments on human health, comfort, and performance*. C.R: C. press.

Parvathamma, G. L. (2016). Farmers suicide and response of the Government in India-An Analysis. *IOSR Journal of Economics and Finance, 7*(3), 1–6.

PAU. (2019). *Progress Report 2018–19 of the All India coordinated research project on ergonomics and safety in agriculture*. Ludhiana, India: Punjab Agricultural University.

Pawar, J.G. (1978). Investigation of Human Energy Requirements for Power Tiller Operators. M.Tech. Thesis, Department of Agricultural Engineering, Punjab Agricultural University, Ludhiana, India.

Pawar, J. G. (1979). *Ergonomic studies ofpower tiller - A review*. Ludhiana, India: Report on Human Engineering, Punjab Agricultural University.

Pessina, D., & Guerretti, M. (2000). Effectiveness of hearing protection devices in the hazard reduction of noise from used tractors. *Journal of Agricultural Engineering Research, 75*(1), 73–80.

Pranav, P. K., & Biswas, M. (2016). Mechanical intervention for reducing dust concentration in traditional rice mills. *Industrial Health, 54*(4), 315–323.

Prasad, N. (1995). *Some Investigations on tractor seat design with regard to Indian Operators (Doctoral dissertation*. Kharagpur, India): I.I.T.

Ragni, L., Vassalini, G., Xu, F., & Zhang, L. B. (1999). Vibration and noise of small implements for soil tillage. *Journal of Agricultural Engineering Research, 74*(4), 403–409.

Raines, N., González, M., Wyatt, C., Kurzrok, M., Pool, C., Lemma, T., et al. (2014). Risk factors for reduced glomerular filtration rate in a Nicaraguan community affected by Mesoamerican nephropathy. *MEDICC Review, 16*, 16–22.

Reeves, M., & Schafer, K. S. (2003). Greater risks, fewer rights: U.S. farmworkers and pesticides. *International Journal of Occupational and Environmental Health, 9*(1), 30–39.

Roncal-Jimenez, C., García-Trabanino, R., Barregard, L., Lanaspa, M. A., Wesseling, C., Harra, T., et al. (2016). Heat stress nephropathy from exercise-induced uric acid crystalluria: a perspective on Mesoamerican nephropathy. *American Journal of Kidney Diseases, 67*(1), 20–30.

Rylander, R. (1986). Lung diseases caused by organic dusts in the farm environment. *American Journal of Industrial Medicine, 10*(3), 221–227.

Sahu, S., Sett, M., & Kjellstrom, T. (2013). Heat exposure, cardiovascular stress and work productivity in rice harvesters in India: implications for a climate change future. *Industrial Health.*

Sam, B., & Kathirvel, K. (2006). Vibration characteristics of walking and riding type power tillers. *Biosystems Engineering, 95*(4), 517–528.

Santos, U. P., Zanetta, D. M. T., Terra-Filho, M., & Burdmann, E. A. (2015). Burnt sugarcane harvesting is associated with acute renal dysfunction. *Kidney International, 87*(4), 792–799.

Sharma, P. K., & Shyam, M. (1993). *Noise and vibration levels of indigenous tractor – A status report*. Bhopal, India: Paper presented at XXVIII convention of ISAE held at CIAE.

Shutske, J. M., Field, W. E., & Chaplin, J. (1994). Grain combine fires: A loss reduction approach. *Applied Engineering in Agriculture, 10*(2), 175–182.

Simon-Nobbe, B., Denk, U., Pöll, V., Rid, R., & Breitenbach, M. (2008). The spectrum of fungal allergy. *International Archives of Allergy and Immunology, 145*(1), 58–86.

Singh, H., & Kaul, R. N. (1972). Human energy requirements of selected farm operations. *Journal of Agricultural Engineering., 11*(3), 44–52.

Spector, J. T., Bonauto, D. K., Sheppard, L., Busch-Isaksen, T., Calkins, M., Adams, D., & Fenske, R. A. (2016). A case-crossover study of heat exposure and injury risk in outdoor agricultural workers. *PLoS one, 11*(10).

Tak, S., Davis, R. R., & Calvert, G. M. (2009). Exposure to hazardous workplace noise and use of hearing protection devices among U.S. workers—NHANES, 1999–2004. *American Journal of Industrial Medicine, 52*(5), 358–371.

Tandon. S. K. (1991). *Studies on multi-functional brush cutter*. Postdoc Report, Silsoe Research Institute, Silsoe, U.K.

Tarlo, S. M., Balmes, J., Balkissoon, R., Beach, J., Beckett, W., Bernstein, D., et al. (2008). Diagnosis and management of work-related asthma: American College of chest physicians consensus statement. *Chest, 134*(3), 1S–41S.

Tewari V. K. (2003). Reduction of vibrations and noise in self-propelled reaper and power tiller. Progress report of the Adhoc project of ICAR, I.I.T., Kharagpur, India.

Thapar, S. (1995). Studies on ride vibrations of tractor and improvements in suspension system of seat. M. Tech. Thesis, Punjab Agricultural University Ludhiana, India.

TNAU. (2002). *Progress Report of the All India Coordinated Research Project on Human Engineering and Safety in Agriculture, 2000-02.* Coimbatore, India: Tamilnadu Agricultural University.

Van den Bogart, H. G. G., Van den Ende, G., Van Loon, P. C. C., & Van Griensven, L. J. L. D. (1993). Mushroom worker's lung: serologic reactions to thermophilic actinomycetes present in the air of compost tunnels. *Mycopathologia, 122*(1), 21–28.

Verma, M. R. (1970). Tractor seat design, fabrication and testing. M. Tech. thesis. Agricultural Engineering Department, I.I.T., Kharagpur, India.

Watts, N., Amann, M., Ayeb-Karlsson, S., Belesova, K., Bouley, T., Boykoff, M., et al. (2018). The Lancet Countdown on health and climate change: from 25 years of inaction to a global transformation for public health. *The Lancet, 391*(10120), 581–630.

Weaver, V. M., Fadrowski, J. J., & Jaar, B. G. (2015). Global dimensions of chronic kidney disease of unknown etiology (CKDu): a modern era environmental and/or occupational nephropathy? *B.M.C. Nephrology, 16*(1), 145.

Wijetunge, S., Ratnatunga, N. V. I., Abeysekera, D. T. D. J., Wazil, A. W. M., Selvarajah, M., & Ratnatunga, C. N. (2013). Retrospective analysis of renal histology in asymptomatic patients with probable chronic kidney disease of unknown aetiology in Sri Lanka. *Ceylon Medical Journal, 58*(4), 142–147.

Xiang, J., Bi, P., Pisaniello, D., Hansen, A., & Sullivan, T. (2014). Association between high temperature and work-related injuries in Adelaide, South Australia, 2001–2010. *Occupational and Environmental Medicine, 71*(4), 246–252.

Yadava, S. B .S, Suneja, Y. L., Mehta, M. L. & Patil, V. A. (1990). Studies on vibration level comparison of a few selected tractors. In *Proceedings of XXVI convention of ISAE held at H.A.U.*, Hisar, India.

Ye, M., Beach, J., Martin, J. W., & Senthilselvan, A. (2013). Occupational pesticide exposures and respiratory health. *International Journal of Environmental Research and Public Health, 10*(12), 6442–6471.

Zander, K. K., Botzen, W. J., Oppermann, E., Kjellstrom, T., & Garnett, S. T. (2015). Heat stress causes substantial labour productivity loss in Australia. *Nature Climate Change, 5*(7), 647–651.

Chapter 10
Pesticides and Chemical Toxicity—Challenges in Farming

History of Pesticide Use

Chemicals used in agriculture primarily include pesticides, fertilizers, fuels, and solvents. The pesticides cover insecticides, fungicides, herbicides, rodenticides, molluscicides, nematocides, and other plant growth regulators. For example, insecticides apply against insects, herbicides for plants, and fungicides against fungi. These complex chemical compounds are in use since the beginning of cropping in order to protect food crops from pests and diseases (Peshin 2002). Ancient writings are indicative of the use of insecticides (sulphur compounds) and other methods to control plant diseases, insects, and pests that perhaps began nearly 4500 years ago by Sumerians. The burning of straw, chaff, dung, and other elements of animal sources were used for smoke formation in the farmyards to dispel insects and mites. The Persians used the Pyrethrum powder (made of the dried flowers of Chrysanthemum cinerariaefolium) as an insecticide to protect stored grains. Many inorganic chemicals (e.g. sodium chlorate, sulphuric acid, Bordeaux mixture based on copper sulphate and lime) and organic substances derived from natural sources were used against fungal diseases and pest control. The organic compounds (e.g. nitrophenols, chlorophenols, petroleum oils, and by-products of coal) were applied against the fungal and insect pests. Ammonium sulphate and sodium arsenate were used as herbicides.

Since the 1950s, there has been an astounding rate of introduction of synthetic pesticides, such as DDT, BHC, Aldrin, dieldrin, endrin, chlordane, parathion, captan, and 2,4-D for various agricultural applications. The organochlorine (OC) and organophosphate (OP) insecticides were introduced in the 1960s. Carbamates, pyrethroids, herbicides, and fungicides appeared in the 1970s–1980s. More than 900 different pesticides have been registered, and nearly 2/3rd of them are for agricultural use. Many of the agrochemicals had a single mode of action. By the 1990s, the emphasis was on identifying new agrochemicals that provide higher selectivity, more user-friendly, and environmentally safe formulations (Morton and Staub 2008). Today, the newer approach of the pest management is emerging, particularly in the

P. K. Nag and L. P. Gite, *Human-Centered Agriculture*, Design Science and Innovation,
https://doi.org/10.1007/978-981-15-7269-2_10

case of genetically engineered crops (e.g. soybeans, corn, canola, and cotton) to produce its self-insecticides or exhibit resistance to pests.

Health and Safety Concerns

Worldwide, agrochemicals protect crops and livestock from insects and diseases and help produce good farm yields. The credits of pesticides have been perceived for enhanced economic potential through increased production of food and fibre, and amelioration of vector-borne diseases. Several diseases, such as malaria, filariasis, dengue, Japanese encephalitis, cholera, and louse-borne typhus, were controlled in India by the systematic application of pesticides in the affected regions. On the other hand, the evidence is overwhelming of the debits of pesticides in health implications to man and his environment. The WHO (2004) puts forward a global estimate for pesticide poisoning at about three million cases every year, resulting in 250,000 fatalities. A report submitted to the United Nations Human Rights Council in 2017 mentioned about 200,000 deaths per year from the toxic exposure of pesticides across the world, and 99% of these deaths being from low-income countries (Foodtank 2017).

Generally, pesticides, e.g. fungicides may be applied by a high-volume technique such as by spray lorries in large land areas, or with low-volume techniques such as misters, fumigation, knapsack spraying. Pesticide exposure can result from off-gassing, dispersion by the wind, dermal/clothing contact with the plants, inhalation of airborne substances, and ingestion by swallowing. The rampant use of pesticides, under the adage, *if little is good, a lot more will be better* played havoc with humans and other life forms. Rachel Carson's eye beckoning book *Silent Spring* brought home the message about the menace of indiscriminate use of pesticides (Carson 1962/2002). It warned that OC compounds could pollute the tissues of virtually every life form on the earth, the air, the lakes and oceans, fishes, and birds. Perhaps Carson's alarm had a resounding effect on the minds of people in search of safer and more environmentally friendly products.

The pesticides are toxic chemicals to cause both acute and chronic health effects. These chemicals bear severe risks to cause carcinogenic, immunologic, neurotoxic, and reproductive effects. The high-risk groups exposed to pesticides include the production workers, formulators, sprayers, mixers, loaders, and farmworkers. In India, the first report of poisoning due to pesticides was from Kerala (Karunakaran 1958)—over 100 people died from consuming parathion contaminated wheat flour. Health hazards associated with the use of pesticides have been reported among farmers in Australia (MacFarlane et al. 2008), Brazil (Recena et al. 2006), cocoa farmers in Ondo State, Nigeria (Tijani 2006), Ekiti State, Nigeria (Oluwole and Cheke 2009), vegetable farmers in the Philippines and Nepal (Lu and Cosca 2011; Ghimire 2014), cotton farmers and fruit farm workers in Pakistan (Khan et al. 2000; Azmi et al. 2006), cotton farmers (Srinivas Rao et al. 2005) and cashew nut farmers (Embrandiri et al. 2012) in India. Quoting India's National Crime Record Bureau data, Taneja (2017) mentioned that, in 2015, there were 7,672 accidental pesticides poisoning cases (other than suicides), resulting in 7,060 deaths.

The symptoms of poisoning of agrochemicals (Table 10.1) depend on the concentration and toxicity of the product, and the amount absorbed (either skin absorbed, inhaled, or ingested). The organophosphorus (OP) compounds are responsible for the maximum number of poisoning among all agricultural pesticides. Whereas all OP compounds are considered within a single group entity, each compound has unique

Table 10.1 Symptoms of poisoning due to agrochemicals

Agrochemical group	Symptoms of poisoning
Arsenicals, e.g. copper aceto-arsenite (Paris green), sodium arsenite	Arsenicals are all-body poisons. Early symptoms are stomach pain, vomiting, and diarrhoea followed by severe hypertension and muscle cramps. Poisoning may be fatal
Bipyridilium compounds, paraquat, and diquat herbicides	These compounds can irritate skin, lips, and eyes; cracking and shredding of fingernails may occur depending on the concentration. If ingested, toxicity may cause vomiting, stomach ache, diarrhoea and also lung, kidney, and liver damage. High dose exposure may cause multi-organ failure and may be fatal in 2 to 3 weeks
Carbamates, e.g. aldicarb, carbofuran, methomyl, propoxur (Baygon)	These insecticides depress acetylcholinesterase activity to produce symptoms similar to those of OP compounds
Carbon tetrachloride	Used as a fumigant along with other hazardous substances. Low-level exposure causes damage to the skin and internal organs. At high exposure, the person may witness CNS symptoms, narcotic effects, and headache, dizziness, mental confusion, nausea, and vomiting
Chlorophenoxy-acetates, e.g. MCPA, MCPB, 2,4-DEP, herbicides	Poisoning by ingestion can cause burning, mouth-watering, stomach cramps, and diarrhoea. A person may show convulsions, mental confusion, and dizziness
Chloropicrin	A fumigant and a powerful irritant to all-body surfaces. Inhalation may cause breathing difficulties and excessive vomiting
Dinitro compounds	These are used as herbicides, and they produce yellow stains wherever contact occurs. The compounds are toxic to the liver, kidney, and brain, and may be fatal. Early symptoms are fatigue, excessive sweating, and thirst. Continued exposure may lead to increasing level of anxiety, restlessness, elevated breathing and heart rate, and an increase in body temperature

(continued)

Table 10.1 (continued)

Agrochemical group	Symptoms of poisoning
Mercurials—organic and inorganic compounds	Mercurials are all-body poisons. They accumulate in the body, leading to a breakdown in essential body functions
Methyl bromide	Inhalation of this hazardous fumigant, even in small quantities, may produce headaches, weakness, irritation of the eyes, and stomach ache. The compound in low exposure may seriously affect the CNS and other body organs, leading to fatality
Organochlorine (OC) compounds, e.g. aldrin, dieldrin, DDT	OC compounds act on the CNS and show multiple symptoms, such as excitement, tremors, convulsions, and coma; nausea and vomiting are common
Organophosphorus (OP) compounds, e.g. azinphos/methyl, phensulphothion, ethyl and methyl parathion	OP compounds suppress the body acetylcholinesterase activity that is responsible for neuro-muscular transmission. Continued exposure causes muscle twitching and weakness, blurred vision, dizziness, convulsions, nausea, vomiting, diarrhoea; the affected may suddenly stop breathing. Atropine and pralidoxime are the usual antidotes that are injected into the affected individual
Phosphine	A highly poisonous fumigant affects the stomach and the CNS. Symptoms are nausea, stomach pains, vomiting and diarrhoea, convulsions, and loss of consciousness; fatality may occur within 24 h of exposure
Pyrethrins and synthetic pyrethroids, e.g. decamethrin, cypermethrin, permethrin	Natural pyrethrins are generally of low toxicity. Synthetic pyrethroids are more toxic. A high dose of exposure may act on the CNS and cause convulsions
Ethylene dichloride	A fumigant has the potential to damage the kidney and liver. Symptoms similar to sea sickness, such as dizziness and vomiting, may appear after a delay of some hours. The compound effects the skin and manifest as dermatitis
Hydrogen cyanide	A fumigant variety affects respiratory enzymes. Low exposure may produce irritation of nose and throat, dizziness, nausea, headache, tightness of the chest. High exposure may produce unconsciousness, convulsions, and may be fatal

biochemical characteristics and receptor-specific outcomes in acute OP poisoning (Eddleston et al. 2005; Peter et al. 2010). The muscarinic symptoms manifest as salivation, lacrimation, urination, defecation, gastric cramps, and emesis. The nicotinic responses are fasciculation, muscle weakness, paralysis, tachycardia and hypertension. Besides, the CNS receptor responses (such as anxiety, convulsions, and respiratory depression) have been vividly reflected in acute OP poisoning. Lee and Tai (2001) noted that the muscarinic symptoms are the most frequent, followed by CNS and nicotinic. Some toxicological syndrome may appear within minutes or hours of acute exposure (Peter et al. 2014), and some are delayed after initial mild symptoms. Evidence shows that the symptoms vary as an organ-specific, such as cardiovascular (Anand et al. 2009; Karki et al. 2004), respiratory (Eddleston et al. 2006; Noshad et al. 2007), or neurological (Singh and Khurana 2009) manifestations. Workers engaged in the production of dust, and liquid formulations of malathion, methyl parathion, DDT and lindane may also manifest a high occurrence of psychological, neurological, cardiorespiratory, and gastrointestinal symptoms (Bhatnagar et al. 2002). Some pesticides are known skin irritants.

Certain agrochemicals, including pesticides, are termed as endocrine disruptors. They elicit adverse effects by mimicking or antagonizing natural hormones in the body. Their long-term, low-dose exposure might be linked to health effects such as immunosuppression, hormone disruption, diminished intelligence, reproductive abnormalities, and cancer. A review of the literature (Blair and Zahm 1995) appears to indicate that farmworkers across the world experience elevated incidences of cancer, including leukaemia, non-Hodgkin's lymphoma, multiple myeloma, soft-tissue sarcoma, and cancers of the skin, lip, stomach, brain, and prostate. No affirmative etiologic factors explain the cancer excesses, since the farmers are exposed to many toxic substances, including pesticides, dust, engine exhausts, and zoonotic microbes.

Pesticide Residue in Food Items and Human Tissues

The presence of pesticide residues in food samples varies significantly from region to region, year to year, and also from one specific food item to another within the same food group. The pesticide residue level in human tissues is an index of exposure (Kashyap et al. 2002). In occupational exposures, the residue level merits an insight reflective of prolonged work environment exposure. Accidental exposures in the general population may appear through the food chain or intended self-poisoning by ingestion. Human blood is the most accessible body fluid to ascertain pesticide residue levels (Bhatnagar 2001). Since pesticides tend to accumulate in the fat, a higher level of its presence in mother's milk is a reflection of increased burden and their translactational passage (Banerjee et al. 1997). The storage and bioaccumulation of these chemicals are influenced by the compound intensity, efficiency of absorption, species, age, nutritional status, and integrity of the organs. Preventive measures are warranted to reduce the mother's body burden and avoid any potential health effects.

Safe Use of Pesticides

Human exposure to pesticides is widespread, and unsafe application of pesticides can put people and the environment at risk. For agricultural application, it is essential to identify the specific weed, disease, or pest affecting the area and accordingly to apply a particular pesticide(s) or use alternative methods of control. Manufacturers are required to provide mandatory labelling on the product regarding what the pesticide is to be used for, the rate of application, re-entry times, clothing and PPE needed, the symptoms of poisoning, and the antidote. The key issues under consideration are (a) the registration of pesticide for specific usage, (b) the weather conditions, site and timing of application, (c) use of appropriately calibrated application equipment, and (d) avoidance of contamination of the vicinity of application and waterways (Cooper and Dobson 2007; Damalas and Eleftherohorinos 2011). This chapter embodies concise guidance for the safe and effective use of pesticides concerning farm applications.

Pesticide Evaluation for Approval and Registration

For decades now, the practice of the registration of pesticides and their formulations has been formalized. However, the guidance and procedure for pesticide registration vary across countries. The pesticide chemistry, supplemented by experimental data of environmental and toxicology studies, form the foundation in evaluation and approval for the registration of a given pesticide. The International Code of Conduct on the Distribution and Use of Pesticides (FAO 2003) recognizes that … *In the absence of an effective pesticide registration process and governmental infrastructure for controlling the availability of pesticides, some pesticides importing countries rely on the industry to promote the safe and proper distribution and use of pesticides. In these circumstances, foreign manufacturers, exporters, and importers, as well as local formulators, distributors, repackers, advisers, and users, must accept a share of the responsibility for safety and efficiency in distribution and use.*

FAO/IAEA project on International Food Contaminant and Residue Information System (INFOCRIS) releases comprehensive physicochemical, toxicological, and ecotoxicological profiles of active ingredients and metabolites of pesticides (Slorach 2006). The Codex Alimentarius Commission (2000) of WHO/FAO attempts to harmonize food regulations, develop food standards relating to food commodity characteristics, food labelling, hygiene, additives and pesticide residues, and the safety of foods derived from biotechnology (http://www.codexalimentarius.net/web/standard_list.jsp). The Codex Alimentarius Commission reviews residue and analytical aspects of the pesticides, including data on their metabolism, fate in the environment, use patterns, and estimating the maximum residue levels (MRLs). Currently, there are over 400 methods for pesticide residue analysis (M-series), and about 200 methods for physical and chemical test methods for formulations (MT-series) and reagents (R-series).

FAO (2016) undertakes evaluations of pesticides, taking into account the good agricultural practice (GAP), analytical methods for residue analysis and experimental studies. These evaluations have been based on data supplied by the manufacturer as part of the process of establishing MRLs. The WHO, coordinating through the international programme on chemical safety (IPCS) establishes toxicological endpoints, such as NOEL, and recommends technical endpoints, such as acceptable daily intake and acute reference dose of pesticides (http://www.inchem.org/pages/jmpr.html). The US EPA (https://www.epa.gov/pesticide-reevaluation) evaluates three categories of pesticides, namely, conventional, biopesticides, and antimicrobials, and ensures that the pesticides have minimal adverse effects on humans, the environment, and non-target species.

The process of pesticide registration covers the evaluation of ingredients of a pesticide, the crop and the amount, frequency, and timing of its use, including storage and disposal practices. The evaluations contain data summaries and risk assessments, regulatory decisions, fact sheets with a synopsis of critical regulatory information, and recommendations (http://www.epa.gov/pesticides/factsheets/chemical_fs.html). The European Commission 91/414 Plant Protection Directives (http://ec.europa.eu/food/plant/protection/evaluation/exist_subs_rep_en.html) refer to a centralized evaluation system including human health and environmental data, risk assessments, the regulatory decision on active substances, and the endpoints. The Directive 91/414/EEC regulates the national product approval requirements among the member countries (Klein et al. 1993). Only the listed active substances (pesticides, growth regulators, pheromones) can be used as plant protection products.

The pesticide action network (PAN) maintains a database of hundreds of active ingredients, including summaries related to chemistry, poisoning symptoms, toxicology, environmental fate, and ecotoxicology. The hazard classification, countries of registration, restrictions or bans, and international treaty listings (PIC, POP) are also included in the database. Several countries, including India, China, Australia, U.K., Japan, and Canada, have national regulatory authorities that maintain data compilations or technical summaries for pesticides and its formulations. The Insecticide Act of India, promulgated in 1968 and enforced in August 1971 and amended from time to time (Govt. of India 2020), regulates the import, manufacture, sale, transport, distribution, and use of insecticides, intending to prevent risks to human beings or animals. The pesticides registered for use in India and other restrictions are listed in Table 10.2.

Table 10.2 The pesticides registered for use in India, and the list of ban pesticides listed in PAN International

India*						PAN international
Insecticides/ Pesticides Registered for use (Listed in two columns in alphabetical order)		Pesticides restricted for use	Pesticides Banned for manufacture, import, and use	Pesticide formulations banned for import, manufacture, and use	Pesticides refused registration	Banned pesticides
Column 1	Column 2					
2,4-D Amine salt, 2,4-Dichloro-phenoxy Acetic Acid, Abamectin, Acephate, Acetamiprid, Afidopyropen, Allethrin, Alphacypermethrin, Alphanaphthyl Acetic Acid, Aluminium Phosphide, Ametroctradin, Ametryn, Ampelomycesquisqualis, Anilophos, Atrazine, Aureofungin, Azadirachtin (Neem Products), Azimsulfuron, Azoxystrobin, Bacillus sphaericus, Bacillus subtillus, Bacillus thuringiensis var. galleriae, Bacillus thuringiensis var. israelensis, Bacillus thuringiensis var. kurstaki Barium carbonate, Beauveriabassiana, Benalaxyl (TIM), Benalaxyl M, Bendiocarb, Benfuracarb, Bensulfuron Methyl, Bentazone TI, Beta Cyfluthrin, Bifenazate, Bifenthrin, Bispyribac sodium, Bitertanol, Boscalid+ Pyraclostrobin, Bromadiolone, Buprimate (FI-WRT),	Hexythiazox, Hydrogen cyanamide, Imazamox, Imazethapyr, Imidacloprid, Imiprothrin, Indoxacarb, Indaziflam+ Glyphosate ammonium (FI), Iprobenfos (Kitazin), Iprodione, Iprovalicarb, Isoprothiolane, Isoproturon, Kasugamycin, Kresoxim methyl, Lambdacyhalothrin, Lime Sulphur, Lufenuron, Magnesium phosphide plates, Malathion, Mancozeb, Mandipropamid, Mepiquate chloride, Meptyldiinocop, Mesosulfuron methyl + Iodosulfuron methyl sodium, Metaflumizone, Metalaxyl, Metalaxyl-M, Metaldehyde, Metamifop TI, Metamitron (TIM), Metarhiziumanisopliae, Methabenzthi azuron, Methomyl(12.5% & 24% formulations), Methoxyfenazide (FI- WRT), Methyl bromide, Methyl chlorophenoxy acetic acid (MCPA), Metiram, Metofluthrin, Metolachlor, Metrafenone,	Aluminium phosphide, Captafol, Cypermethrin, Dazomet, Dichloro diphenyl, Trichloroethane (DDT) Fenitrothion, Methyl bromide, Monocrotophos, Trifluralin	Aldicarb, Aldrin, Benzene hexachloride, Benomyl, Calcium cyanide, Carbaryl, Chlorbenzilate, Chlordane, Chlorofenvinphos, Copper acetoarsenite, Diazinon, Dibromochloropropane (DBCP), Dieldrin, Endosulfan (order of the Supreme Court of India, 13 May 2011; disposed of 10 Jan. 2017), Endrin, Ethyl mercury chloride, Ethyl parathion, Ethylene dibromide (EDB), Fenarimol, Fenthion, Heptachlor, Lindane (Gamma-HCH),	Carbofuran 50% SP, Methomyl 12.5% L, Methomyl 24% formulation, Phosphamidon 85% SL, Captafol 80% Powder (ban for use), Nicotine sulfate (ban for use), Dalapon (WD), Ferbam (WD), Formothion (WD), Nickel chloride (WD), Paradichlorobenzene (PDCB) (WD), Simazine (WD), Sirmate (WD), Warfarin (WD)	2,4, 5-T, Ammonium sulphamate, Azinphos ethyl, Azinphos methyl, Binapacryl, Calcium arsenate Carbophenothion, Chinomethionate (Morestan), Dicrotophos, EPN, Fentin acetate, Fentin hydroxide, Lead arsenate, Leptophos (Phosvel), Mephosfolan, Mevinphos (Phosdrin), Thiodemeton / disulfoton, Vamidothion	4-Chlorophenol, Aldicarb, Azinphos-Methyl, Calcium Cyanamide, Captafol, Chlordane, Chlorophene/2-benzyl, Chlorotoluron, Cyhexatin, Cypermethrin (Beta), DDT-Dichlorodiphenyltrichloroethane, DEET, Demeton, Demeton-O, Dicofol, Di-Nitro-Ortho-Cresol/DNOC, Dinotefuran, Endosulfan, Ethylene dibromide / EDB / 1,2-dibromoethane, Ethylene dichloride / 1,2-Dichloroethane, Ethylene oxide, Fluoroacetamide, Glufosinate (Including

(continued)

Table 10.2 (continued)

India*						PAN international
Insecticides/ Pesticides Registered for use (Listed in two columns in alphabetical order)		Pesticides restricted for use	Pesticides Banned for manufacture, import, and use	Pesticide formulations banned for import, manufacture, and use	Pesticides refused registration	Banned pesticides
Column 1	Column 2					
Buprofezin, Butachlor, Captan, Carbendazim, Carbofuran, Carbosulfan, Carboxin, Carfentrazone ethyl, Carpropamid, Cartap hydrochloride, Chlorantraniliprole, Chlorfenopyr, Chlorfluazuron, Chlorimuron ethyl, Chlormequat chloride (CCC), Chlorothalonil, Chlorpropham (TI),TIM, Chlorpyriphos, Chlorpyriphos methyl, Chlothianidin, Chromafenozide, Cinmethylene, Clodinafop-propargyl, Clodinafop-propargyl+sodium acifluorfen, Clomazone, Clothianidin, Copper hydroxide, Copper oxychloride, Copper sulphate, Coumachlor, Coumatetralyl, Cuprous oxide, Cyantraniliprole, Cyazofamid, Cyenopyrafen (FI-WRT), Cyflumetofen, Cyfluthrin, Cyhalofop-butyl, Cymoxanil, Cypermethrin, Cyphenothrin, Cyproconazole (TI), Dazomet, Deltamethrin (Decamethrin),	Metribuzin, Metsulfuron methyl, Milbemectin, Myclobutanil Novaluron, Nuclear polyhyderosis virus of Helicoverpaarmigera and Spodopteralitura, Orthosulfamuron, Oxadiargyl, Oxadiazon, Oxathiapipron, Oxycarboxin, Oxydemeton-methyl, Oxyfluorfen, Paclobutrazol, Paraquat dichloride, Penoxasulam, Penconazole, Pencycuron, Pendimethalin, Penflufen*, Penoxsulam, Permethrin, Phenthoate, Phosalone, Picoxystrobin TIM, Pinoxaden, Prallethrin, Pretilachlor, Primiphos-methyl, Prochloraz TI, Profenophos, Prohexadione calcium, Propamocarb hydrochloride 66% w/w min (aqueous concentrate), Propanil, Propaquizafop, Propergite, Propetamphos, Propiconazole, Propineb, Propoxur, Pseudomonas fluorescens, Pymetrozin (FI), TIM, Pyraclostrobin, Pyraclostrobin+epoxiconazole,		Linuron, Maleic hydrazide, Menazon, Methoxy ethyl, mercury chloride, Methyl parathion, Metoxuron, Nitrofen, Paraquat dimethyl sulfate, Pentachloro nitrobenzene (PCNB), Pentachlorophenol, Phenyl mercury acetate, Sodium cyanide (banned for Insecticidal purpose), Sodium methane arsonate, Tetradifon Thiometon (vide S.O 3951(E), 8th August, 2018), Toxaphene(Camphe chlor), Tridemorph, Trichloro acetic acid (TCA)			Glufosintae-ammonium), Hexachloro-benzene/ Benzene hexachloride (HCB/BHC), Hexachlorocyclohexane (HCH), Hydrogen cyanamide, Lindane, Mercury compounds, Metaldehyde, Methamidophos, Methoxyethyl mercury chloride (MEMC), Methyl eugenol, Methyl parathion, MGK repellent, Monocrotophos, Paraquat dimethyl, Parathion (Ethyl), Pentachlorophenol (PCP) and salts, Phosphamidon, Prochloraz, Propoxycarbazone sodium,

(continued)

Table 10.2 (continued)

India*						PAN international
Insecticides/ Pesticides Registered for use (Listed in two columns in alphabetical order)		Pesticides restricted for use	Pesticides Banned for manufacture, import, and use	Pesticide formulations banned for import, manufacture, and use	Pesticides refused registration	Banned pesticides
Column 1	Column 2					
Diafenthiuron, Dichloro Diphenyl Trichloroethane (DDT), Dichloropropene and Dichloropropane (DD mix), Diclofop-Methyl, Diclosulam, Dicofol, Difenoconazole, Diflubenzuron, Dimethoate, Dimethomorph, Dinocap, Dinotefuron, Dithianon, Diuron, Dodine, D-trans Allethrin, Edifenphos, Emamectin benzoate, Epoxyconazole, Ethephon, Ethion, Ethiprole, Ethofenprox (Etofenprox), Ethoxysulfuron, Ethylene dichloride and carbon tetrachloride mix (EDCT 3:1), Etoxazole(FI), Famoxadone, Fenamidone, Fenazaquin, Fenitrothion, Fenobucarb (BPMC), Fenoxaprop-p-ethyl, Fenpropathrin, Fenpyroximate, Fenvalerate, Fipronil, Flonicamid, Fluazifop-p-butyl, Flubendiamide, Flucetosulfuron,	Pyrazosulfuron ethyl, Pyrethrin (pyrethrum), Pyridaben (FI- WRT), Pyridalyl, Pyriproxyfen (TI), Pyrithiobac sodium, Pyroxasulfon(FI- WRT), Quinalphos, Quizalofop ethyl, Quizalofop-P-tefuryl, S-bioallethrin, Sodium paranitrophinolate, Spinetoram, Spirotetramat, Spinosad, Spiromesifen, Streptomycin + tetracycline, Sulfentrazone TIM, Sulfosulfuron, Sulfoxaflor, Sulphur Tebuconazole, Tembotrione, Temephos, Tetraconazole (FI), Thiacloprid, Thifluzamide, Thiobencarb (Benthiocarb), Thiocyclam hydrogen oxalate, Thiodicarb, Thiomethoxain, Thiophanate-methyl, Thiram, Tolfenpyrad (TIM), Topramezone,Transfluthrin, Triacontanol, Triadimefon, Triafamone (Triafamone 20% w/w + Ethoxysulfuron 10% WG % w/w SC FI), Triallate, Triasulfuron,					Safrole, Sodium dichromate, Sulfluramid, Trichlorfon, Ziram

(continued)

Table 10.2 (continued)

India*						PAN international
Insecticides/ Pesticides Registered for use (Listed in two columns in alphabetical order)		**Pesticides restricted for use**	**Pesticides Banned for manufacture, import, and use**	**Pesticide formulations banned for import, manufacture, and use**	**Pesticides refused registration**	**Banned pesticides**
Column 1	**Column 2**					
Fluchloralin, Fluen sulfone 47% TC (MUP) (FI), Flufenacet, Flufenoxuron, Flufenzine, Flumioxazin, Fluopicolide, Fluopyram and its metabolite, Flupyradifurone, Flusilazole (TI), Fluthiacet methyl (TIM), Fluvalinate, Fluxapyroxad 167 g/L + Pyraclostrobin 333g/l SC (FI - WRT), Fomesafen, Forchlorfenuron, Fosetyl-Al Gibberellic acid, Glufosinate ammonium, Glyphosate, Haloxyfop-R-methyl 10.55%.EC(FI), Helosulfuron methyl, Hexaconazole, Hexazinone,	Trichodermaharzianum, TrichodermaViride, Tricyclazole, Triflumezopyrim (TIM), Trifloxistrobin, Trifluralin (use in wheat shall be banned), Validamycin, Verticilliumlecanii, Zinc phosphide, Zineb, Ziram					

*Insecticides to phase out by India, 31st December 2020 are Alachlor, Dichlorovos, Phorate, Phosphamidon, Triazophos, Trichlorfon

Pesticide Specifications

Developing specifications for pesticides and its formulations follow an internationally defined mechanism on (a) quality standards for pesticides, (b) approval and acceptance of pesticides, (c) protection against inferior products, and (d) biological efficacy linked to specification requirements (http://npic.orst.edu/factsheets/formulations.pdf). After manufacturing a pesticide technical material (active ingredient, active substance), a formulation is prepared by mixing the technical material with other inert ingredients, like solvents, mineral clays, wetting agents or other adjuvants (e.g. surfactants, crop oils, antifoaming agents). The pesticide formulations may be formed from one or more active ingredients in different variations, whereby to improve its effectiveness and specific use (Unruh et al. 2003). The international coding system for pesticide formulations (CropLife International 2008) covers (a) water-dispersible granules (Code WG), (b) emulsifiable concentrate (Code EC)—a homogeneous formulation applied as an emulsion after dilution in water, and (c) vapour releasing product (Code VP) that contains one or more volatile active ingredients.

Prior Informed Consent

With the dramatic growth in the production of pesticides and its consumption for agricultural and other applications, global concerns are vivid on exchanging information on hazards associated with these chemicals. The joint initiatives of UNEP and FAO resulted into introducing the voluntary prior informed consent (PIC) procedure under the Rotterdam Convention 1998, to help national governments to assess the risks of hazardous chemicals to the human health and the environment, and to make informed decisions on their import (UNEP/FAO 2017; Kummer 1999). This multilateral environmental agreement promotes shared responsibility to notify the importing countries about the international trade of any agrochemicals (e.g. pesticides), which may be banned or severely restricted in the exporting country. A total of 41 chemicals (twenty-four pesticides, formulations, and industrial chemicals) are currently subject to the PIC procedure (http://www.pic.int/en/table7.htm).

Pesticides: 2,4,5-T and its salts and esters, Aldrin, binapacryl, captafol, chlordane, chlordimeform, chlorobenzilate, DDT, dieldrin, DNOC, Dinoseb, EDB, ethylene dichloride, ethylene oxide, fluoroacetamide, HCH, heptachlor, hexachlorobenzene, mercury compounds, monocrotophos, parathion, pentachlorophenol, toxaphene.

Formulations: Benomyl, carbofuran, thiram, monocrotophos, methamidophos, phosphamidon, methyl parathion, and parathion.

Obsolete Pesticides

Obsolescence of a pesticide may arise when a product has been de-registered nationally or banned internationally. Commonly, a stock of pesticides becomes obsolete with the degradation of product and packaging, due to long-term storage. Old and deteriorated stocks of pesticides, often remain stored in poor conditions and pose a threat to human health and the environment. It is a national imperative to introduce adequate facilities to dispose of such stocks in a safe and environmentally sound manner. FAO outlined guidelines for the safe disposal of bulk and small quantities of pesticides (http://www.fao.org/WAICENT/FAOINFO/AGRICULT/AGP/AGPP/Pesticid/Disposal/common/ecg/103811_en_w1604e.pdf); (http://www.fao.org/WAICENT/FAOINFO/AGRICULT/AGP/AGPP/Pesticid/Disposal/common/ecg/103825_en_No_7Small_quantities_stocks.pdf).

Toxicity Factor Value

The composite pesticide risk assessment in agriculture varies as a function of cropping systems, soil, and climatic conditions in a geographic region. A measurement approach should ideally take account of the quantity, type, and the toxicity of active ingredients, when and how they are applied, and the extent of non-target organisms exposed. Accordingly, the toxicity factor value is derived as an index value to assess pesticides, in terms of the risks associated with applying different active ingredients. The multi-attribute index (Benbrook et al. 2002) comprises four components, as acute mammalian toxicity (AM), chronic mammalian toxicity (CM), Ecotoxicity (ECO), and impacts on biointensive IPM systems (BioIPM) in a generalized form of the equation:

$$\text{Value for Pesticide x} = (a)\text{AMx} + (b)\text{CMx} + (c)\text{ECOx} + (d)\text{BioIPMx},$$

where (a), (b), (c), and (d) are the weights assigned to each component index. The Wisconsin potato production risk index for pesticide (x) was weighted as

$$(0.5) * \text{AMx} + \text{CMx} + \text{ECOx} + (1.5) * \text{BioIPMx}$$

In potato cropping seasons, the insecticides which triggered the acute toxicity are the OP compounds (azinphos-methyl and methamidophos) and the carbamates (carbofuran and oxamyl). The chronic risk criterion includes seven active ingredients, i.e. the metribuzin (a herbicide), the permethrin and endosulfan (insecticides), and the maneb, mancozeb, chlorothalonil, and triphenyltin hydroxide (fungicides). A lower weight was assigned to acute mammalian toxicity because of relatively lower risk to applicator exposure and workers handling the harvested potatoes. No adjustment was applied to the chronic mammalian toxicity index given the potential for low-level pesticide exposures in drinking water, the air, and occupational exposure. No adjustment was made for the ecotoxicity index, giving importance to impacts on birds, fish, and other aquatic organisms. In a geographic region where human exposure might be higher due to intensive manual labour in the production process, the weights to acute toxicity risk might be higher. A higher weightage was given to the BioIPM component since a pesticide has an impact on soil microorganisms and arthropods.

Pesticide Resistance

Most pest species bear the ability to circumvent the toxic effects of specific types of pesticides. Pesticide resistance is a genetic phenomenon that develops on a pest population. Repeated application of pesticides can kill pests, but the resistant survivors pass-on the resistance genes to the next generation (Bellinger 1996). When a pest develops resistance to a pesticide, switching on a different class of pesticide having different modes of action is the likely alternative option. In the case of the insecticide resistance, the preferred approach is to minimize insecticide use about pest/insect-resistant varieties. With IPM measures of the removal of crop residues and water sources from the land, one can eliminate food sources for pests and rodents. For fungicide resistance, one may apply a mixture of different fungicide families, whereas, in the case of the herbicide resistance, crop rotation allows using herbicides of a different mode of action and managing weed problems. The use of non-chemical weed control methods is preferable. For the rodenticide resistance, one may ensure that all baiting points are inspected weekly, and the treatment is given for the elimination of the infestation. The strategy for anti-coagulant resistance management may not be taken as permanent baits.

Integrated Pest Management

Indiscriminate use of agrochemicals may cause health hazards to the exposed population. It also damages crops, livestock, wildlife, and the environment. The level at which pests become a viable threat to the crop yield is critical in deciding pest control preventive action. The IPM strategy aims at the appropriate intervention of physical, biological, and chemical methods to preserve the quality of the crop and minimal effects on the environment. In other words, IPM targets to minimize reliance on chemical pesticides and to use alternatives or the pesticides of reduced risk and to prevent pest menace through better crop management.

Agenda 21 of the United Nations Conference on Environment and Development at Rio de Janeiro (Weiss 1992) identified IPM as one of the requirements in promoting sustainable agriculture and rural development. The concept of

IPM (http://www.fao.org/WAICENT/FAOINFO/AGRICULT/AGP/AGPP/Pesticid/Code/Download/code.pdf) calls for *the careful consideration of different pest control techniques and integration of measures that discourage the growth of pest populations and keep the application of synthetic pesticides at an economically justified level, and minimize risks to human health and the environment.* In general, the IPM control actions include (a) cultural controls, i.e. crop rotation to replenish soil nutrients, (b) biological controls, i.e. insect predators to destroy pests, and (c) breeding of pest-resistant varieties. Besides, the methods include highly targeted spraying of pesticides or broadcast spraying of non-specific pesticides, or mechanical control by deploying traps, electronic triggering devices or the electricity as a light source to trap pests.

Despite that a great deal of IPM programme details is available (Peshin 2013; Berg 2004; Gallagher et al. 2009), the core issue remains in percolating the knowledge to grass-root farmers and making them aware about agricultural pests and ecosystems. By adoption and effective implementation of IPM, the farmers can manage pests and also minimize environmental, health, and economic risks.

Biopesticides

Biopesticides, such as microbial, plant-incorporated-protectants, and biochemical pesticides (Table 10.3), provide useful pest management tools in place of synthetic pesticides and against chemical pesticide resistance. Microbial pesticides consist of a microorganism (e.g. a bacterium, fungus, virus, or protozoan) as the active ingredient to control target specific pests. Plant-incorporated-protectants that plants produce from genetic material, e.g. the gene for the *Bacillus thuringiensis* (Bt) pesticidal protein, is introduced into the plant's genetic material, and the plant produces a substance to destroy the pest. Biochemical pesticides are naturally occurring substances, such as insect sex pheromones that interfere with mating, and other scented plant that extracts attract insect pests to traps.

The Bt biopesticide is a naturally occurring bacterium with several strains. The target insect species are determined about the particular Bt that produces a protein having the ability to bind to a larval gut receptor, thereby causing the insect larvae to deprive of survival resources. Several products of Bt-based insecticides have been developed. Plants have been modified with short sequences of genes from *Bt* to express the crystal protein *Bt,* and thereby plants can produce the proteins to protect themselves from insects. Bt GM crops are explicitly protected against corn borers, tobacco budworm, cotton bollworm, pink bollworm, and potato beetle (http://www.bt.ucsd.edu/bt_crop.html). During the drum priming process of seed preparation of carrot and onion seeds, mixtures of fungi (*Clonostachys rosea* IK726 or *Trichoderma harzianum* T22) and bacteria (*Pseudomonas chlororaphis* MA342 or *P.fluorescens* CHA0) can be applied as biocontrol agents (http://ec.europa.eu/environment/integration/research/newsalert/pdf/95na1.pdf).

Table 10.3 Examples of biopesticides

Use	Bacteria	Viruses	Fungi	Protozoa	Nematode	Others
Insect control	*Bacillus thuringiensis, B. sphaericus, Paenibacillus popilliae, Serratia entomophila*	nuclear polyhedrosis viruses, granulosis viruses, non-occluded baculoviruses	*Beauveria spp, Metarhizium, Entomophaga, Zoopthora, Paecilomyces fumosoroseus, Normuraea, Lecanicillium lecanii*	*Nosema, Thelohania, Vairimorpha*	Steinernema *spp, Heterorhabditis spp*	Pheromones, parasitoids, predators, microbial by-products
Weed control	*Xanthomonas campestris pv. Poannua*		*Colletotrichum gloeosporioides, Chondrostereum purpureum, Cylindrobasidium laeve*			
Plant disease control	Competitive and Soil Inoculants - *Bacillus pumilus, B. subtilis, Pseudomonas* spp, *Streptomyces griseoviridis Burkholderia cepacia*		*Ampelomyces quisqualis, Candida spp., Clonostachys rosea f. catenulate, Coniothyrium minitans, Pseudozyma flocculosa, Trichoderma spp*			

(continued)

Table 10.3 (continued)

Use	Bacteria	Viruses	Fungi	Protozoa	Nematode	Others
Nematicides	*Bacillus firmus*, *Pasteuria penetrans*		Nematode Trapping Fungi - *Myrothecium verrucaria*, *Paecilomyces lilacinus*		Mollusc parasitic nematode - *Phasmarhabditis hermaphrodita*	

Source: Society for Invertebrate Pathology—Microbial Control Division—Biopesticides, http://www.dropdata.net/SIP_micontrol/biopesticides.htm

Like in chemical pesticides, biopesticides also have to undergo a registration process. Several guidance documents of the European Commission, OCED, US EPA, and other country-specific standards are available on registration and Re-registration of biological pesticides, including microbial, pheromones and semiochemicals, plant-incorporated protectants, aiming at ensuring safe food and environmental health and safety. Most biopesticides are selective in their mode of action and thereby reduces adverse impacts on non-target organisms. Because of its selectivity, biopesticides likely carry the potential risk of resistance.

Pesticide Residue Trial

In order to determine the presence of pesticide residues in farm crops and foodstuffs, the analytical methods are chosen to define the critical parameters explicitly for its robustness. The residue trials aim at quantifying the highest likely residue level in treated crops at harvest or on removal from the store following the good agricultural practice (GAP). Guidance on the pesticide residue trials (Codex Alimentarius Commission 2000) covers (a) the design of residue trials, (b) application of the plant protection product, (c) records of plant growth and development data, (d) records of weather and soil data, and (e) sampling. The readers may refer to various guidance on study protocols, and methods of pesticide residue analysis. The US FDA has compiled pesticide residue analysis methods (PAM I and PAM II)—PAM I on multi-residue methods for fatty and non-fatty foodstuffs, and PAM II on methods for individual pesticides (http://www.fda.gov/Food/ScienceResearch/LaboratoryMethods/PesticideAnalysisManualPAM/default.htm); (http://www.fda.gov/Food/ScienceResearch/LaboratoryMethods/PesticideAnalysisManualPAM/ucm113710.htm).

A multi-residue method based on LC-MS/MS (Greulich and Alder 2008) covers 300 pesticides in drinking water. A QuEChERS method (**Qu**ick, **E**asy, **Ch**eap, **E**ffective, **R**ugged, **S**afe) has a broad acceptance for rapid multi-residue analysis of pesticide (polar and pH-dependent compounds) (Anastassiades et al. 2003). Method validation and quality control for pesticide residues analysis depend on the acceptable values of certain analytical parameters (Buschmann 2013). These are the mean recovery of an analyte (range 70–110%), selectivity (specificity) of the method, and the analytical calibration extends to the lowest and highest nominal concentration of the analyte (±at least 20%). Repeatability refers to the closeness of agreement between mutually independent test results obtained with the same method on identical test material, equipment, laboratory setup, and the operator with a short interval of time. Reproducibility refers to the closeness of agreement between independent results obtained by the same method on identical test material under different conditions. Limit of detection (LOD) of a procedure is the lowest amount of an analyte in a sample that can be detected. Limit of quantitation (LOQ)/determination (LOD) refers to the regulatory perspective as the lowest concentration tested at which an unambiguous identification of the analyte is proven and an acceptable mean recovery with an acceptable relative standard deviation (RSD) is obtained.

The analytical methods demand an excellent laboratory set up and trained personnel. The emphasis has been given to theoretical modelling of the physico-chemical and biological properties of agrochemicals. The readers may refer to the initiatives on the use of quantitative structure-activity relationships (QSARs) for theoretical analysis of chemicals, including pesticides (Sheridan et al. 2016; Willett 2009).

Perspective and Recommendations

Society accrues a great benefit from the use of pesticides and other agrochemicals. However, the cost-benefit scenario from pesticide use differs between developed and developing countries. In the developed world, the users of toxic chemicals get more excellent protection through (a) the enforcement of regulations, (b) better awareness and education among the farm operators, (c) awareness on PPE and its better availability, and (d) access to healthcare facilities and social security benefits. On the contrary, the developing countries are confronted with multiple constraints. Gross indifference of the enforcement machinery resulted in the indiscriminate use of even banned or restricted pesticides. Education and training of the exposed population is a significant vehicle to ensure the safe use of pesticides. However, poor literacy of farmworkers and their economic constraints deter the use of proper PPE. Also, the tropical climatic conditions hinder people from working with protective clothing and equipment. Often the poorly maintained spray equipment or other applicators are in use.

Research from different countries, e.g. Nigeria (Oluwole and Cheke 2009), the Philippines (Lu and Cosca 2011), North Carolina, USA (Heiberger 2015) indicates that the awareness of farmers and farmworkers need to be substantially raised on health hazards associated with the handling of hazardous pesticides. In a questionnaire-based investigation on vegetable producing farmers in the Philippines, Lu and Cosca (2011) noted the risk factors, such as the use of incomplete PPE, dermal contact while mixing pesticides, and spills of backpack sprayers, and re-entering the recently sprayed land area.

The extent of manifestation of symptoms of pesticide exposure, as mentioned earlier, such as headache, vomiting, skin rash, respiratory problems, and convulsions, depends on the toxicity of the compound, dosage and exposure time. The depression in the Plasma Cholinesterase (PChE) level has been identified as a reliable indicator to monitor pesticide intoxication in farmworkers (Dasgupta et al. 2007). According to the US Washington State Guidelines (Furman 2006), the agricultural pesticide handlers should not work with pesticides until their PChE levels reach within 20% of baseline values.

The economic impact of exposure of pesticides in non-target species (including humans) amounts to about $8 billion annually in developing countries. What is required is to weigh all the risks against the benefits to ensure a maximum margin of safety. In the poorer countries, the imperatives to use pesticides stand against the challenges to overcome hunger of teeming millions and communicable diseases, like malaria, at the cost of reasonable risks. The approach of using pesticides must be pragmatic and should be based on the judgment that human health and safety are not in jeopardy. A large number of human variables, such as age, sex, race, socio-economic status, diet, and state of health in combination, may affect human health against exposure to pesticides. Also, the long term effects of low-level exposure to one pesticide may be influenced by concomitant exposure to other pesticides and environmental contaminants.

There is a general lack of epidemiological database on the incidence of acute poisoning to pesticides of different classes. Surveillance studies of outbreaks and accidental exposure with pesticides, cohort analyses, randomized trials of intervention procedures add to the generation of valuable information. Monitoring the end product of human exposure in the form of residue levels in the body fluids and tissues of the general population is an alternative avenue to examine the impacts of exposure to pesticides. Notification of pesticide poisoning by all hospitals and medical practitioners would go a long way for nation-wide monitoring.

Guidance on Safety and Health

The wide geographical distribution of farmworkers of different sizes of land holdings, as well as self-employed and landless labourers demand closer attention to the health and safety concerns as regards to practices of agrochemicals and pesticides. The local government, agricultural associations, and agrochemical manufacturers are responsible in educating community leaders, healthcare workers, and farmworkers about the health hazards associated with pesticides and other agrochemicals. Dissemination of information may be made through easy-to-read material, mass media advertisements and direct access to the farmers. Understandably, the illiterate farmers in rural settings are grossly constrained of self-educating on the correct handling of agrochemicals and pesticides. This contribution summarizes guidelines on the good practices in pesticide handling, distribution, formulation, storage, and disposal, as well as the record-keeping of events and incidents. The guidance will make farm managers, rural development authorities, health workers, and community leaders aware of the harmful effects of agrochemicals and their role in minimizing the risks.

General Information

Farm owners, agricultural extension workers, block development officers, primary healthcare workers should generally be aware of the site location (*village, farm*) and crops where the agrochemicals are applied. The responsibility may be assigned:

- To identify the agrochemicals in use in the locality *(quantity of use per unit area; indicate other methods to control pests in the locality);*
- *To k*eep records of the retailers who sell the products *(brand and chemical name);*
- To check that the products supplied by the retail shop are correctly labelled *(packaged or repackaged containers);* the classification category in text and symbols, such as very toxic, toxic, harmful, corrosive, irritant, flammable, oxidizing, and explosive;
- To educate farmworkers and others about the route of entry of chemicals into the body *(inhalation, ingestion, or skin absorption). B*reathing of agrochemicals might happen if they are in the form of gases, fine spray droplets, dust, fumes, and smoke. Pesticides easily penetrate the skin surface. Rapid skin absorption may take place in a hot climate since the skin pores are wide open. Pesticide formulations containing penetrative solvents (e.g. kerosene, petroleum products) might be more hazardous. A minute quantity of toxic agrochemicals could be fatal if ingested, contaminated through the lips and mouth.
- To attach responsibility to retailers if agrochemicals are sold or stored without a manufacturers' labelling on the package or authorized re-labelling of containers.

Chemical Safety Data Sheet

Manufacturers, retailers, standards bodies, product approval, and registration authority provide the *chemical safety data sheet (identification and classification, hazards of the agrochemical product, measures of health and safety, and emergency procedures*);

Identification

- how and when to use the product safely and effectively; dose/application rates, timing, and method of treatment or application;
- warnings to prevent incorrect or inappropriate use; expiry date of the product, as labelled in the container;
- safety interval between the application of agrochemical and harvesting of crop or consumption of its product;
- safety precautions in storage, mixing, and application of pesticides, and disposal of used containers; protective clothing; emergency measures, first aid and medical facility in the event of contamination and poisoning;

Labelling and Re-Labelling

- hazard symbol; trade name, manufacturer or distributor; batch identification of the product; active ingredient;
- purpose and directions of use; regulations (e.g. registration, compliance); safety precautions; warnings; first aid and advice to health personnel.

Toxicity Classification

The amount of the substance that kills up to 50% of small sample animals (e.g. rats) within a specified period, when the substance ingested by animals (lethal dose of ingestion 50, LD 50) or inhaled (lethal inhaled concentration 50, LC 50). The hazard classification symbol identifies the agrochemical container.

- *Very toxic*—a substance inhaled/ingested or penetrated the skin, maybe extremely serious, acute (immediate) or chronic (long-term) health risks and even death. A symbol denoting *Toxic* is also a highly hazardous substance. Very toxic and toxic substances are labelled by the hazard class (class Ia and Ib).
- *Harmful*—a substance inhaled/ingested or penetrated the skin, may cause moderate health risks (class II).

Irritancy

The *harmful* symbo*l* wording with *irritant* indicates a non-corrosive substance; prolonged or repeated contact with the skin or mucous membrane can cause inflammation. Toxicity class III, IV, or V is according to the country classification.

Corrosivity

A *corrosive* substance can destroy living tissues on its contact. Severe skin and flesh burns might occur from splashes of the substances.

Flammability

- Symbol *extremely flammable* indicates that the liquid would boil at body temperature and would catch fire if exposed to a flame; *Highly flammable* denotes a

substance that may become hot and catch fire in contact with air at ambient temperature. Symbol *flammable* denotes a substance that would catch fire if allowed to warm above ambient temperature.
– *Oxidizing* denotes a substance that releases substantial heat when reacts with other flammable substances.

Explosivity

- Symbol *explosive* indicates that a substance may explode under the effect of a flame or if subjected to shocks or friction.

Operation and Handling of Pesticides

The local authority and retailers ensure that agrochemicals are transported safely. If contamination takes place by spillage or container damage during transport, preventive measures are required so that other items transported are not contaminated.

Packaging

Generally, every country has national regulations that apply to the packaging of agrochemicals. Packages of agrochemicals are made of glass, metal, plastic, or paper of different sizes. The types of the package used as containers of agrochemicals should be so designed that the contents do not escape during handling, storage, stacking, loading, and unloading.

Awareness among concerned is necessary that packages should be resistant to the pressure, atmospheric conditions, or the corrosive action of chemicals. Since the users may not be adequately trained to understand the packaging requirements, the agrochemicals should never be repackaged at the users' level.

Transport

Safe transport would mean that products in good-quality containers are transported, where the modes of transport may be cart, manual carrying, open truck, or lorry. The driver must have the necessary training to transport agrochemicals. The driver of the vehicle should take necessary care that the vehicle avoids collisions or violent falls. Any burst or weakening of the container can cause spillage of contents. The chemical safety data sheets of the agrochemicals must be available at the transporting vehicle.

Transfer

Retail suppliers and spray operators must be aware that agrochemicals are not to be transferred from one container to another. The practice of transferring between containers has been banned in many countries. Only in exceptional cases, transfer of agrochemical products may be allowed. One must ensure that the receiving container is the manufacturer's container for the same product, like the original container. The container should not be overfilled with a liquid to avoid spillage due to any expansion in volume by temperature variation.

Storage

The FAO Pesticide Storage and Stock Control Manual (http://www.fao.org/docrep/V8966E/V8966E00.htm#2) provided details about the precaution necessary for pesticide storage and disposal. The pesticide storage should typically be away from the dwelling area or water bodies. Areas liable to flooding or pollution of underground water supply sources have not preferred for the location of storage. The storage area should be sufficiently ventilated and adequately illuminated. No direct sunlight should fall on to agrochemicals, since ultra-violet light may deteriorate containers and contents. Floors of the storage area should be non-slippery and impermeable for easy cleaning in case spillage or leakage. The shelves and floor area should be demarcated by colour and space for storing and stacking products in an orderly manner and delivering quickly for transportation. Over stacking of drums and bags should be discouraged for easy inspection and handling of materials. A register of the agrochemical stock should be kept updated, mentioning details of incoming pesticides, arrival details, and also outgoing pesticides mentioning the date and destination. The pesticide store should be equipped with (i) PPE (nitrile rubber or neoprene gloves, rubber boots, overalls, goggles, masks or half-face respirators), (ii) water fountain to wash hands and face, and eyewash, (iii) fire extinguisher, and (iv) absorbents to absorb leaked pesticides. Concerned local authorities, fire brigade, and agricultural inspectors may be well informed about the siting, quantities, and products stored.

Dispensing

Fertilizers, dust, and granules are supplied ready for use. Pesticides require dilution from concentrated formulations. Dispensing agrochemicals require an understanding of the instruction given in the package/container:

- the correct dilutions and doses; method of dispensing agrochemical to the applicator to prevent accidental splash-back of the concentrated substance;
- wearing protective clothing and gloves, as specified;

- carefully dispensing packs of agrochemical dust and powders into applicators to avoid particles of being airborne and inhaled.

Disposal of Pesticides Containers

The supplier, retailer, local authority, and community leader are the stakeholders for the safe disposal of pesticides, as per the recommended guidance (CropLife International 2007). Agrochemical wastes and empty containers should never be dumped indiscriminately. Preferably there should be a designated disposal site in the locality away from land drains or watercourses. Wastes should be disposed of by persons trained to handle waste disposal.

Empty containers must not be used to store food or water since a tiny amount of agrochemical residue left in the container could lead to severe illness or death. Containers of highly flammable pesticide formulations must not be burnt and buried. Lightly contaminated packaging may be burned as a means of disposal. Containers of hydrogen cyanide gassing powders or group of phosphides carry a risk of reaction with water. Therefore, these containers should not be cleaned with water when empty.

Pesticide Application

When agrochemicals are sprayed under pressure, the liquid emerges through the nozzles and breaks up into minuscule aerosol particles. For the period, a particle remains airborne depending on the particle characteristics, gravitational forces, the viscosity of air, and the rate of evaporation of the liquid constituting the aerosol. The smaller spray particles remain in the air for a more extended period, and are likely to drift away from its intended target. There is always a likelihood that the pesticide particles move away beyond the target to non-target receptors (water bodies, plants, and animals in the vicinity of application (Felsot et al. 2010). The sprayer design (e.g. knapsack sprayer) has a significant role in optimizing the spray drift, by suitably modifying the nozzle type, nozzle spray angle, and nozzle spacing (Ganzelmeier 2002). The spray drift may also be affected by climatic factors (e.g. wind, ambient temperature, and humidity) and the height of released spray relative to the crop area.

Therefore, the person engaged in pesticide spraying should be trained and certified. Training covers multiple aspects of the application:

- the choice of the equipment and checking for its proper functioning; the application techniques; filling the applicator with the agrochemical, calibrating, operating, maintenance and repair in the event of malfunction;
- knowledge of safety precautions required, before, during and after application of agrochemicals; use of protective clothing and other safety equipment including breathing apparatus, as required;

- risks of pesticide application to people, animals, and the environment; safe disposal of empty containers, tank washings, and surplus pesticides;
- record-keeping of the date, place, and name of the user of the agrochemicals.

Spraying Precautions

Since agrochemicals are a potential threat to areas adjoining the application site, the spray men should know the primary precautions of avoiding spray drift and observing separation distances between the site of application and any sensitive areas, such as sources of food or water supplies. The use of PPE for people working with pesticides is an essential safety requisite. However, for a variety of reasons, the adoption of PPE/safety kit by the workers in many countries is dismal. The cost consideration of a spraying safety kit is a credible deterrent for the poor farmers to avoid having a kit.

Working garments, such as shirts, overalls, trousers, socks, shoes/boots, should be kept clean and free of agrochemical deposits by washing immediately after use. In general, working clothes should be comfortable for use in hot tropical weather. The garments should not be unduly burdensome and restrict body movements of the spraying operators. A pesticide formulation with organic solvents may penetrate protective garments, and therefore, suitable material may be chosen. If exposure is limited to occasional liquid spills or to dry powders or granules, a coverall made of cotton or polyester fabric may be sufficient.

Head protection covers all parts above the shoulders, excepting the face. Eye and face protection covers the forehead and face to a point below the jaw. Non-fogging goggles may be worn when handling dust or granules. The protective gloves and garments made from neoprene, nitrile or butyl material of at least 0.4 mm in thickness are resistant to most agrochemical formulations. Gloves of wrist length may be required for manual spraying of toxic pesticides, elbow-length for handling granules, and shoulder-length for dipping hands in pesticides. Gloves, boots, goggles, and face masks should be washed thoroughly and allowed to dry.

Respiratory protection includes respirators of different types, such as (a) disposable for protection against dust, fumes, and mist; (b) chemical cartridge types; (c) canister types; (d) fully self-contained types; and (e) airline breathing units. Respirators may be half-faced, covering nose and mouth, or full-faced, covering nose, mouth, and eyes. In the case of highly hazardous agrochemicals, the air is filtered through a cartridge or canister containing chemical absorbents that, in turn, absorb or adsorb the hazardous substances. The masks and cartridges should be replaced periodically to ensure protection.

The BIS has made an amendment to IS 3906 on lever-operated knapsack sprayer, making mandatory to provide a mask, hand gloves and safety goggles with each sprayer (BIS 1995/2001). The quality of the kit, its efficacy in preventing pesticide exposure to workers has not been specified. The essential components of a knapsack spraying safety kit are illustrated in Fig. 10.1. The design and development of a spraying safety kit (CIAE 2014) illustrated herewith includes a face mask, hand

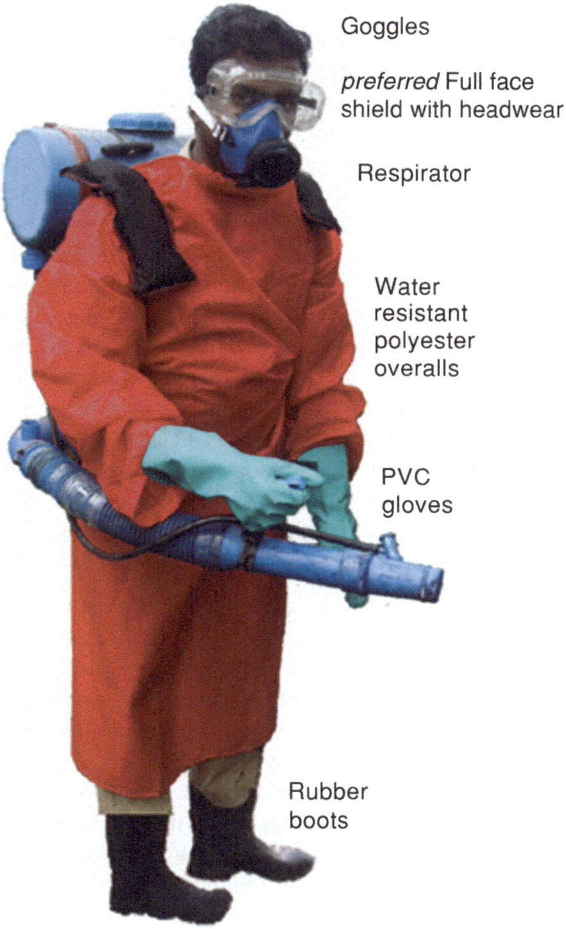

Fig. 10.1 Spraying safety kit

gloves, eye protector, footwear and an overall. The double-layered polypropylene of water repellent quality was used in the filtering material of the face mask. For chlorpyrifos pesticide, the mask filtering efficiency was found to be 84%. For comfortable wear by the workers, the mask was made of a flexible plastic body with an exhale valve. The temperature rise inside the mask was about 1.6 °C, and the breathing resistance was 0.68 m bar (i.e. 68 Pa or 7 mm of water). The PVC hand gloves were neither acidic nor alkaline, and therefore, was safe for hand and skin. The gloves had excellent gripping comfort, and the temperature rise inside the gloves was about 2.6 °C. The eye protector weighed 66 g. The overall was made of water-resistant polyester and was comfortable to wear from a thermal point of view. The pesticide (chlorpyrifos and cypermethrin) penetration through the fabrics was only about 1.5%.

The general precautions are described below:

- The sprayer operators should use appropriate personal protective equipment, including protective clothing as prescribed.
- Mix the correct application/dilution rate of agrochemicals. For mixing two or more agrochemicals, ensure that there is no risk of a chemical reaction.
- Handle containers carefully to prevent spillage during pouring into an applicator. In the case of spillage, people should be away until the area is cleaned up and disposed of safely (washing or using absorbent material by soil or sawdust to soak up the agrochemical).
- Consider whether weather conditions are satisfactory to avoid excessive wind speeds and spray drift towards sensitive areas (e.g. drinking-water supply). Do not allow other workers or children to enter into the field when pesticides are sprayed. Place warning signs on the field, as necessary.
- Post-spraying, thoroughly wash hands, face and neck, and other parts of the body that might become contaminated.
- Never blow the sprayer nozzle to clear the block. Safely dispose of any surplus in the application equipment. Wash the application equipment and the washings should be drained into a soak-away pit without any risk to the environment.
- Decontaminate protective clothing (apron, boots, face shield, gloves, respiratory protection equipment). Work clothing should be washed every day after spraying.

The knapsack sprayers are primarily in use in small land holdings for localized application. Routine maintenance of knapsack sprayers may ensure that the equipment is in good condition, with no potential to leak or spill. The tractor-drawn sprayer or aerial spraying from helicopters and fixed-wing aircraft have been adopted in extensive farm holdings and sizes, having similar crops over a large area. In tractor-drawn sprayer, the closed systems allow regulating the transfer of agrochemicals from a container into the spray tank automatically. The driver smoothly operates the sprayer controls fitted on to the tractor. It is necessary to ensure that only the approved pesticides and fertilizers are sprayed and that had minimal contamination to the locality and water bodies. Indiscriminate aerial spraying of endosulfan (OC compound) over cashew nut plantations in Kasargod (Kerala, India) caused disastrous health consequences (NIOH 2002; Embrandiri et al. 2012). Hundreds of newborns suffered congenital anomalies, mental retardation, and health complications. Many people lost their lives due to exposure to endosulfan, as reported from Sri Lanka (Roberts et al. 2003). Therefore, the local command and control (police, enforcement agencies, hospital authorities) must be well informed of the potential risks of aerial sparying, taking into view of likely contamination of potable water, water bodies used for rearing aquatic life, and irrigation of crops, and habitants.

Other Agrochemical Applications

Agrochemicals are also applied as veterinary products, fertilizers, and commodity chemicals. In animal injection treatments, prevention requires to use a needle set in a retractable spring-loaded sleeve to avoid self-injection of the handler. For treatment against disease-causing organisms in livestock buildings, a measured amount of potassium permanganate and formalin are sprayed as a fumigant. Fertilizers, as granules or rock powders, may be applied by using hand tools and wearing suitable gloves. Proper precaution may be taken in handling liquid fertilizers or silage additives containing concentrated acids.

Re-Entry

The re-entry period is the time interval between the agrochemical application and entering the treated area for safety reasons. The harvesting interval is the time between the agrochemical application and the harvesting of a crop for eating. A hazard warning sign may be displayed on the sprayed field, indicating the minimum re-entry periods. For entry into a treated area before the re-entry period, appropriate protective clothing should be worn.

Agrochemical Spillage and Fire

Agrochemical spillage in the farm locality or greenhouses may take place from the leaks of the defective container, while transferring from container to applicator, or worn pipe couplings/hose lines of equipment. When two or more substances come in contact, and those may react and produce gaseous vapours, heat, and fire. Also, the climatic conditions such as high humidity and direct sunlight may speed up the rate of damage. Essential interventions in the event of spillage are to contain the spread of spillage and contamination, keeping people and animals away from the area, soaking up the liquid spillage with absorbent (dry sand, soil or wood dust), and clearing up the contaminated matter for safe disposal. The people engaged in the cleaning activity must wear protective clothing and thoroughly wash immediately afterwards.

In the event of agrochemical fire, there is an apparent likelihood of toxic gas emission proportional to the number and type of the products involved. Firefighting would need precautions for applying the appropriate type of extinguishing agent. Portable extinguishers should be available at the storage site; however, more massive fires should be fought by professional firefighters.

First aid is the first available knowledge to treat a condition of poisoning or injury until the casualty is taken to a medical practitioner. In confirmed case of poisoning with agrochemicals, the casualty may be moved to an open, fresh, shaded area and contaminated clothing removed to avoid further contamination. The contaminated skin and eyes may be flushed with clean water, and cover the casualty with warm clothing. Injuries may result from chemical burns, and therefore, the person aware of the first aid treatment takes measures depending on whether the casualty is in a conscious or unconscious state. The medical practitioner may be informed about the type of agrochemical exposure. Local leaders of the villages must maintain a case history of any pesticide-related incidents and educate all concerned to prevent a recurrence of similar events in the locality.

References

Anand, S., Singh, S., Nahar Saikia, U., Bhalla, A., Paul Sharma, Y., & Singh, D. (2009). Cardiac abnormalities in acute organophosphate poisoning. *Clinical Toxicology, 47*(3), 230–235.

Anastassiades, M., Scherbaum, E., & Bertsch, D. (2003, May). Validation of a simple and rapid multi-residue method (QuEChERS) and its implementation in routine pesticide analysis. MGPR Symposium, Aix en Provence, France (pp. 1–7).

Azmi, M. A., Naqvi, S. N. H., Azmi, M. A., & Aslam, M. (2006). Effect of pesticide residues on health and different enzyme levels in the blood of farm workers from Gadap (rural area) Karachi—Pakistan. *Chemosphere, 64*(10), 1739–1744.

Banerjee, B. D., Zaidi, S. S. A., Pasha, S. T., Rawat, D. S., Koner, B. C., & Hussain, Q. Z. (1997). Levels of HCH residues in human milk samples from Delhi, India. *Bulletin of Environmental Contamination and Toxicology, 59*(3), 403–406.

Bellinger, R. G. (1996). *Pest resistance to pesticides*, publication, Department for Entomology, Clemson University.

Berg, H. V. D. (2004). *IPM Farmer Schools: A synthesis of 25 impact evaluations* (53 pp). Wageningen University.

Bhatnagar, V. K. (2001). Pesticides pollution: trends and perspectives. *ICMR Bulletin, 31*(9), 87–88.

Bhatnagar, V. K., Karnik, A. B., Suthar, A. M., Zaidi, S. S. A., Kashyap, R., Shah, M. P., et al. (2002). Biological indices in formulators exposed to a combination of pesticides. *Bulletin of Environmental Contamination and Toxicology, 68*(1), 22–28.

BIS. (1995/2001). IS 3906, Crop Protection Equipment - Hand operated knapsack sprayer, Piston type, - Specification. Bureau of Indian Standartds, New Delhi.

Blair, A., & Zahm, S. H. (1995). Agricultural exposures and cancer. *Environmental Health Perspectives, 103*(suppl 8), 205–208.

Benbrook, C. M., Sexson, D. L., Wyman, J. A., Stevenson, W. R., Lynch, S., Wallendal, J., et al. (2002). Developing a pesticide risk assessment tool to monitor progress in reducing reliance on high-risk pesticides. *American Journal of Potato Research, 79*(3), 183–199.

Buschmann, J. (2013). The OECD guidelines for the testing of chemicals and pesticides. In *Teratogenicity testing* (pp. 37–56). Totowa, NJ: Humana Press.

Carson, R. (1962/2002). *Silent spring*. Boston, MA, USA: Houghton Mifflin.

CIAE. (2014). Research Highlights 2012–14. All India coordinated research project on ergonomics and safety in agriculture. Technical Bulletin No. CIAE/2014/176. Central Institute of Agricultural Engineering, Bhopal, India.

Codex Alimentarius Commission. (2000). http://www.codexalimentarius.net/web/standard_list.jsp.

Cooper, J., & Dobson, H. (2007). The benefits of pesticides to mankind and the environment. *Crop Protection, 26*(9), 1337–1348.

CropLife International. (2007). *Guidelines for the safe warehousing of crop protection products*. Brussels, Belgium: CropLife International aisbl.

CropLife International. (2008). Catalogue of pesticide formulation types and international coding system, CropLife technical monograph No. 2.

Damalas, C. A., & Eleftherohorinos, I. G. (2011). Pesticide exposure, safety issues, and risk assessment indicators. *International Journal of Environmental Research and Public Health, 8*(5), 1402–1419.

Dasgupta, S., Meisner, C., Wheeler, D., Xuyen, K., & Lam, N. T. (2007). Pesticide poisoning of farm workers–implications of blood test results from Vietnam. *International Journal of Hygiene and Environmental Health, 210*(2), 121–132.

Eddleston, M., Eyer, P., Worek, F., Mohamed, F., Senarathna, L., von Meyer, L., et al. (2005). Differences between organophosphorus insecticides in human self-poisoning: a prospective cohort study. *The Lancet, 366*(9495), 1452–1459.

Eddleston, M., Mohamed, F., Davies, J. O., Eyer, P., Worek, F., Sheriff, M. R., et al. (2006). Respiratory failure in acute organophosphorus pesticide self-poisoning. *Journal of the Association of Physicians, 99*(8), 513–522.

Embrandiri, A., Singh, R. P., Ibrahim, H. M., & Khan, A. B. (2012). An epidemiological study on the health effects of endosulfan spraying on cashew plantations in Kasaragod District, Kerala, India. *Asian Journal of Epidemiology, 5*(1), 22–31.

FAO. (2003). *International code of conduct on the distribution and use of pesticides (Revised version, 123rd session of the FAO Council, November 2002)*. Rome: FAO.

FAO. (2016). *Evaluation of pesticide residues for estimation of for estimation of maximum residue levels and calculation of dietary intake*, Plant Production and Protection paper 224, Food and Agriculture Organization, Rome.

Felsot, A. S., Unsworth, J. B., Linders, J. B., Roberts, G., Rautman, D., Harris, C., et al. (2010). Agrochemical spray drift; assessment and mitigation—A review. *Journal of Environmental Science and Health Part B, 46*(1), 1–23.

Foodtank. (2017). United Nations: Agroecology, not Pesticides, is the Future for Food. https://foodtank.com/news/2017/03/united-nations-report-rejects-pesticide-myth/.

Furman, J. (2006). *Cholinesterase Monitoring for Agriculture Pesticide Handlers: Guidelines for Health Care Providers in Washington State*. Department of Labor & Industries, WISHA Policy & Technical Services. (http://www.lni.wa.gov/Safety/Topics/AtoZ/Cholinesterase/files/ProvidersGuidelines1).

Gallagher, K. D., Ooi, P. A., & Kenmore, P. E. (2009). Impact of IPM programs in Asian agriculture. In: *Integrated pest management: dissemination and impact* (pp. 347–358). Dordrecht: Springer.

Ganzelmeier, H. (2002). The prospect of European harmonisation-plant protection equipment under test. *Journal of Central European Agriculture, 3*(4).

Ghimire, S. (2014). *Practice related to pesticide use and health hazards amongst vegetable farmers in gotikhel*. Lalitpur: Nepal Health Research Council, Govt. of Nepal.

Govt. of India. (2020). The Insecticides Act, 1968. India Code, Digital Repository of all Central and State Acts. https://indiacode.nic.in/handle/123456789/1551?sam_handle = 123456789/1362.

Greulich, K., & Alder, L. (2008). Fast multiresidue screening of 300 pesticides in water for human consumption by LC-MS/MS. *Analytical and Bioanalytical Chemistry, 391*(1), 183–197.

Heiberger, S. (2015). Journal of Agromedicine "Leader in the Field" 2015: John J. May, MD. *Journal of Agromedicine, 20*(1), 1–2.

http://www.inchem.org/pages/jmpr.html. International Programme on Chemical Safety (IPCS), INCHEM Internationally Peer Reviewed Chemical Safety Information.

https://www.epa.gov/pesticide-reevaluation. Reevaluation: Review of Registered Pesticides.

http://www.epa.gov/pesticides/factsheets/chemical_fs.html.

http://ec.europa.eu/food/plant/protection/evaluation/exist_subs_rep_en.html, Council Directive 91/414/EEC, concerning the placing of plant protection products on the market. European Commission, 1991.

http://www.pic.int/en/Table7.htm.

http://www.fao.org/WAICENT/FAOINFO/AGRICULT/AGP/AGPP/Pesticid/Disposal/common/ecg/103811_en_w1604e.pdf.

http://www.fao.org/WAICENT/FAOINFO/AGRICULT/AGP/AGPP/Pesticid/Disposal/common/ecg/103825_en_No_7Small_quantities_stocks.pdf.

http://www.fao.org/WAICENT/FAOINFO/AGRICULT/AGP/AGPP/Pesticid/Code/Download/code.pdf.

http://www.dropdata.net/SIP_micontrol/biopesticides.htm, Biopesticides, Microbial Control Division, Society for Invertebrate Pathology.

http://www.bt.ucsd.edu/bt_crop.html.

http://ec.europa.eu/environment/integration/research/newsalert/pdf/95na1.pdf.

http://www.fda.gov/Food/ScienceResearch/LaboratoryMethods/PesticideAnalysisManualPAM/default.htm.

http://www.fda.gov/Food/ScienceResearch/LaboratoryMethods/PesticideAnalysisManualPAM/ucm113710.htm.

http://www.fao.org/docrep/V8966E/V8966E00.htm#2; FAO Pesticide Storage and Stock Control Manual.

Karki, P., Ansari, J. A., Bhandary, S., & Koirala, S. (2004). Cardiac and electrocardiographical manifestations of acute organophosphate poisoning. *Singapore Medical Journal, 45,* 385–389.

Karunakaran, C. O. (1958). The Kerala food poisoning. *Journal of the Indian Medical Association, 31*(5), 204–207.

Kashyap, R., Bhatnagar, V. K., & Saiyed, H. N. (2002). Integrated pest management and residue levels of dichlorodiphenyltrichloroethane (DDT) and hexachlorocyclohexane (HCH) in water samples from rural areas in Gujarat State, India. *Archives of Environmental Health: An International Journal, 57*(4), 337–339.

Khan, M. F., Khan, M. I., Aslam, M., & Naqvi, S. N. H. (2000). Study of cholinesterase level in blood of cotton field workers exposed to pesticides. *Journal of Baqhdad Medical University, 3,* 12–16.

Klein, A. W., Goedicke, J., Klein, W., Herrchen, M., & Kördel, W. (1993). Environmental assessment of pesticides under directive 91/414/EEC. *Chemosphere, 26*(5), 979–1001.

Kummer, K. (1999). Prior informed consent for chemicals in international trade: The 1998 Rotterdam Convention. *Review of European, Comparative & International Law, 8,* 323.

Lee, P., & Tai, D. Y. H. (2001). Clinical features of patients with acute organophosphate poisoning requiring intensive care. *Intensive Care Medicine, 27*(4), 694–699.

Lu, J. L., & Cosca, K. (2011). Pesticide application and health hazards: implications for farmers and the environment. *International Journal of Environmental Studies, 68*(2), 197–208.

MacFarlane, E., Chapman, A., Benke, G., Meaklim, J., Sim, M., & McNeil, J. (2008). Training and other predictors of personal protective equipment use in Australian grain farmers using pesticides. *Occupational and Environmental Medicine, 65*(2), 141–146.

Morton, V., & Staub, T. (2008). A short history of fungicides. *Online, APSnet Features, 10*(1094), 0308.

NIOH. (2002). *The Investigation of Unusual Illnesses Allegedly Produced by Endosulfan Exposure in Padre Village of Kasargod District (N.* Kerala): National Institute of Occupational Health, Ahmedabad, India.

Noshad, H., Ansarin, K., Ardalan, M. R., Ghaffari, A. R., Safa, J., & Nezami, N. (2007). Respiratory failure in organophosphate insecticide poisoning. *Saudi Medical Journal, 28*(3), 405.

Oluwole, O., & Cheke, R. A. (2009). Health and environmental impacts of pesticide use practices: a case study of farmers in Ekiti State. *Nigeria. International Journal of Agricultural Sustainability, 7*(3), 153–163.

Peshin, R. (2002). Economic benefits of pest management. *Encyclopedia of pest management.* Boca Raton: Marcel and Dekker Inc.

Peshin, R. (2013). Farmers' adoptability of integrated pest management of cotton revealed by a new methodology. *Agronomy for Sustainable Development, 33*(3), 563–572.

Peter, J. V., Jerobin, J., Nair, A., & Bennett, A. (2010). Is there a relationship between the WHO hazard classification of organophosphate pesticide and outcomes in suicidal human poisoning with commercial organophosphate formulations? *Regulatory Toxicology and Pharmacology, 57*(1), 99–102.

Peter, J. V., Sudarsan, T. I., & Moran, J. L. (2014). Clinical features of organophosphate poisoning: A review of different classification systems and approaches. *Indian Journal of Critical Care Medicine, 18*(11), 735.

Recena, M. C. P., Caldas, E. D., Pires, D. X., & Pontes, E. R. J. (2006). Pesticides exposure in Culturama, Brazil—knowledge, attitudes, and practices. *Environmental Research, 102*(2), 230–236.

Roberts, D. M., Karunarathna, A., Buckley, N. A., Manuweera, G., Sheriff, M. H., & Eddleston, M. (2003). Influence of pesticide regulation on acute poisoning deaths in Sri Lanka. *Bulletin of the World Health Organization, 81,* 789–798.

Sheridan, R. P., Wang, W. M., Liaw, A., Ma, J., & Gifford, E. M. (2016). Extreme gradient boosting as a method for quantitative structure–activity relationships. *Journal of Chemical Information and Modeling, 56*(12), 2353–2360.

Singh, G., & Khurana, D. (2009). Neurology of acute organophosphate poisoning. *Neurology India, 57*(2), 119.

Slorach, S. A. (2006). Assuring food safety: The complementary tasks and standards of the World Organisation for Animal Health and the Codex Alimentarius Commission. *OIE Scientific and Technical Review, 25,* 813–821.

Srinivas Rao, C. H., Venkateswarlu, V., Surender, T., Eddleston, M., & Buckley, N. A. (2005). Pesticide poisoning in south India: opportunities for prevention and improved medical management. *Tropical Medicine & International Health, 10*(6), 581–588.

Taneja, S. (2017). Why India continues to use lethal pesticides. DownToEarth. https://www.downtoearth.org.in/news/agriculture/vidarbha-s-toxic-trail-59173.

Tijani, A. A. (2006). Pesticide use practices and safety issues: the case of cocoa farmers in Ondo State, Nigeria. *Journal of Human Ecology, 19*(3), 183–190.

UNEP/FAO. (2017). The Rotterdam Convention on the prior informed consent- Procedure for certain hazardous chemicals and pesticides in international trade, 1998 (revised in 2017). (http://www.pic.int/en/Table7.htm).

Unruh, J. B., Brecke, B. J., & Trenholm, L. E. (2003). *Weed management in home lawns.* Institute of Food and Agricultural Sciences, EDIS: University of Florida Cooperative Extension Service.

WHO. (2004). Pesticides and Health-The Impact of Pesticides on Health: Preventing Intentional and Unintentional Deaths from Pesticide Poisoning. https://www.who.int/mental_health/prevention/suicide/en/PesticidesHealth2.pdf.

Weiss, E. B. (1992). United Nations conference on environment and development. *International Legal Materials, 31*(4), 814–817.

Willett, P. (2009). A Bibliometric Study of Quantitative Structure-Activity Relationships and QSAR & Combinatorial Science. *QSAR & Combinatorial Science, 28*(11–12), 1231–1236.

Part V
Ergonomics Application in Design

Chapter 11
Ergo-Design Criteria for Farm Tools and Machinery

Introduction

The history of farming hand tools can be traced back to 7000 BC, and perhaps since the very beginning, the man took up cultivation. The traditional hand tools include axe, pickaxe, spade, crowbar, sickle, hand hoe, mallet, and these tools have been used for cultivation of crops, digging soil, cutting firewood, and hunting forest animals. With superior technology and metallurgy, there have been continuing evolution of improved versions of hand tools and devices to use in functions of farming and allied activities. The farmers and all other rural inhabitants possess some farm tools to meet their daily household and farming needs. Inevitably, looking at the drudgery proneness and productivity requirements, all-round advances in farming methods, practices, tools, and machinery are visible. The science and technology of ergonomics and human factors look into the core issues and criteria for better design and use of farm machinery and tools. Primarily emphasis is on the productivity, safety, man–machine compatibility, and biological, motivational, and climatic requirements. The user acceptance of tools and machinery is based on product engineering, and aesthetics, such as shape, colour, and form. The socio-economic contexts, user expectations and requirements determine the kind of technology, machinery, and methods that come to stay in an agricultural community. Understanding of the user requirements in vastly different crops and farming activities and acceptability of the farm tools and machinery by the farming community are the apparent driving force of the concerned professionals for their design innovation. With due recognition of the multitude of issues that go into the user functions in farming activities, this chapter primarily highlights some of the criteria utilized in the ergo-design of farm tools and machinery.

P. K. Nag and L. P. Gite, *Human-Centered Agriculture*, Design Science and Innovation,
https://doi.org/10.1007/978-981-15-7269-2_11

General Ergonomics Criteria

Anthropometry

As regard to deciding functionality, dimensions of any tool/equipment, and man–machine hardware compatibility, concerns primarily focus on the body dimensions of the users. Since natural dynamic human movements are a necessary part of farm work, the equipment and workplace should match the body size of the operator. Therefore, the anthropometric data are principally useful in determining the size and shape of the tools and setting the dimensions of workspace. The critical factor to take into account is to define the user population, about their age, race, gender, and occupation. A substantial variation exists in human body dimensions. Thus, *one design fits all* is not practically feasible to generalize the equipment and workplaces to suit 100% of the users. The design is made in such a way to satisfy about 90% of the users. This is achieved through the use of the 5th and 95th percentile limits. It means that those people who fall outside these limits may not match to the criteria concerned. The anthropometric measures fall into three main categories and deal with issues, such as clearance, reach, and posture.

1. The clearance criteria deal with concerns like headroom and legroom. Access problems between and around obstacles also fall into this category. Here, the limiting user will be a significant member of the population, generally, one who is 95th percentile in the relevant aspect.
2. The reach criteria concern the location of controls, the storage of materials, and the situations where these are necessary to reach to perform a task. The limiting user will be a small member of the population, usually the 5th percentile in the relevant aspect.
3. The postural criteria include those concerned with the posture to be adopted at the workplace for optimal operation of various controls and carrying out activities. In many cases, the operator takes a working posture based on the situations of the working surface given practical limitations in the farm field. Here a limiting user will be identified, keeping into consideration the job requirement.

The readers may refer to Chap. 6, Table 6.2 that provides 79 body dimensions of Indian male and female farmworkers (Gite 2017). These dimensions seemed representative to use in farm equipment design. The limitation exists in the availability of similar data of farming population from other countries.

ISO 7250 parts 1 to 3 standard (ISO 1996/2017, ISO 2010, ISO 2015), and NASA (1978a, b) are useful sourcebooks for anthropometric data. Data of 56 body dimensions from 10 countries, namely, Austria, Germany, Italy, Japan, Kenya, Korea, the Netherland, Thailand, the United States, and China are available in ISO 7250 part 2-2010. There are regional variations in different body dimensions. Data on essential body dimension parameters of men and women from 13 countries have been compiled in Chap. 6, Tables 6.3 and 6.4. It is necessary to have population-specific and occupation-specific anthropometric data of each country, for use in equipment design. The data of Indian farmworkers (Gite 2017) have been used for application in the design of farming tools and equipment, as elaborated in Chaps. 12 and 13.

Muscular Strength

Most open-field farming jobs are manual. The muscle strength criteria are applicable where a worker has to apply force to do the work. Table 11.1 includes safe force limits or the isometric strength values given by various researchers. Since the safe force limits vary with gender, age-range, and population characteristics, it is apparent to have the country-specific muscle strength data of farmworkers. The ISO 11228/part 2 (ISO 2007/2016) has been developed based on the studies of European and western countries, and those are not directly applicable to people in Southeast Asian countries.

Table 11.1 Safe force limits as given by different researchers/standards

Source	Population	Force	Value (N)	Remarks
Davis and Stubbs (1977)	British male workers	Push Pull	245 343	Force applied at the shoulder level. Limits suitable for 99% of young males only (age below 35 years)
Snook and Ciriello (1991)	American male workers	Push Pull	216 235	Limits suitable for 90% of the population. Handle height 1440 mm for push and 950 mm for the pull
	American female workers	Push Pull	137 137	Limits suitable for 90% of the population. Handle height 1350 mm for push and 890 mm for the pull
Health Council of Netherlands (2012)	Dutch workers	Push Pull	200 145	Limits suitable for 85% of the population
ISO 11228/part2 (ISO 2007/2016)	Male workers	Push Pull	230 240	Limits suitable for 90% of the population, Handle height 950 mm
	Female workers	Push Pull	130 140	Limits suitable for 90% of the population. Handle height 890 mm
Gite (2017)	Indian male farmworkers	Push Pull	148 153	Limits suitable for 90% of the population. Handle height—Acromion height 1360 mm
	Indian female farmworkers	Push Pull	93 109	Limits suitable for 90% of the population. Handle height—Acromion height 1260 mm

The stated ISO standard gives a procedure to determine whole-body pushing and pulling force limits according to specific characteristics of the population and the task. That is, one needs to have data on maximum isometric strength, age, gender, and stature as well as the requirements of the task (i.e. frequency, duration, and distance of push/pull task).

Davis and Stubbs (1977) mentioned that when the frequency of force application in pushing/pulling activities is higher than once per minute, then the safe load needs to be adjusted to 70% of the maximum safe force limit. In the US industries, manual material handling tasks (DARCOR 2001) are designed in such a way that at least 75% of the female population and 99% of the male population can safely perform those tasks. The Health Council of Netherlands (2012) recommended that the exerted forces must be less than 50% of the maximum force values, and thereby to keep health risk as low as possible. However, it has been viewed that dynamic effort of a repetitive nature should not exceed 30% of the maximum, and only about 50% is satisfactory for work up to 5 min. In the case of static loading, only less than 15% of the maximum force can be exerted by the muscles for a prolonged duration (Nag 1991). Understandably, muscle exertion in farming activities are combined nature of isotonic and isometric contractions, and the optimization of the force limit would demand characterization of the extent of the load from either of the components of muscle contraction.

If the farm tools or machinery is to be made suitable for the 5th percentile worker (in terms of muscle strength), then the force required in the operation of the equipment in push or pull mode should not exceed the 5th percentile muscle strength values. The force limit maybe 30% of the maximum strength of the 5th percentile worker in the case of repetitive movements. For example, the 5th percentile strength values for push and pull of Indian farmworkers were 133 and 140 N for males, and 84 and 101 N for females, respectively. Taking 30% of these values, the force limits for continuous or repetitive work arrive at 40 N push and 42 N pull for males, and 25 N push and 30 N pull for females, respectively. In the case of torque application with both hands in standing posture, the 5th percentile values are 31 and 25 Nm respectively. For sustained force application in rotary mode, the corresponding recommended limits are 9 Nm and 8 Nm respectively. If the force required for the operation of a tool is more than these values, the worker can operate that tool or device intermittently, by taking frequent rest pauses.

Safe Limits for Lifting and Carrying of Loads

Agricultural operations involve various manual material handling (MMH) activities. There are different forms of MMH activities, such as lifting, lowering, pushing, pulling, holding, turning and carrying of loads of different size and shape. The safe load-lifting limits (Table 11.2) vary between 23 to 25 kg for males and 14 to 16 kg for female workers, respectively. It has been viewed that a worker should not carry a load exceeding 40% of one's body weight. In carrying loads, the person also carries ones'

Table 11.2 Safe load-lifting limit (kg)

Reference	Male	Female
Snook (1978)	23	14
Maiti and Ray (2004)	–	15
NIOSH (2007)	23	–
ISO 11228 Part I (ISO 2014)	25	15
Hunt (2017)	25	16

body weight. The mode of load carrying should be such that the static loadings of muscles of hands and arms are avoided. Also, the limits are applicable for plains, and for the hilly region, however, the limit requires downwards adjustment depending on slope and the terrain. To design the load-carrying activity for 90% of the workers, the 10th percentile values of body weight can be taken for setting the threshold. For example, in the case of Indian farmworkers, the 10th percentile values of body weight are 45.0 kg and 37.5 kg for male and female workers, respectively, and the limits for safe load-carrying works out at 18 kg for males and 15 kg for females (Gite 2017).

Working Postures

Human working posture refers to the relative orientation of the parts of the body in space (Pheasant 2003). In a workplace situation, each person develops unique postural dynamics about one's physiology, habits, the use of tools and devices, and the circumstances of work. The postural orientation exerts a load on the musculoskeletal and spinal structures. The strain due to awkward posture carries the risk of spinal and paraspinal discomfort and damage, such as joint impairments (arthritis), inflammation of tendons and tendon sheaths, chronic muscle pain, and joint degeneration (arthroses). Work- and posture-related musculoskeletal disorders (WMSDs) constitute a significant occupational health problem in all spheres of activity. The rising costs of medical expenses, reduced productivity, wage compensation, and lower quality of life are the alarming scenario in both developed and developing countries.

Refer to farm activities described in Chap. 2, the agricultural activities demand a diverse form of primary and secondary postures, including standing, sitting, stooping, bending, or squatting. In many a time, the body posture depends on the working conditions and environment where the work is carried out. A good working position is the one that brings a minimum of static muscular effort, and the given task is performed more effectively and with least muscular discomfort. Any operation in squatting or bending posture involves drudgery, and that reflects in pain and discomfort among the workers. As far as possible, stooping or squatting positions should be avoided. Preferably, work may be carried out in a sitting posture, avoiding the vibratory surface. When the job demands a standing position, provision should be made for a change of posture to avoid discomfort in lower limbs and back muscles.

Various postural observational techniques apply to assess posture-related musculoskeletal disorders and discomforts. Ergonomics toolkits and methods, such as ovako working posture analysing system (OWAS) (Louhevaara et al. 1992), rapid upper limb assessment (RULA) (McAtamney and Corlett 1993), and rapid entire body assessment (REBA) (Hignett and McAtamney 2000) apply to body posture analysis, workplace evaluation and design intervention. Generally, these methods attach numbers and scores to different body orientations, joint angles, and movements, along with the handling of physical loads. From the accrued ranking, suggestions may be put forward for intervention and remedial measures (Kee and Karwowski 2007).

Handle Grip and Handle Height for Manually Operated Equipment

The hand is primarily adapted for reaching, grasping and manipulating, and other associated activities, such as pushing, adjusting objects, striking blows, and supporting the body in space. Accordingly, the type of handgrip varies with the function, as a cylinder, ball, ring, pliers and pinch, as the variants of power and precision grips (Nemeth 1985; Freivalds 1987; Nag et al. 2003). Whereas several deeper and external muscles are associated with hand movements, four muscles principally involved in elbow joint action are brachialis, biceps, brachioradialis, and pronator teres. The rotation of the forearm affects the efficiency of biceps, brachioradialis, and pronator teres muscles. In pronated hand position, these muscles exert the least contractile force while being at their combined best in the mid-position. In a supinated position, the increased contractile force of the biceps offsets the force levels in other muscles. Thus, the power produced in the supinated position is somewhat higher than that in a pronated position, although both are less than in a neutral position. In the sagittal plane, pulling and pushing force is nearly the same, whether the arms are held sideways or forwards (Grandjean 1982). Therefore, the positioning of hands should be such that they are close to their neutral position. That is, the hand placement on handle grip in mid-position is desirable. The length of the crossbar will depend on the elbow-elbow breadth, and it can be taken as 40 to 60 cm.

The optimum handle height for push–pull type manually operated equipment was found to be 0.7 to 0.8 times of the shoulder (acromion) height (Gite and Yadav 1990). For a hand-operated rotary maize sheller operated by females, Kharade et al. (2010) reported the optimum crankshaft height as 0.9 times of elbow height, and the optimum crank length as 0.45 times of shoulder grip length.

Physiological Cost and Efficiency

The aerobic capacity of a person, i.e. the maximum oxygen consumption rate, $VO_{2\,max}$, (l/min), sets the limit of physical work capacity. For American, European, and African workers, the values of aerobic capacity for male workers range from 3 to 4 l/min whereas for southeast Asian workers, e.g. Indian workers, these values range from 2 to 2.5 l/min. For women, the aerobic capacity is generally about 70 to 75% of that of men.

The physiological cost of any operation is expressed in terms of heart rate and O_2 consumption rate. For an 8 h work period, a workload requiring O_2 at a rate of 35 to 40% of $VO_{2\,max}$. The heart rate for such a workload may be about 110 to 120 beats/min. In most individuals, a heart rate of around 120–130 beats/min corresponds to a workload of about 50% of the individual's $VO_{2\,max}$. Incidentally, this is also the work rate when most persons begin to be slightly out of breath.

The human energy expenditure of a person at rest or while doing work can be calculated from the measurement of O_2 consumed by the person in the specified time, and multiplying it by the calorific value of O_2, i.e. about 20.9 kJ per litre of O_2 (Nag et al. 1980). The efficiency of work is the ratio of useful work to the energy consumed to produce that work under favourable conditions, and the value may go up to 20 to 30%. However, in-field activities, the efficiency level generally varies between 3 and 30% (Table 11.3), as reported in different unit operations (Grandjean 1982).

Table 11.3 Maximum efficiency in various physical tasks

Activity	Efficiency (%)
Shovelling in a stooped posture	3
Screw driving	5
Shovelling in normal posture	6
Lifting weights	9
Turning a handwheel	13
Using a heavy hammer	15
Carrying a load on the back on the level, returning without load	17
Carrying a load on the back up an incline, returning without load	20
Going up and down ladders, with and without load	19
Turning a handle or crank	21
Going up and downstairs, without load	23
Pulling a cart	24
Cycling	25
Pushing a cart	27
Walking on the level, without load	27
Walking uphill on a 5^0 slope, without load	30

Cardiorespiratory Responses in Different Force Application Modes

Cardiorespiratory responses, such as heart rate and O_2 consumption rate of a person in a given activity depend on the workload, the amount of force applied by a person, and the walking speed. That is, the heart rate and the O_2 consumption, for example, vary with the speed of walking, pedalling or cranking, and the power applied during the operation of farm equipment. The data given herewith in Tables 11.4, 11.5, 11.6 may be used to estimate the heart rate and O_2 consumption in operating farm tools and machinery by the Indian farmworkers. These cardiorespiratory responses may vary among the farmworkers in different regions owing to ethnicity and their ergonomics characteristics.

Discomfort rating/score—As stated above, a worker spends physiological energy in operating tools and machinery. Besides, one may experience discomfort due to

Table 11.4 Heart rate (HR) and O_2 consumption during push and pull force application by farmworkers (TNAU 2014)

Type of force applied	Force application (N)	HR (beats/min)				O_2 consumption (l/min)			
Working speed, km/h		0.5	1.0	1.5	2.0	0.5	1.0	1.5	2.0
Push	50	84	88	92	96	0.44	0.50	0.53	0.58
	100	89	93	100	104	0.50	0.54	0.62	0.67
	150	92	99	103	106	0.54	0.61	0.66	0.69
	200	95	101	105	112	0.57	0.63	0.68	0.76
	250	98	103	112	118	0.61	0.66	0.77	0.83
Pull	50	93	94	95	98	0.49	0.51	0.52	0.57
	100	98	99	102	102	0.57	0.59	0.64	0.65
	150	101	103	104	105	0.63	0.66	0.68	0.70
	200	105	108	109	110	0.70	0.74	0.76	0.78

Table 11.5 Heart rate (HR) and O_2 consumption during pedalling force application by farmworkers (MPUAT 2014)

	Power output (W)	HR (beats/min)				O_2 consumption (l/min)			
Pedalling speed (rpm)		40	50	60	70	40	50	60	70
Male workers	60	120	117	115	117	0.84	0.77	0.79	0.82
	70	131	127	124	128	1.17	1.11	1.08	1.14
	80	143	137	136	138	1.35	1.25	1.19	1.29
	90	155	148	147	148	1.44	1.40	1.34	1.42
Female workers	50	128	124	121	124	0.84	0.77	0.79	0.82
	60	137	134	132	136	1.03	0.97	0.94	1.00
	70	150	145	143	143	1.21	1.11	1.05	1.15
	80	161	160	154	154	1.30	1.26	1.20	1.28

Table 11.6 Heart rate (HR) and O_2 consumption during hand-cranking force application by farmworkers (IITKGP 2016; PAU 2016)

Force applied in cranking mode (40 rpm)	HR (beats/min)					O_2 consumption (l/min)				
Power output (W)	21	34	47	60	75	21	34	47	60	75
Male workers	88	102	114	129	146	0.60	0.80	1.00	1.20	1.60
Female workers	83	94	109	123	138	0.50	0.70	1.00	1.20	1.60

postural stress or muscular fatigue. Different techniques for assessing discomfort has been described in Chap. 5. Essentially, therefore, a farm tool or a device needs to be designed and operated in such a way that it involves minimum discomfort and fatigue.

Design of Workplace of Tractors and Self-propelled Machines

Optimum Body Joint Angles for Driving

The optimum values of body joint angles for comfortable work posture during driving have been proposed by various researchers, as shown in Fig. 11.1 and Table 11.7.

Hand and Leg Reach Envelopes, and Location of Controls

Defining the hand and leg reach envelopes are critical components to form layout design of the personal workspace, and optimize areas for placement of controls of workstations, tractors, and other self-propelled vehicles. About designing workplace layout on self-propelled farm machinery (e.g. a tractor), it is necessary to have various controls within optimum reach of the operator (Fathallah et al. 2008; Mehta et al. 2008). The reach envelopes give unbiased feedback concerning a point in a workspace that is reachable for hand and leg controls (Yang and Abdel-Malek 2009). Generally, the reach envelopes are determined using (a) experiment-based methods visualizing human reach from a large number of volunteers (Matthews and Knight 1971; Chaffin 2002), (b) the voxel-based method (Troy and Guerin 2004), and (c) the closed-form solution (Yang et al. 2005).

Experimental-based methods have been applied in ergonomic design of tractor controls (Kumar et al. 2009; Eltawil and Hegazy 2011; Khadatkar et al. 2017) with the use of human anthropometry, and boundary conditions. For layout design of controls in various vehicles, including farm machinery, two terms widely used are seat reference points (SRP) and seat index point (SIP). As per IS: 11806 (BIS 1994),

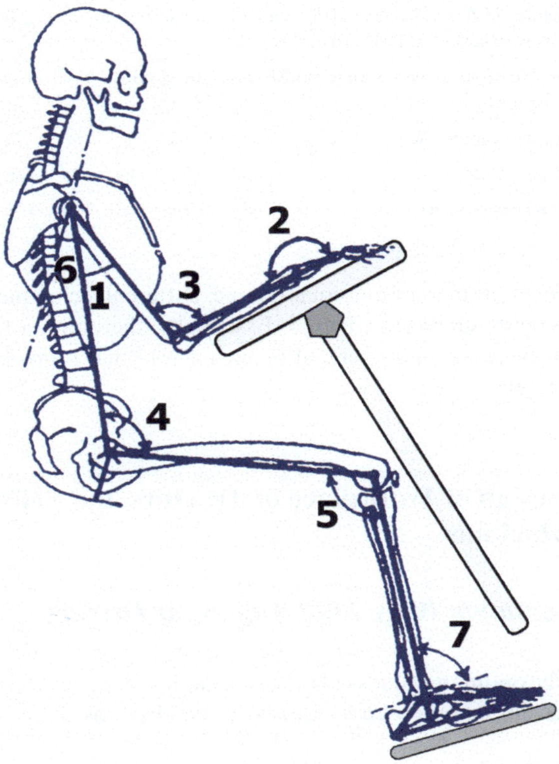

Fig. 11.1 Body joint angles (in degree) for comfortable posture during driving (Angle numbers correspond to the Sr. No. given in Table 11.7)

the SRP is a point in the central longitudinal plane of the seat where the tangential plane of the lower backrest and a horizontal plane intersect. The SIP is the intersection on the central vertical plane through the seat centerline of the theoretical pivot axis between the human torso and thighs, with reference to IS: 11113 (BIS 1999/2004) and ISO 5353 (ISO 1995). Generally, the SRP is used in checking the machine or measuring the distances between the seat to controls (Gellerstedt et al. 2006). Figure 11.2 shows the relative locations of SRP and SIP. The SIP usually located 90 mm above and 140 mm in front of SRP, according to IS 12343 (BIS 1998, Mehta et al. 2008).

Potdar et al. (2017) developed a procedure for outlining the optimum area for placement of tractor controls, and hand and leg reach envelopes for men and women, and also combined population. Figures 11.3a, b and 11.4a, b depict the sample hand and leg reach envelopes derived from the 5th and 95th percentile anthropometric data Indian male farmworkers, as given in Chap. 6.

The limiting values of X-, Y-, and Z-coordinates derived from the hand and leg reach envelopes (Table 11.8) for locating hand, and leg controls are based on the 5th and 95th percentile values of male workers. That is, the hand controls should not be located nearer than 318 mm and farther than 485 mm in the horizontal direction

Table 11.7 Optimum body joint angles (in degree) for comfortable posture during driving

Sr. No.	Body joint	Angles for comfortable postures						
		Dupuis (1959)	Morrison and Harrigton (1962)	Babbs (1979)	Hansson and Pettersson (1980)	Courtney and Evans (1987)	Göbel and Luczak (1996)	Tilley (2002)
1	Angle of upper arm to vertical	31	<35	15–35	16–21	30	5–25	0–35
2	Wrist angle	–	–	170–190	–	–	–	–
3	Elbow angle	104	NG	80–110	105–115	100	95–135	80–165
4	Hip angle	90	90	85–115	100	92.5	100–115	95–100
5	Knee angle	123	135	95–120	105–120	100	110–120	110–120
6	Angle of trunk to the vertical	5	–	15–25	–	10	20	–
7	Ankle angle	89	75–105	85–95	90–100	–	90–100	90–100

from SIP (Fig. 11.3a). Also, these controls should be placed between 186 mm and 410 mm in a vertical direction above SIP. In the lateral direction, the controls should be located within ± 528 mm from a centerline passing through SIP (Fig. 11.3b).

Accordingly, the leg controls in self-propelled machines should not be located nearer than 410 mm and farther than 493 mm in the horizontal direction from SIP (Fig. 11.4a). These controls should not be placed nearer than 261 mm and farther than 396 mm downwards from SIP. In lateral direction, the leg-operated controls, these should preferably be placed between 90 and 215 mm on either side from the centerline of the tractor (Fig. 11.4b). Limiting values for the workplace of other sets of users like 5th and 95th percentile females can easily be worked out using a similar approach.

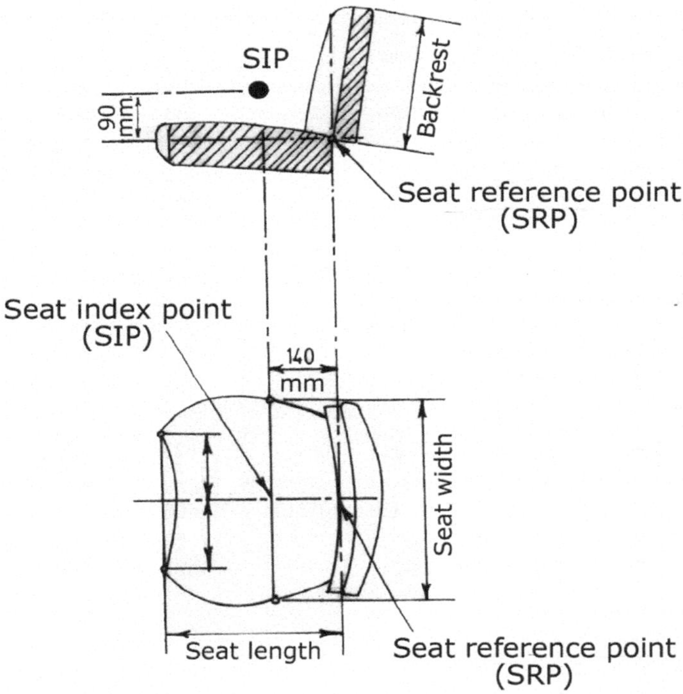

Fig. 11.2 Seat reference point and seat index point on a tractor

Actuating Force Limits for Controls of Farm Machinery

Limits for actuating forces for controls on self-propelled machines, such as tractors/other farm machines, are summarized in Table 11.9.

Environmental Aspects

Setting criteria for the working environment in the farming would demand to ascertain exposure conditions of the workers to a multitude of stress agents, such as vibration, noise, dust, exhaust emissions, chemicals, adverse thermal conditions, and poor lighting.

a

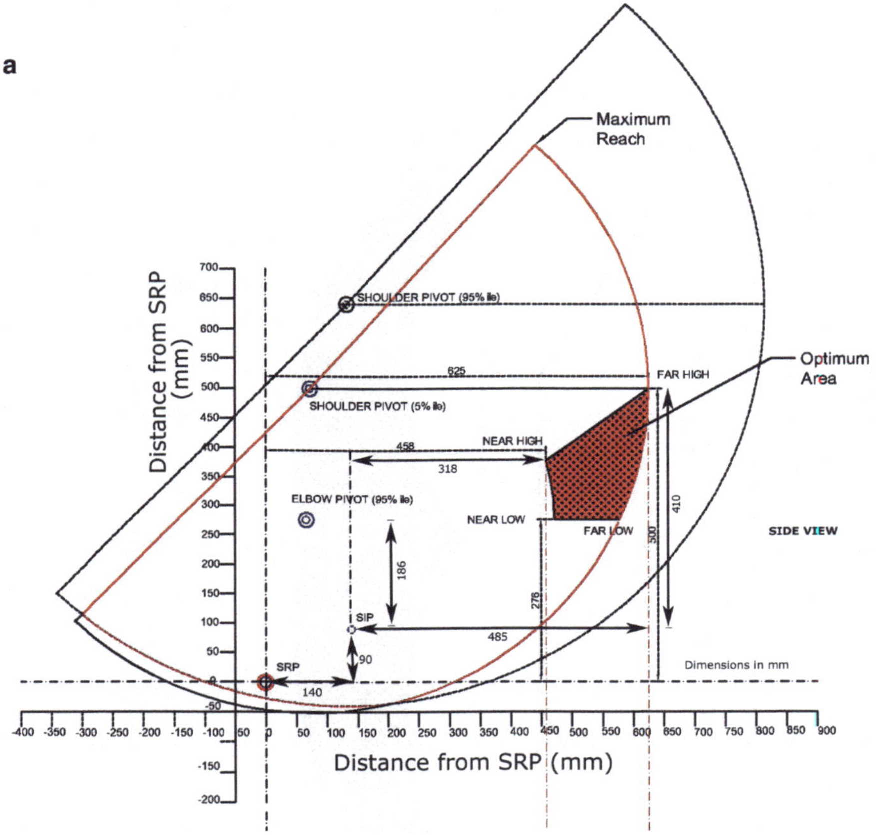

Fig. 11.3 **a** Hand reach envelopes (side view) for 5th and 95th percentile Indian male farmworkers (all dimensions are in mm). **b** Hand reach envelopes (plan view) for 5th and 95th percentile Indian male farmworkers (all dimensions are in mm)

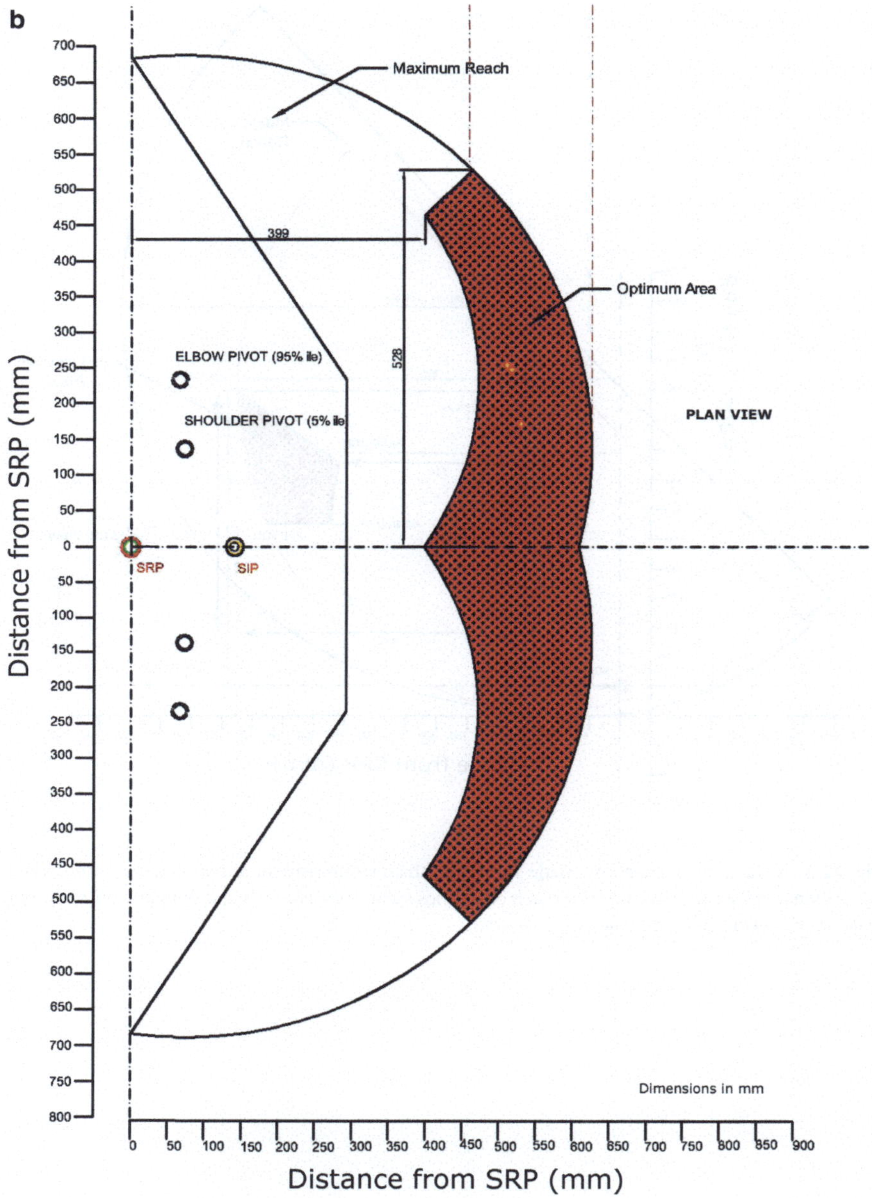

Fig. 11.3 (continued)

Vibrations

During the operation of farm machines, the operators are exposed to the whole body as well as hand-arm vibrations, transmitted through the seat and steering/handle, respectively. In the case of tractors, the whole body vibration arises mainly from the travel over uneven surfaces. The most crucial vibration component is the vertical oscillations of the vehicle, and below 2 Hz, the human body responds to vertical vibration as a dead weight. Above this frequency, relative motion occurs between regions and body organs due to resonance effects within the skeletal structures and connecting tissues. The primary body resonance occurs in the range up to 8 Hz. A considerable amplification of vibration occurs between the buttocks and the upper body regions, resulting in vibration-related discomfort. In practice, the dominant frequencies are in the range of 2.5–4.0 Hz. The vertical components of the vibration (a_z) are more severe, as compared to the horizontal (a_x and a_y) as well as rotational modes.

The high-frequency vibrations arise from the internal combustion engines or the mechanisms of the machinery. The transmission of high-frequency vibration through the body parts to the head is limited. The resonance of the head is encountered at

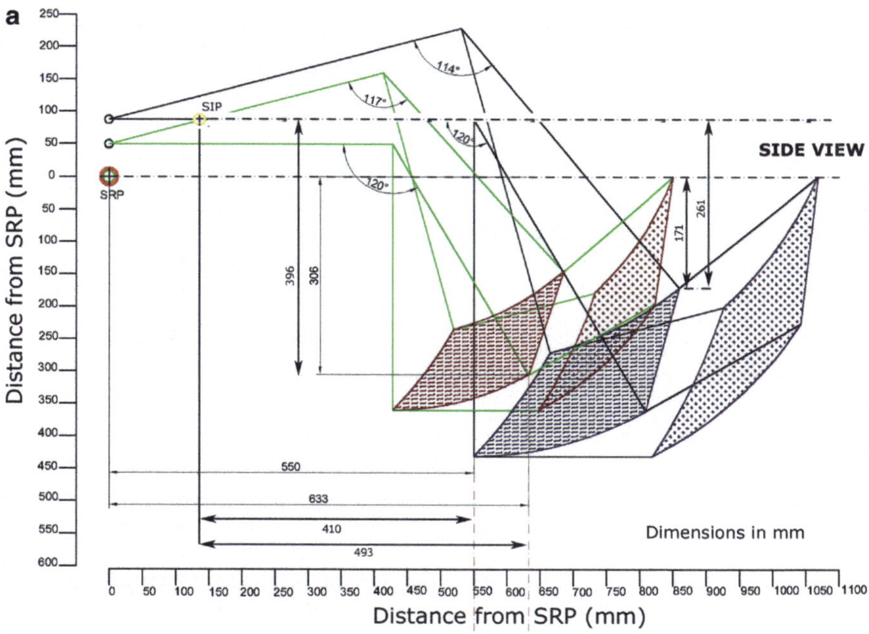

Fig. 11.4 **a** Leg reach envelopes (side view) for 5th and 95th percentile Indian male farmworkers (all dimensions are in mm). **b** Leg reach envelopes (plan view) for 5th and 95th percentile Indian male farmworkers (all dimensions are in mm)

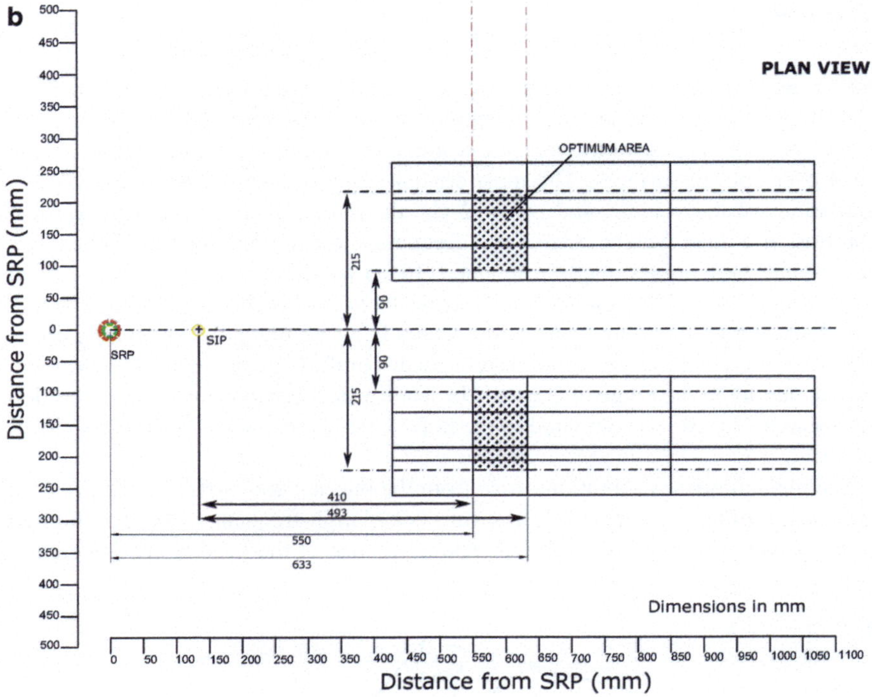

Fig. 11.4 (continued)

Table 11.8 X, Y, and Z coordinates from SIP for optimum location of hand- and leg-operated controls based on hand, and leg reach envelopes of male farmworkers

Particulars	Limiting values (mm)					
	X		Y		Z	
	Min	Max	Min	Max	Max	
Hand reach envelope, based on 5th and 95th percentile values	318	485	186	410	±528	
Leg reach envelope, based on 5th and 95th percentile values	410	493	261	396	±90	±215

Note the limiting values of coordinates given here are for the location of the centre of controls

20 to 30 Hz, with the amplitude of head vibration exceeding shoulders about three times. The eyeball resonance may be experienced at 60 to 90 Hz.

In the case of power tillers, vibration arises due to engine as well as the operating tool, i.e. rotary tiller or other equipment. Here the vibrations are in the frequency range of 20–250 Hz during various field operations (Dewangan and Tewari 2009), and are most likely to cause cumulative upper arm discomfort and disorders. In the case of a power-operated knapsack sprayer, the vibrations mainly arise from the engine.

Table 11.9 Actuating force limits (N) for controls on tractors/other farm machines

Control	Type of control	Force applied	Maximum operating force limit					
			IS 10703 (BIS 1992/2012)	ISO 15077 (ISO 2008)	Zander (1972)	Hansson (1991)	Mehta et al. (2011)	Gite (2017)
Clutch	Pedal	Push	350	450	200	150	125	115
Brakes	Pedal	Push	600	450	200	150	260	251
Parking brake	Hand lever	Pull	400	100	100	80	60	46
PTO coupling	Hand lever	Pull	200	100	100	80	60	46
Steering	Wheel	Torque	250	–	50	50	51	52
Hydraulic controls position	Hand lever	Pull	70	20	20	24	25	24 (push) 27 (pull)
Hand accelerator	Hand lever	Pull	–	100	100	40	25	32
Foot accelerator	Pedal	Push	–	90	–	50	50	47
Gear selection lever	Hand lever	Pull & Push	–	100	100	40	46	46
Speed selection lever	Hand lever	Pull & Push	–	100	100	40	46	46
Differential lock	Pedal	Push	600	450	200	150	130	251

The amplitude of vibration reaches peak between 50 and 130 Hz, with engine speeds in the range of 3000–8000 rpm.

For hand-arm vibrations, the ISO 5349-1 (ISO 2016) may be referred. The human body is most sensitive to frequency ranges of 4–8 Hz for vertical vibration and below 2 Hz for horizontal (a_x and a_y) vibration. The reduction in the transmission of vibration through the seat and steering/handles may be partly effective through the provisions of suspension, isolators and cushions. The other way to protect the operator from debilitating vibrations is to optimize the exposure time.

The safe exposure limits for whole-body vibration as per ISO 2631/1 (ISO 2017) and hand-arm vibration as per ISO 5349-1 (ISO 2016) are given in Table 11.10. The ISO 2631/1 specifies safe limits for human exposure to vibration in three different

Table 11.10 Exposure limits for whole-body and hand-arm vibration

Daily exposure time (h)	Whole-body vibration Acceleration, m/s^2 (r.m.s)		Hand-arm vibration Acceleration, m/s^2 (r.m.s)
	Vertical (a_z)	Horizontal (a_x, a_y)	x-, y-, and z-axes
8	0.630	0.448	2.1
4	1.060	0.710	2.9

ways, namely, (a) the preservation of health and safety (exposure limit), (b) the preservation of working efficiency (fatigue decreased proficiency boundary), and (c) the preservation of comfort (reduced comfort boundary). In the case of whole-body vibration, the fatigue decreased proficiency limit and reduced comfort boundary limits are obtained by dividing these values by 2.0 and 6.3, respectively. As per IS 12207 (BIS 1999), the mechanical vibration (amplitude of vibration) at footrest (left and right), steering wheel, and seat (with the driver seated) of a tractor should not exceed 100 μm.

Noise

The frequency range of hearing for a healthy young person extends from 20 to 20,000 Hz. The audible sound pressure levels range from the threshold of hearing at 0 dB(A) to the limit of pain, which can be more than 130 dB(A). The most critical effect of noise on the workers is the temporary and permanent damage to the hearing organ. Chronic exposure to noise leads to hearing loss depending on the source intensity, frequency content, and time duration.

The farmworkers, especially tractor drivers, suffer from hearing loss due to tractor noise (Kumar et al. 2005). Similarly, noise levels in threshers, power tillers, and other machines may exceed the safe exposure limit. Therefore, farmworkers experience one of the highest rates of hearing loss among all occupations. The exposure to high noise levels can be reduced either by modifying the machines or by changing the operator's workplace. International agencies, namely, OSHA (2013), ISO (2013/2018), specify 90 dB(A) as the permissible level for an 8 h/day exposure (40 h/week). NIOSH (2018) prescribes 90 dB(A) as the permissible exposure limits but suggests 85 dB(A) as the recommended exposure limit. When the workers are exposed to noise exceeding 90 dB(A), feasible administrative or engineering controls are the available options to mitigate exposure levels. Besides, the use of personal protective equipment (e.g. earmuffs, earplugs) partly contributes to protecting the operators from the noise.

Extreme Weather Conditions

Most of the agricultural operations are open fields under sun, rain, or cold. During these operations, the workers often subject to extreme heat in the summer season and severe weather in the winter season. During the rainy season, workers get wholly drenched in doing operations, like transplanting. For tractors without cabs, the tractor drivers encounter extreme weather during farming operations. When the weather extremes are beyond the tolerance of the human body, the situation may lead to errors and accidents. Wherever feasible, the work schedules may be modified depending on the weather situation.

Humans dynamically respond to as one's sense of warmth and comfort against the climatic and behavioural conditions of the environment (Pantavou et al. 2011). In essence, humans subject to the influence of both intrinsic and extrinsic factors in expressing one's thermal warmth and comfort. Therefore, thermal comfort is often viewed as the condition of mind about ones' satisfaction with the thermal environment, which is assessed mainly by subjective evaluation (ANSI/ASHRAE 2010/2013). Six factors that determine comfort are ambient air temperature, humidity, radiation, air movement, intrinsic clothing, and level of activity. The zone of thermal comfort is defined to the conditions where 4/5th of sedentary or slightly active persons find the environment thermally acceptable. The acceptable ambient temperature of comfort would be somewhat higher in the summer than in the winter, being 23 to 27 °C and 20 to 25 °C, respectively (humidity ranged between 30 and 60%). In dry tropics, the upper limit of ambient temperature for the comfort zone can be 30 °C. An air velocity of 0.2 to 0.5 m/s is considered optimum for a comfortable environment.

Based on a cross-sectional study on a sample 1144 farmworkers in eastern India, Sen and Nag (2019a) explored the human susceptibility to the tropical hot outdoor environment during paddy and potato cropping activities. The study evaluated the efficacy of human thermal indices, such as WBGT, HI, Humidex, UTCI, PET, SET*, PMV, in human susceptibility categorization. Sen and Nag (2019a, b) emphasized the utility of the E_{sk} (evaporative heat exchange through skin surface) as a useful thermoregulatory parameter to assess the thermal strain of farmworkers. The defined level of E_{sk} at ~ 200 W/m^2 corresponded to the comfortable temperature range within 19.5 to 22.5 °C for WBGT, PET, and T_a. Beyond this specific range of warmth, a proportionate increase in E_{sk} would result in cumulative heat-related symptoms of stress and strain. Three principal components of heat-related responses, such as (a) physical fatigue and responses, (b) neural stressors, and (c) behavioural effects yielded to a generalized quantitative climatic susceptibility indicator.

Lighting

Sufficient light is necessary for a better and safe performance by the workers. The illumination level in sunlight is 107527 lx, whereas during full daylight, it is 10752 lx (NOAO 2020). For agricultural work in the open space during day time, there is always sufficient light in regular working hours. However, in closed areas or while working during the night, the view is often less, bearing the risk of accidents. Recommendations for proper light intensity have been specified by various organizations (OSHA 2011; NOAO 2020; BIS 2015). It is suggested to provide light intensity at work areas in the range of about 100 to 300 lx, depending on work requirement. For more delicate work, the intensity requirement can be as high as 500 to 1000 lx.

Dust, Exhaust Emissions, and Agrochemicals

During various agricultural operations, the workers face problems of exposure to dust originating from crop residues or soil dust. In land development and tillage operations, the machinery operators often suffer from respiratory illness due to exposure to dust. The dust also irritates the eyes and the skin. The problem of dust is very much excruciating in post-harvest activities, such as cleaning, milling, and grinding, ginning, where the dust concentration may be as high as 500 mg/m^3 of air. In acute cases where the persons are exposed to dust from mouldy materials, they may suffer from pulmonary diseases. The level of dust in the air is determined by its mass content and the size of dust particles. From the occupational health perspectives, the dust is classified into total dust, inhalable dust, and respirable dust (particle size lesser than 5 μm). As per the US OSHA (2019), the limit values of exposure to dust concentrations are 10 mg/m^3 and 5 mg/m^3 for total dust and respirable dust, respectively. When the dust in the work environment exceeds the limits, corrective dust prevention measures for workplace control and personal protective measures are necessary to safeguard the health of workers.

The critical pollutants in the exhaust emissions of internal combustion engines are CO, NO_x, SO_2, and unburnt hydrocarbons. The CO reacts with haemoglobin in the bloodstream to form carboxy-haemoglobin and may cause headaches, nausea, weakness, and dizziness. Nitric oxide (NO) also has a high affinity for haemoglobin in the blood, producing an O_2 deficiency, which may cause fatigue. The nitrogen oxide may produce nitric acid in the lungs causing acute irritation and damage to tissues. As smoke emanates from the engine exhaust, the unburnt hydrocarbons are responsible for irritation of the eyes and lungs. The permissible 8-h exposure limits in the cab environment, as given by OSHA (2019) are 25 ppm for CO and one ppm for NO_2.

While operating the plant protection equipment, the workers are exposed to pesticides and other chemicals. The workers who do not use any protective wears are at a high risk of contamination. These chemicals enter into the human body through inhalation, ingestion, as well as dermal routes. The dermal pathway predominates in agriculture. The permissible toxicity depends upon the degree and duration of exposure and is also affected by pesticide formulation, concentration, method of application, operating conditions (*refer to Chap. 10 on pesticide toxicity*). The mixing of chemicals with water and spraying the mixture in the field are significant points of contamination. Irrespective of how a pesticide enters into the body, acute poisoning may occur after one exposure, and repeated exposures can cause chronic poisoning. Contamination of the human body is assessed through visible symptoms, like skin irritation, eye irritation, headache, difficulty in breathing, vomiting, and dizziness. The objective data on contamination and toxicity can be obtained through the analysis of urine and blood samples. All spray operators should be provided with safety kits to minimize the health hazards associated with the agrochemicals. A typical spraying kit includes a goggle, a pair of gloves, body overalls, a mask or respirator.

Use of Ergo-Design Criteria

The design of farm tools and equipment has often been viewed in different contexts. For an engineer, the design starts from the power and capacity of the tool/equipment to be fabricated. The ergonomist would like to know the user of the tool/equipment and how it will be operated, i.e. by muscle power or by a power source. The users can be both men and women workers. The user perspectives as regards requirements, aesthetics, and preference of use are central to the success of any ergo-design intervention. Accordingly, one must determine the capabilities and limitations of the user population, i.e. anthropometric, muscle strength, and physiological data. This information becomes essential in deciding the functional parameters of the tool/equipment. Therefore, once the researcher/engineer decides the type of tool/equipment to be designed, for example, a hand tool or a manually operated equipment or power-operated equipment, one has to identify the user population. To work out the details on operational performance and engineering aspects of the tool/equipment, the researchers may consult the reference sources of agricultural machinery design (Liljedahl et al. 1989; Kepner et al. 2017; Ojha and Michael 2008) 2008 as well as mechanical engineering handbooks. The criteria/information given above and the examples embodied in Chaps. 12 and 13 may be helpful to all concerned in their design work from ergonomics perspectives.

References

ANSI/ASHRAE. (2010/2013). Thermal Environmental Conditions for Human Occupancy - ASHRAE 55-2013. American Society of Heating, Refrigerating and Air-Conditioning Engineers, Inc. (www.ashrae.org) p. 37.

Babbs, F. W. (1979). A design layout method for relating seating to the occupant and vehicle. *Ergonomics, 22*(2), 227–234.

BIS. (1992/2012). IS 10703. Agricultural tractors – Maximum actuating forces required to operate controls. Bureau of Indian Standards, New Delhi.

BIS. (1994). IS 11806. Earth-moving machinery, and tractors and machinery for agriculture and forestry – seat reference point. (Now Withdrawn). Bureau of Indian Standards, New Delhi.

BIS. (1998). IS 12343. Agricultural tractors -Operators seat - Technical requirements. Bureau of Indian Standards, New Delhi.

BIS. (1999/2004). IS 11113. Earth-moving machinery, and tractors and machinery for agriculture and forestry – seat index point. Bureau of Indian Standards, New Delhi.

BIS. (1999). IS 12207. Recommendations on selected performance characteristics of agricultural tractors. Bureau of Indian Standards, New Delhi.

BIS. (2015). IS 6665- Code of practice for industrial lighting. Bureau of Indian Standards, New Delhi.

Chaffin, D. B. (2002). On simulating human reach motions for ergonomics analyses. *Human Factors and Ergonomics in Manufacturing & Service Industries, 12*(3), 235–247.

Courtney, A. J., & Evans, W. A. (1987). A preliminary investigation of bus cab design for Cantonese drivers. *Journal of Human Ergology, 16*(2), 163–171.

DARCOR. (2001). *The ergonomics of manual material handling- Pushing and pulling tasks.* DARCOR limited Toronto Canada.

Davis, P. R., & Stubbs, D. A. (1977). Safe levels of manual forces for young males (2). *Applied Ergonomics, 8*(4), 219–228.

Dewangan, K. N., & Tewari, V. K. (2009). Characteristics of hand-transmitted vibration of a hand tractor used in three operational modes. *International Journal of Industrial Ergonomics, 39*(1), 239–245.

Dupuis, H. (1959). Effects of tractor operation on human stresses. *Agricultural Engineering, 40*(9), 510–525.

Eltawil, M. A., & Hegazy, R. A. (2011). Ergonomic Studies on Controls Layout, Drivers Anthropometric and Noise Exposure for Egyptian Farm Tractors. *Misr Journal of Agricultural Engineering, 28*(4), 898–916.

Fathallah, F. A., Chang, J. H., Berg, R. L., Pickett, W., & Marlenga, B. (2008). Forces required to operate controls on farm tractors: Implications for young operators. *Ergonomics, 51*(7), 1096–1108.

Freivalds, A. (1987). The ergonomics of tools. *International Review of Ergonomics, 1,* 43–75.

Gellerstedt, S., Erikson, G., Frisk, S., Hultåker, O., Synwoldt, U., Tobish, R., & Weise, G. (2006). European Ergonomic and Safety Guidelines for Forest Machines 2006. *A handbook produced by Ergowood, a project co-financed by the European Commission.* European Commission contract No. QLK5-CT-2002-01190.

Gite, L. P., & Yadav, B. G. (1990). Optimum handle height for a push-pull type manually-operated dryland weeder. *Ergonomics, 33*(12), 1487–1494.

Gite, L. P. (2017). Development of Ergonomical Design Guidelines for Agricultural Tools, Equipment and Work Places, Final Report of the Emeritus Scientist Project, Central Institute of Agricultural Engineering, Bhopal, India.

Göbel, M., & Luczak, H. (2000). *Ergonomic design of urban bus driver workplaces. Advances in Ergonomics, Occupational Health and Safety.* New Age International (P) Limited, New Delhi, 67–73.

Grandjean, E. (1982). *Fitting the task to the man.* Taylor and Francis London.

Hansson, J. E. (1991). *An ergonomic checklist for tractors and agricultural machinery.* National Swedish Institute of Occupational Health (Arbetsmiljöinstitutet).

Hansson, J. E., & Pettersson, B. (1980). Ergonomic checklist for transport and materials handling machinery. Arbetsmedicinska Institute (The National Institute of Occupational Health), Byggnadsindustrins Arbetsforskningsstiftelse (The Building Industry Labour Research Foundation), Forskningsstiftelsen Skogsarbeten (The Logging Research Foundation).

Health Council of the Netherlands. (2012). *Pushing, pulling and applying force in work situations.* Publication no. 2012/37E. The Hague: Health Council of the Netherlands.

Hignett, S., & McAtamney, L. (2000). Rapid entire body assessment (REBA). *Applied Ergonomics, 31*(2), 201–205.

Hunt, S. (2017). *A guide to manual handling and lifting techniques.* U.K.: Work Safety Advice Cheshire.

IITKGP. (2016). Progress report 2015-16. All India coordinated research project on ergonomics and safety in agriculture. Indian Institute of Technology, Kharagpur, India.

ISO. (1995). ISO 5353. Earth-moving machinery, and tractors and machinery for agriculture and forestry—Seat index point. International Organization for Standardization, Geneva.

ISO. (1996/2017). ISO 7250-1. Basic human body measurements for technological design—Part 1: Body measurement definitions and landmarks. International Organization for Standardization, Geneva.

ISO. (2007/2016). ISO 11228- Part 2. Ergonomics -Manual handling - Part 2: Pushing and pulling. International Organization for Standardization, Geneva.

ISO. (2008). ISO 15077. Tractors and self-propelled machinery for agriculture- operator's controls-actuating forces, displacement, location and method of operation. International Organization for Standardization, Geneva.

ISO. (2010). ISO 7250-Part 2. Basic Human Body Measurements for Technological Design - Statistical summaries of body measurements from national populations. International Standard Organization, Genava.

ISO. (2013/2018). ISO 1999. Acoustics—Determination of occupational noise exposure and estimation of noise-induced hearing impairment. International Organization for Standardization, Geneva.

ISO. (2014). ISO 11228- Part 1. Ergonomics - Manual handling - Part 1: Lifting and carrying. International Organization for Standardization, Geneva.

ISO. (2015). ISO 7250-Part 3. Basic Human Body Measurements for Technological Design - World-wide and regional design ranges for use in product standards. International Standard Organization, Genava.

ISO. (2016). ISO 5349-1.Mechanical vibration—Measurement and evaluation of human exposure to hand-transmitted vibration—Part 1: General requirements. International Organization for Standardization, Geneva.

ISO. (2017). ISO 2631-1. Mechanical vibration and shock—Evaluation of human exposure to whole-body vibration—Part 1: General requirements. International Organization for Standardization, Geneva.

Kee, D., & Karwowski, W. (2007). A comparison of three observational techniques for assessing postural loads in industry. *International Journal of Occupational Safety and Ergonomics, 13*(1), 3–14.

Kepner, R. A., Bainer, R., & Barger, E. L. (2017). *Principles of farm machinery*. New Delhi: CBS Publishers and Distributors.

Khadatkar, A., Mehta, C. R., & Gite, L. P. (2017). Development of reach envelopes for optimum location of tractor controls based on central Indian male agricultural workers. *Agricultural Engineering Today, 41*(2), 34–39.

Kharade, H. S., Deshmukh, M., Tiwari, P. S. & Gite, L.P. (2010). Optimum crankshaft height and crank length for hand-operated rotary maize seller. *PKV Research Journal, 34*(2), 177–180.

Kumar, A., Bhaskar, G., & Singh, J. K. (2009). Assessment of controls layout of Indian tractors. *Applied Ergonomics, 40*(1), 91–102.

Kumar, A., Mathur, N. N., Varghese, M., Mohan, D., Singh, J. K., & Mahajan, P. (2005). Effect of tractor driving on hearing loss in farmers in India. *American Journal of Industrial Medicine, 47*(4), 341–348.

Liljedahl, J. B., Turnquist, P. K., Smith, D. W., & Hoki, M. (1989). *Tractors and their power units* (4th ed.). New York: Van Nostrand Reinhold.

Louhevaara, V., Suurnäkki, T., Hinkkanen, S., & Helminen, P. (1992). *OWAS: A method for the evaluation of postural load during work*. Helsinki: Finish Institute of Occupational Health.

Maiti, R., & Ray, G. G. (2004). Manual lifting load limit equation for adult Indian women workers based on physiological criteria. *Ergonomics, 47*(1), 59–74.

Matthews, J., & Knight, A. A. (1971). *Ergonomics in agricultural equipment design*. Silsoe, Bedford, UK: National Institute of Agricultural Engineering Wrest Park.

McAtamney, L., & Corlett, E. N. (1993). RULA: A survey method for the investigation of work-related upper limb disorders. *Applied Ergonomics, 24*(2), 91–99.

Mehta, C. R., Gite, L. P., Pharade, S. C., Majumder, J., & Pandey, M. M. (2008). Review of anthropometric considerations for tractor seat design. *International Journal of Industrial Ergonomics, 38*(5–6), 546–554.

Mehta, C. R., Pandey, M. M., Tiwari, P. S., Gite, L. P., & Khadatkar, A. (2011). Tractor controls actuating force limits for Indian operators. *Industrial Health, 49*(4), 523–533.

Morrison, C. S., & Harrington, R. E. (1962). Tractor seating for operator comfort. *Agricultural Engineering, 43*(11), 632–635.

MPUAT. (2014). Biennial report 2012-14. All India coordinated research project on ergonomics and safety in agriculture. Maharana Pratap University of Agriculture and Technology, Udaipur, India.

Nag, P. K. (1991). Endurance limits in different modes of load holding. *Applied Ergonomics, 22*(3), 185–188.
Nag, A., Nag, P. K., & Desai, H. (2003). Hand anthropometry of Indian women. *Indian Journal of Medical Research, 117,* 260–269.
Nag, P. K., Sebastian, N. C., & Mavlankar, M. G. (1980). Occupational workload of Indian agricultural workers. *Ergonomics, 23*(2), 91–102.
NASA. (1978a). *Anthropometric Source Book*, vol. I- Anthropometry for Designers, NASA Reference Publication 1024. National Aeronautics and Space Administration, Washington.
NASA. (1978b). *Anthropometric Source Book*, vol. II – A Handbook for Anthropometric Data, NASA Reference Publication 1024. National Aeronautics and Space Administration, Washington.
Nemeth, S. E. (1985). Hand tool design. In D. C. Alexander & B. M. Pulat (Eds.), *Industrial ergonomics: A Practitioner's Guide* (p. 35). Atlanta, Ga: Industrial Engineering and Management Press.
NIOSH. (2007). *Ergonomics guidelines for manual material handling*. Publication No. 2007-131. Cincinnati, USA: National Institute for Occupational Safety and Health.
NIOSH. (2018). *Noise and hearing loss prevention - Guidance and Regulations*. https://www.cdc.gov/niosh/topics/noise/reducenoiseexposure/regsguidance.html. National Institute for Occupational Safety and Health, Page last reviewed: February 6, 2018.
NOAO. (2020). Recommended Light Levels (Illuminance) for Outdoor and Indoor Venues. https://www.noao.edu/education/Nqltkit/Activity_Documents/Safety. Tucson, USA: The National Optical Astronomy Observatory.
Ojha, T. P., & Michael, A. M. (2008). *Principles of agricultural engineering* (Vol. 1). New Delhi: Jain Brothers.
OSHA. (2011). Safety and health regulations for construction, Illumination. CFR 1926.56. https://ecfr.io/Title-29/pt29.8.1926. United States Department of Labour, Washington DC., U.S.A: Occupational Safety and Health Administration.
OSHA. (2013). OSHA Technical Manual (OTM), Section III: Chapter 5 - Noise. OSHA Directive TED 01-00-015. Occupational Safety and Health Administration. United States Department Of Labour, Washington D,C., U.S.A.
OSHA. (2019). Permissible Exposure Limits – Annotated Tables. https://www.osha.gov/dsg/annotated-pels/index.html. Occupational Safety and Health Adminastration. United States Department Of Labour, Washington DC., U.S.A.
Pantavou, K., Theoharatos, G., Mavrakis, A., & Santamouris, M. (2011). Evaluating thermal comfort conditions and health responses during an extremely hot summer in Athens. *Building and Environment, 46*(2), 339–344.
PAU. (2016). Progress report 2015-16. All India coordinated research project on ergonomics and safety in agriculture. Punjab Agricultural University, Ludhiana, India.
Pheasant, S. (2003). *Bodyspace: Anthropometry, ergonomics and the design of work* (2nd ed.). London: Taylor & Francis.
Potdar, R. R., Mehta, C. R., Gite, L. P., Narwariya, B. S., Gaikwad, B. B., & Agrawal, K. N. (2017). *Hand and Leg Reach Envelopes of Indian Operators for Locating Tractor Controls*. Bhopal, India: Central Institute of Agricultural Engineering.
Sen, J., & Nag, P. K. (2019a). Human susceptibility to outdoor hot environment. *Science of the Total Environment, 649,* 866–875.
Sen, J., & Nag, P. K. (2019b). Effectiveness of human-thermal indices: Spatio–temporal trend of human warmth in tropical India. *Urban Climate, 27,* 351–371.
Snook, S. H. (1978). The ergonomics society the society's lecture. (1978). The design of manual handling tasks. *Ergonomics, 21*(12), 963–985.
Snook, S. H., & Ciriello, V. M. (1991). The design of manual handling tasks: revised tables of maximum acceptable weights and forces. *Ergonomics, 34*(9), 1197–1213.
Tilley, A. R. (2002). The measure of man and woman: human factors in design. John Wiley & Sons.

TNAU. (2014). Progress report 2012-14. All India coordinated research project on ergonomics and safety in agriculture. No. AEC&RI/Pub.14/2014. Tamil Nadu Agricultural University, Coimbatore, India.

Troy, J., & Guerin, J. (2004). Human swept volumes (No. 2004-01-2190). SAE Technical Paper. *SAE Transactions, 113*(1), 249–253.

Yang, J., & Abdel-Malek, K. (2009). Human reach envelope and zone differentiation for ergonomic design. *Human Factors and Ergonomics in Manufacturing & Service Industries, 19*(1), 15–34.

Yang, J., Abdel-Malek, K., & Nebel, K. (2005). On the determination of driver reach and barriers. *International Journal of Vehicle Design, 37*(4), 253–273.

Zander, J. (1972). *Ergonomics in machine design: a case-study of the self-propelled combine harvester*. Wageningen, Netherlands: H. Veeman and Zonen.

Chapter 12
Ergonomics Application in Design of Farm Tools and Equipment

Introduction

Ergonomics considerations in the design of farm tools and equipment are essential for more productivity, reduction in drudgery, and better health and safety of farmworkers. The engineers, scientists, and machinery designers attempt to use anthropometric and muscle strength data of the user population and apply suitable biomechanics principles and other ergo-design guidelines to arrive at appropriate design specifications of tools and equipment. Some examples are embodied herewith to show the design process from an ergonomics perspective. Examples and descriptions added herewith are primarily based on the data of the Indian population, as given in Chap. 6. The readers may use ones' country-specific anthropometric data for the purpose. One can add other socio-dynamic issues, depending on the requirements of the user population of tool and equipment. The analysis given here includes tools and equipment for crop production and processing operations for the field as well as horticulture and plantation crops, as mentioned below:

- Transplanting—Manually operated rice transplanter;
- Fertilizer application—Manually operated fertilizer broadcaster;
- Weeding—Long-handled hand tools, grubber weeder, manually operated wheel hoe and Cono weeder;
- Spraying—Lever-operated and power-operated knapsack sprayers;
- Harvesting—Sickle hand tool;
- Dehusking and shelling—Maize sheller hand tool, pedal-operated maize dehusker sheller;
- Threshing—Manually operated and power-operated threshers;
- Winnowing—Manually operated paddy winnower;
- Decortication—Manually operated groundnut decorticators;
- Tree climbing—coconut tree climber;
- Pruning—Pruning secateurs; and
- Chaff cutting—Manually operated and power operated chaff cutters.

P. K. Nag and L. P. Gite, *Human-Centered Agriculture*, Design Science and Innovation,
https://doi.org/10.1007/978-981-15-7269-2_12

Farm activities require a whole range of tools and equipment, which is out of bound to bring in the discussion in the present contribution. The tools and equipment covered herewith are examples of different modes of operation, e.g. pulling, pushing, cranking, pedalling, treadling, and gripping. Readers can make use of the examples for the design application of farm equipment and machinery involving similar modes of operation.

Transplanting—Three Row Rice Transplanter

Fig. 12.1 **a** Manual rice transplanting in the puddled field; **b** three-row rice transplanter

Transplanting: Manually operated three-row rice transplanter (Fig. 12.1)	
Description	The machine is used for transplanting about 3 weeks old mat type rice seedlings (at three leaves stage) in three rows simultaneously in the puddled field. The components of the transplanter include frame, floats, seedling tray, operating handle, fingers (pickers), tray drive unit, and depth control mechanism. The worker puts the equipment at one end of the puddled field, then pushes the handle to collect the seedling from the tray and to put them into the soil. After that, the worker pulls the handle upwards and then moves backwards to plant seedlings at another place. The output of a person using this equipment is generally 170 m^2/h, as against 60 m^2/h in traditional manual transplanting in bent posture (CIAE 2014)
User	Female workers
Mode of operation	Push and Pull
Design process	
Weight	The rice transplanter is portable equipment, and therefore, the weight of the machine should be less than 15 kg (40% of the 10th percentile body weight of female workers)

(continued)

(continued)

Transplanting: Manually operated three-row rice transplanter (Fig. 12.1)	
Height	Female workers operate the equipment in standing posture. One has to push the handle downwards and forwards to transplant the seedlings. The handle height during work (the handle is about 30° to vertical towards the operator) should be more than 95th percentile metacarpal III height and less than the 5th percentile elbow height of female workers to avoid discomfort during force application in a bent posture. Thus, the handle height should be within the range of 729 mm–890 mm, respectively. Considering 30° angle of the handle with vertical during work, the handle height (handle in vertical position) should be within 841 mm to 1028 mm
Handle crossbar length	The desirable position of holding the handle crossbar should be in the line of the arms. The minimum length of the handle crossbar should be more than 250 mm (2 times handbreadth across thumb plus 30 mm clearance), considering the working position and placement of the worker's hand. The elbow-elbow breadth may be considered for the handle crossbar length. The 5th percentile elbow-elbow breadth of females (280 mm) is the recommended length if the handle crossbar
Handle grip (cylindrical)	Since the 95th percentile value of middle finger, palm grip diameter is the lower limit, and the 5th percentile grip diameter (inside) is the upper limit. These values for female workers are 32 and 35 mm, respectively. Therefore, the handle grip diameter recommended is 32 mm
Force limits	The operation requires pushing the handle to collect the seedling from the tray and then putting them into the soil. After that, the worker pulls the handle upwards and then moves backwards to plant seedlings at another place. Thus, the operation requires a push and pull force with both hands in standing posture. As mentioned in Chap. 6, the 5th percentile value of push and pull strength with both sides are 84 and 101 N, respectively, for female workers. Taking a 30% limit for continuous operation, the force required in the process of transplanting should not exceed 26 N in pushing and 31 N in pulling mode. The power needed in pulling the transplanter with seedling mats placed on it is about 150 N. The power requirement reduces as the seedlings get reduced during transplanting. Thus, the workers may have to take frequent rest pauses during work

Fertilizer Application—Broadcaster

Fig. 12.2 Manual fertilizer broadcaster

Fertilizer application: Manually operated fertilizer broadcaster (Fig. 12.2)	
Description	Fertilizer broadcaster is used for uniform application of granular fertilizer in the field. It consists of a hopper with agitator, spreading disk, gear, cranking handle, cushion pad, and shoulder straps. The broadcaster requires cross-mounting, as it is a belly mounted device. The quantity of fertilizer in the hopper can be observed through a transparent lid. During the operation, the worker walks forwards and rotates the handle to broadcast the fertilizer. The output of a person using this equipment is about 1.15 ha/h (Singh et al. 2007)
User	Female workers
Mode of operation	Hand cranking
Design process	
Total weight and hopper capacity	The equipment is operated by hand cranking while walking. The total weight of the equipment (filled with fertilizer) should not exceed 40% of the bodyweight for comfortable carrying. Therefore, for women to operate the machine, the total weight should not exceed 15 kg (37.5 kg as the 10th percentile body weight). As the empty weight of the unit may be about 5 kg, the hopper capacity would be about 10 kg of fertilizer

(continued)

(continued)

Fertilizer application: Manually operated fertilizer broadcaster (Fig. 12.2)	
The total height of the broadcaster	The height of the broadcaster should be such that it does not touch the chin of the operator during work. There should be adequate clearance between the chin and the hopper lid. When the broadcaster is belly mounted, its lower portion is generally in line with the iliocristale point, i.e. the waist of the operator. The height of the broadcaster is decided based on 5th percentile iliocristale height (820 mm) and 5th percentile acromial height (1173 mm), which is about 40 to 50 mm less than the chin height. Therefore, the total height of broadcaster should not exceed 353 mm (i.e. the difference between the acromial height and iliocristale height of the 5th percentile operator)
Crank length	Ideally, the upper level of the crank should not come above the elbow level to have better comfort. Thus, taking the 5th percentile value of metacarpal III height (which is the lowermost handgrip height in standing posture) as 587 mm and the elbow height as 890 mm, the total diameter of the crank works out to be 303 mm $\approx$ 300 mm. Therefore, the crank length can be taken as 150 mm for comfortable operation
Handle grip diameter (cylindrical)	The diameter of the grip is such that while holding the grip, the operator's most extended finger should not touch the palm. Also, the grip should not exceed the internal grip diameter. The 95th percentile range of middle finger palm grip diameter is the lower limit, and the 5th percentile grip diameter (inside) of the female workers is considered as the upper limit. Therefore, the handle grip diameter recommended is 32 to 35 mm. The handle grip should be free to rotate on its axis for convenient cranking
Force limit for cranking	For continuous operation, the force limit can be taken as 30% of the 5th percentile value of cranking torque in standing posture, which is 21 Nm. The torque required should not exceed 6.3 Nm (i.e. 0.3 $\times$ 21 Nm). Taking the crank length as 0.15 m, the force required in cranking may be about 42 N

Weeding—Hand Tools and Weeders

Weeding – Hand Tools and Weeders: Hand tools (e.g. Dutch hoe, Grubber weeder)	
Weeding is a labour intensive work in farming. Manual weeding involves different types of vintage tools, such as the long-handled Dutch hoe and Grubber weeder. The Dutch hoe, with the steel blade of 100 to 150 mm wide, cuts weeds just under the soil surface. The operator generally pushes forwards or pulls the tool back to break the soil crust and uproot the weeds. Grubber weeder consists of a ferrule, three tynes and sweep type blades of 100–200 mm wide. In the case of Grubber weeder, the operator pulls the weeder to break the soil crust and uproot the weeds. The tools weigh about 1.5 to 2 kg. The output of such a device is about 50 m2/h (Pandey et al. 1997)	
Handle diameter and length	The desirable shape of the tool handle is cylindrical or slightly oval. The 95th percentile value of the middle finger palm grip diameter of females is 32 mm, and the recommended handle grip diameter is 32 mm. As a thumb rule, the total tool height with handle should not be less than the eye height of the worker in standing posture (Hopfen and Biesalski 1953). The 95th percentile value of eye height for male workers is 1646 mm. The angle of the handle during operation should preferably be less than 45°. During the farm operation, the upper limb and the tool are in one line. Taking the acromial height of the 95th percentile male as 1477 mm and the maximum permissible angle as 45°, the total length (upper limb plus tool handle plus tool) works out to be 2088 mm (1477/*sin* 45). Of this, the elbow grip length would be 405 mm. The length of the device plus tool handle would be 2088 minus 405 = 1683 mm (i.e. the minimum length of the device plus tool handle ≈ 1.70 m)
Push/pull force limit	Dutch hoe is operated by both male and female workers. Therefore, 30% of the 5th percentile push or pull force in the standing position may be taken as the desired force limit. The 5th percentile values of push and pull strength with both hands in a standing posture for males are 133 and 140 N, respectively. The corresponding 30% of these values are 40 and 42 N. In the case of female workers, the 5th percentile values of push and pull strength are 84 and 101 N, respectively. The corresponding 30% limits are 26 and 31 N. Therefore, the hoe should be such that the operator exerts less than 26 N push or 31 N pull. Whereas the Grubber weeder is pull-type equipment. Therefore, the operator should not exert a pull force more than 31 N. Wherever more power is needed, the operator may take rest breaks in between the work schedule depending on the severity of work

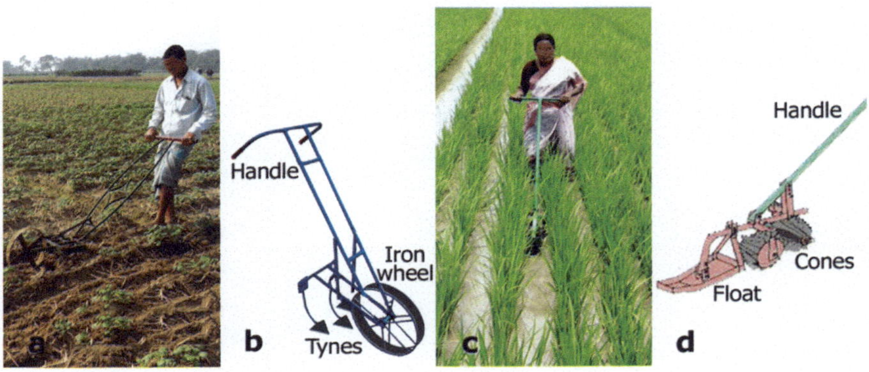

Fig. 12.3 **a**, **b** Wheel hoe weeder, and **c**, **d** Cono weeder

Weeding equipment: Wheel hoe (Fig. 12.3a, b) and Cono weeder (Fig. 12.3c, d)		
Description	The wheel hoe is used for weeding and interculture in upland row crops. It consists of one or two wheels, a frame, a V-blade fixed on a tyne, a U-clamp, a cross handlebar, and a long handle. A worker operates the tool by push–pull action. The weight of this device is about 4 to 6 kg. A person can do weeding work of about 150 m^2 area per hour (Singh et al. 2007)	Cono weeder is used for uprooting and burying of weeds in between standing rows of rice crops in wetlands. It consists of a handle, two truncated conical rollers with straight or serrated blades on the periphery. A worker walks in the field and operates the device in push–pull mode. The weight of a cono weeder is about 5 to 8 kg. A person can cover about 280 m^2/h by using the equipment (TNAU 2012)
User	Male and female workers.	
Mode of operation	Pushing and pulling	
Design process		
Posture	A good working posture is one that can be sustained with a minimum of static muscular effort and in which the given task can be performed more effectively. As such, weeding operation either in squatting or bending posture did not cause a marked difference in energy expenditure. Nevertheless, the drudgery caused due to bent posture gets reflected in terms of musculoskeletal discomfort of the worker. When work can be done in standing posture, it should not be done in a bent position. Use of long-handled tools such as wheel hoe and cono weeder can help to obviate the bent posture of the operators	
Handle height	The height of the handle should be optimized to give the least human fatigue in weeding operation. About 0.7 to 0.8 times of the acromial height is the suggested handle height (Gite and Yadav 1989). Since both men and women operate the equipment, about 0.8 times of the 5th percentile acromial height (women), and 0.7 times of the 95th percentile acromial height (men), may be considered for setting the handle height. The 5th percentile value of the acromial height of Indian female workers is 1173 mm, and the 95th percentile value of acromial height for male workers is 1477 mm. Therefore, the handle height may be adjustable between 938 and 1034 mm. For a fixed handle, the handle height may be kept at 1000 mm	
Handle length	The length of the handle of the tool and the angle of operation are interdependent. The angle of operation is based on functional design and geometry of the tool and generally lies between 30° and 45°. By determining the point of attachment, and the height of the handle, the length of the handle can be calculated	
Handle crossbar length	The weeding operation is a prolonged activity. Therefore, the desirable position of holding the handle crossbar should be in the line of the arms. The elbow-elbow breadth is to be considered for handle crossbar length. Since both men and women operate the equipment, the 95th percentile elbow-elbow breadth of male and female workers may be considered. The 95th percentile values of elbow-elbow breadth for female and male workers are 440 and 465 mm, respectively, which may be taken as the range of the handle crossbar length. For a fixed type of handle, a value of 450 mm can be taken	
Handle grip	The desirable shape of handle grip is cylindrical. The diameter of the grip should be such that while holding the grip, the operator's most extended finger should not touch the palm. At the same time, the grip should not exceed the internal grip diameter. The 95th percentile value of the middle finger palm grip diameter of male workers is the lower limit, and the 5th percentile grip diameter (inside) of the female workers is the upper limit. However, as the wheel hoe and cono weeder are mostly used by women, and their 95th percentile value of middle finger palm grip diameter is 32 mm, the handle grip diameter recommended is 32 mm	
Push and pull force limits	As per the Indian data, the 5th percentile values of push and pull force levels with both hands in a standing posture for female workers are 84 and 101 N, respectively. Therefore, the wheel hoe and cono weeder should be such that the operators do not have to exert more than 26 N push or 31 N pull (i.e. 30% of the 5th percentile push/pull force of female workers). If more power is needed, the operator may take rest breaks in between the work schedule, depending on the severity of work	

Spraying—Knapsack Sprayer

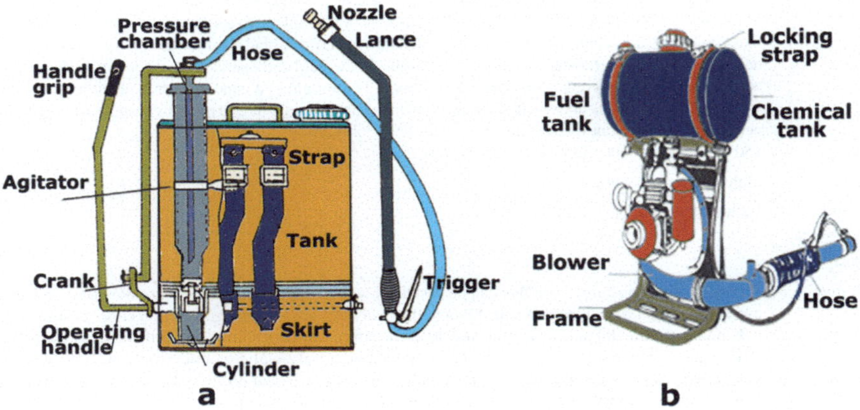

Fig. 12.4 **a** The schematics of lever-operated and **b** powered knapsack sprayers

Spraying: Manually and power operated knapsack sprayers (Fig. 12.4)	
Description	The illustration above (Fig. 12.4) shows the constructs of the engine-operated and manually operated knapsack sprayers. A lever-operated knapsack sprayer (Fig. 12.4a) is a hand-operated hydraulic energy type sprayer. It consists of a plastic or metallic tank of capacity 10 to 16 litres, a hand-operated pump, a pressure chamber, stand, hose, a lance with the trigger, nozzle, and two shoulder straps. The pressure chamber is generally fitted inside the tank. The operating lever may be under-arm or over-arm type. The operator walks at varying speeds (1.5 to 2.5 km/h) and operates the handle with one hand, and holding the lance with the other hand. The operators experience discomfort in prolonged handling of the sprayer. The average working heart rate of the spray operators is 124 beats/min at a walking speed of 2.5 km/h, and the output 0.1 ha/h (Gite 1996). A typical power-operated knapsack sprayer (Fig. 12.4b) consists of a petrol engine of 1 to 2 hp fitted with a blower, a chemical mixture tank, and a delivery hose fitted with air hose, flow regulator knob, and an atomizer. The blower generates a high-velocity air to which the chemical solution is fed, which is then broken into fine droplets at the atomizer. A worker carries out the spraying by guiding the hose on the crop plants
User	Male workers
Mode of operation	In the case of a manual sprayer, the force for handle movement is applied in elbow flexion and extension mode
Total weight and tank capacity	The total weight of the knapsack mounted equipment (filled with chemical and water mixture) should not exceed 40% of the bodyweight. That is, the weight should not exceed 18 kg (45 kg is the 10th percentile body weight), which seems appropriate. In the case of knapsack sprayer, the empty weight is about 5 kg, and the tank capacity maybe about 13 l. In the case of the power-operated unit, as the empty weight is about 9 kg, the tank capacity maybe about 9 l

(continued)

(continued)

Spraying: Manually and power operated knapsack sprayers (Fig. 12.4)	
Size of tank	For a 13-litre capacity sprayer, the total capacity of the sprayer tank maybe 14.2 l, taking 5% freeboard, and the space occupied by pump components inside the tank. For 9-litre sprayer capacity of power sprayer, the total tank capacity will be about 9.5 l, including 5% freeboard. For both types of sprayers, the total height of the sprayer should be such that the operator feels comfortable and minimum body discomfort during spraying. As a thumb rule, the height of the frame of sprayer should not exceed the 5th percentile waist back length, which is 380 mm. Regarding the centre of gravity (CG) location of the loaded unit in the vertical plane, it is recommended that the unit should be about 22.5 cm above the lumbosacral joint. The recommended tank size for a 13-litre capacity sprayer is 300 mm height, 340 mm width, and 140 mm depth. The sprayer tank needs a stand of 100 mm in height. About 20 mm height of the stand might cover-up the container, and thus, the total height of the sprayer may be 380 mm. The horizontal distance between the sprayer surface and handle pivot point depends on the constructional requirement, such as sprayer tank depth. For a 140 mm depth of the tank, it may be about 120 mm (Gite 1996)
Handle dimensions	In the case of the manual sprayer, longer the length of the handle more is the mechanical advantage and less the force required in handle operation. However, the 5th percentile operator should also be able to hold the handle at the extreme position, mainly when the handle is at the end of a downwards stroke. Therefore, for determining the handle length, the 5th percentile length of the acromion to finger-tip (arm reach from a wall—wall to acromion distance) is taken (i.e. 682 mm). The handle length comes to about 554 mm in the acromion to finger-tip length reach envelope. The handle length may be less by 44 mm (half of the palm length), taking into account the placement of the hand on the handle, and therefore, the effective handle length is 510 mm. Regarding handle grip, it should be tapered one, the diameter of the smallest and largest section being within 30 to 39 mm limit
Sprayer operating parameters	The manually operated sprayer operates at a working pressure of 300 kPa (3 kgf/cm^2) for pesticide applications. Typical nozzle discharge is about 0.450 l/min. The handle frequency ranges from 6 to 30 strokes/min depending on spray requirement and other parameters. Taking 20 strokes/min as the optimum handle frequency, the actual swept volume of pump per stroke arrives at 22.5 ml. Further taking a volumetric efficiency of 85%, the theoretical swept volume reaches 26.5 ml. With the likely pressure fluctuations of about 15% of the mean pressure during work (as suggested by FAO 1995), the minimum pressure chamber size should be 20 times that of the swept volume. Thus, the required pressure chamber volume will be 530 ml. Generally, a piston-type pump is used in knapsack sprayers. The diameter of the piston and the suitable length of stroke is selected as 34 and 30 mm, respectively. The pressure chamber, as well as pump (cylinder and piston), maybe inside the tank to avoid leakage. The pump may be mounted at the centre of the tank, and the handle may be fitted on the right or left side of the tank. In the power-operated unit, the blower delivers a minimum of 6 m^3 of air per min with a minimum air velocity of 60 m/s. The sprayer is generally capable of discharging 0.5 to 3.0 l/min
Strap size and position	The straps should be of woven cotton, having a width of 50 mm and thickness 3 mm. The strap length is based on the 95th percentile value of scapula to waist length (730 mm). With a margin, a strap length of 800 mm is recommended. A cushion of a minimum 50 mm width and 20 mm thickness may be provided with each strap, at least on the portion that rests on the shoulder of the operator. These cushions should be covered with a suitable material (e.g. canvas, rexine, and PVC or plastic-coated fabrics) to avoid absorption of liquid

(continued)

(continued)

Spraying: Manually and power operated knapsack sprayers (Fig. 12.4)	
Cut-off device	In manual sprayer, a trigger-type cut-off device is used. The 5th percentile handgrip strength of non-preferred hand may be considered (i.e. 195 N) for the maximum permissible force to operate the trigger. Since the trigger operates only occasionally, 50% of the force level (98 N) can be taken as the limit. The trigger may be provided with an auto-lock so that it remains pressed during spraying, with no strain on the arm muscles. In power sprayer, a lever-type cut-off device is commonly used. The 5th percentile push strength of the non-preferred hand is considered (47 N) to set the maximum permissible force for operating the lever. As the lever is operated occasionally, 50% of this force can be taken as the limit (24 N). The lever is provided with an automatic lock so that it remains pressed during spraying, and the hand is not taxed for the same
Length of lance and type of nozzle	The length of the lance is taken as 600 mm for convenient operation. For spraying, a hollow cone nozzle of A 60450 (i.e. agricultural purpose, cone angle 600 and discharge 450 ml/min) type is used. In power sprayer, a spraying hose is provided. The chemical mixed air stream is broken into fine droplets at the atomizer device fixed at the end of the hose for spraying
Force limit for handle operation	For deciding force limit in handle operation of the manual sprayer, strength values in elbow flexion and extension are to be used. The 5th percentile strength for non-preferred hand in elbow flexion is 127 N. The acceptable force at handle grip works out 43 N, by taking 40% force limit for prolonged work. The pressure chamber, as well as pump (cylinder and piston), maybe inside the tank to avoid leakage. The pump may be mounted at the centre of the tank, and the handle fitted on the right or left side of the tank. The recommended total handle movement as 46° (Gite 1996). From the analysis of muscular strength at different elbow angles, the suggested handle movement is 23° upwards and downwards
Vibrations	In the case of power operated knapsack sprayer, the vibrations mainly arise from the engine. With engine speeds between around 4500 and 5500 rpm, the vibrations may reach a peak at about 75 to 92 Hz. To reduce these vibrations, generally, the engine and pump are mounted on a frame with rubber mountings. Refer to Chap. 11, the vibrations to the back (horizontal direction) and shoulders (vertical direction) should not exceed 0.45 and 0.63 m/s^2, respectively. Also, the vibrations coming to the hand should be less than 2.1 m/s^2. Also, to reduce the transmission of vibrations from the sprayer to back, a cushion pad of size 200 × 200 mm and 20 mm thickness between the machine frame and the operator's back is provided. Also, cushions of 40 mm width and 20 mm thickness are given with the straps (as described elsewhere) that rests on the operator's shoulders
Safety	As the speed of the engine increases from 4000 rpm and above, the noise level at the operator's ear level may exceed 90 dB (A) (Dixit et al. 2006). Therefore, it may be necessary to reduce the work duration or to provide ear muffs to protect the operator from excessive noise exposure. During spraying, the worker is exposed to toxic agrochemicals. Therefore, the worker should use a spraying safety kit (face mask, eye protector, gloves and apron) (refer to Chap. 10 on spraying precautions)

Harvesting—Sickle

Fig. 12.5 **a** to **c** Harvesting hand tool (sickle) of different types

Harvesting: Hand tool (Improved Sickle) (Fig. 12.5)	
Description	There has been a large-scale introduction of mechanization in the harvesting of crops. Among the resource-poor small farmers, the use of traditional tools and implements are still very much relevant. The sickle is a simple and common harvesting tool, that is readily available, easy to use, and a low-cost device, to harvest crops, such as wheat, rice, soybean, chickpea (Nag et al. 1988). A sickle is a curved, serrated or non-serrated edge blade attached to a wooden handle. The tool with serrated blade cuts the crop stalk by sawing action, and the pulling motion is applied in the case of the non-serrated blade. The performance of serrated sickle was found to be better than the non-serrated type (Sutjana 2000). The potential demand for improved sickle is more than 2.27 million in India (Singh 2012). The great need for simple hand tool led to the development of different varieties of sickles in different regions and countries. A survey in different regions of India identified twelve distinctive sickle types (Mondal 2019). A more detailed description of the use of sickle is given elsewhere (*refer to* Chap. 2, Fig. 2.10). Both female and male workers perform repetitive work to cut crops, with a stroke rate of about 30 to 50 per min. A lighter sickle causes less discomfort in the wrist and hand for prolonged use, and preferably, the weight of sickle should be less than 200 g (Nag et al. 1988). The output of a person with this type of sickle is about 150 m2/h (Singh et al. 2007)
Design process	
Dimensions and weight	The field capacity of a sickle is dependent upon the geometry of the blade, and size, shape, and diameter of the handle. The weight of the sickle may vary from nearly 75 to about 270 g. The length of the sickle ranges from 300 to 400 mm, and the handle length varies with 120 to 130 mm

(continued)

(continued)

Harvesting: Hand tool (Improved Sickle) (Fig. 12.5)	
Handle diameter	The diameter of the handle should be such that while holding the sickle, the operator's most extended finger should not touch the palm. At the same time, the grip should not exceed the internal grip diameter. The desirable shape of the handle is cylindrical. Since both male and female workers operate the hand tool, the 95th percentile value of the middle finger palm grip diameter of male workers would be the lower limit, and the 5th percentile value of the grip diameter (inside) of the female workers may be considered as the upper limit. Accordingly, for the Indian population, these values are 38 and 35 mm, respectively. In certain geographic regions, the sickles are used mainly by women in harvesting work. The handle grip diameter of the sickle for women workers may be taken as 32 mm, this being their 95th percentile value of middle finger palm grip diameter
Handle length	The minimum handle length should be 95th percentile value of handbreadth across thumb for male workers, that is, about 110 mm. Considering that a person holds the handle at different positions, and the 15 mm margin for the ferrule, a length of 125 mm is recommended for a sickle handle
Push/pull force limit	As far as the muscle exertion in sickle operation is concerned, the 5th percentile pull force with the right hand in sitting posture of women is 42 N. The acceptable pulling force for sickle arrives at 13 N by applying the limit of 30%

Dehusking and Shelling—Maize Sheller

Fig. 12.6 Tubular maize sheller—torque application by hand

Dehusking and shelling: Tubular Maize Sheller (Fig. 12.6)	
Description	A tubular maize sheller is used for shelling grains from dehusked cobs. The device, octagonal in shape, is made of mild steel sheet and weighs about 200 to 250 gm. Four tapered fins in the sheller help in shelling maize grains. A cob is inserted into it, and shelling is done by twisting action. Output with a tubular maize sheller is about 25 to 30 kg/h (Singh et al. 2007)
User	Female workers
Mode of operation	Torque application by hand.
Design process	
Sheller diameter	A handheld maize sheller should be such that while holding the sheller, the operator's thumb and the fingers should not touch each other. Also, the diameter of the sheller should not exceed the maximum grip length. The 95th percentile value of inside grip diameter is the lower limit, and the 5th percentile value of maximum grip length of the female workers may be considered as the upper limit (i.e. 57 and 84 mm, respectively). That is, the sheller diameter in the above-stated range may be useful for better functional performance and comfort. The weight of the tool may be as low as possible, and therefore, minimum possible dimensions may be used, keeping technical limitations in consideration. It should be cylindrical or octagonal with smooth edges.
Sheller length	The length of the sheller should be such that the palm can have a full grip on it, and the sheller extends beyond the hand. The lower limit of the length of the sheller should be the 95th percentile value of the handbreadth of female workers (i.e. 85 mm should be the minimum length of the sheller)
Push/pull force limit	The 5th percentile value of handgrip torque for female workers is 2 Nm. Considering a 30% factor, the limit for torque application can be taken as 0.6 Nm in designing a hand maize sheller

Fig. 12.7 Pedal-operated maize dehusker sheller

Dehusking and shelling: Pedal-operated maize dehusker sheller (Fig. 12.7)	
Description	The structure of the pedal-operated maize dehusker sheller includes threshing drum, concave, drum cover, flywheel, feeding trough, saddle for pedalling, power transmission, and transportation wheels. Two male workers operate the machine, i.e. one for pedalling and another to feed the unhusked cobs in the hopper. A male worker operating a pedal-operated maize dehusker sheller can sustain a power output of 60 W for long-duration work, and accordingly, the feeding rate may be adjusted. The output of the device is about 130 kg of maize grains with a feed rate of 175 kg of maize cobs per hour (MPUAT 2012)
User	Male workers.
Mode of operation	Pedalling
Design process	
Height of feeding hopper	The height of the feeding hopper of pedal-operated maize dehusker sheller should be lower than the 5th percentile value of acromial height (shoulder height), and more than the 95th percentile iliocristale height (waist height). These values are 1255 and 1050 mm, respectively, which should be the range of the hopper height for the machine operated by male workers. In the likelihood of design considerations that the suggested hopper heights cannot be achieved, a provision of a stable platform may alternatively meet the requirement

(continued)

(continued)

Dehusking and shelling: Pedal-operated maize dehusker sheller (Fig. 12.7)	
Cranking/pedalling details	*Saddle height* The saddle height from the pedal should be equal to the trochanteric height of the workers (Tiwari et al. 2011). This height should be adjustable to accommodate the majority of the user population. The lower and higher values of saddle height may be based on the 5th, and 95th percentile values of trochanteric height (i.e. 700 mm and 920 mm, respectively). *Pedalling speed and power transmission system* The peripheral velocity of the threshing drum required for dehusking and shelling of maize cobs have been reported as 4.5 to 5.5 m/s (Singh 2012; MPUAT 2012). For about 250 mm diameter of the threshing drum, the required threshing drum speed would be 343 rpm to get the desired peripheral velocity. The recommended optimal speed for pedalling is 50 to 60 rpm (Tiwari et al. 2011). Accordingly, to get smooth operation of the equipment, provision of suitable flywheel may be necessary. MPUAT (2012) suggested that a flywheel weighing 14 kg on the drum shaft of the machine may be useful. Tiwari et al. (2011) suggested a 180 mm crank length for pedalling. The commonly available adult bicycle crank can be used, as it is of the same size. All rotating parts of the devices should be provided with safety covers. No parts of the human body should come in contact with moving machine parts
Handle size (cylindrical)	Here, the handle is used for holding support during pedalling operation. The suggested specifications of the handle grip diameter (38 mm) and handle length (120 mm) are the same as the dimensions of other pedal-operated machines
Force required in cranking/pedalling	The force limit for pedalling operation is based on the 5th percentile values of right and left leg strengths, which are 251 and 231 N for male workers. Taking 30% of left leg strength, the force limit for pedalling, as applies to a pedal-operated device, would be 70 N for prolonged work.
Dust	During dehusking and shelling operation, the total, as well as respirable dust in the working zone, are often higher than the recommended limits of 10 mg/m^3 and 5 mg/m^3. In such a situation, workers may be provided with suitable dust masks to protect their health

Threshing—Paddy Thresher

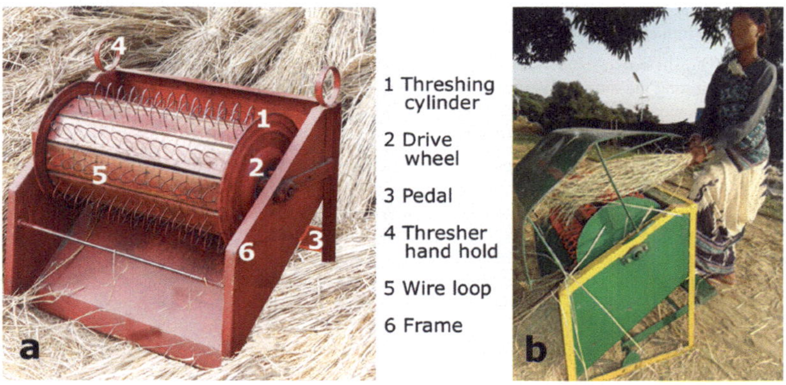

Fig. 12.8 **a** Pedal-operated paddy thresher and **b** thresher with safety guard

Threshing: Pedal-operated paddy thresher (Fig. 12.8)	
Description	The pedal-operated paddy thresher is hold-on type equipment for separating paddy grains from the straw. It consists of mainframe, threshing cylinder mounted on a shaft, gear mechanism, pedal, levers, and a safety cover. Over the decades, there has been substantial reform in rice and wheat threshing methods among the small farmers (Zakiuddin et al. 2012). Chapter 2 includes some description of different crop threshing methods. The pedal-operated thresher is fitted with a set of spur gears to develop a speed of 280 to 320 rpm from a treadling stroke rate of 70 to 80/min. A person in standing posture operates the pedal by one leg in treadling mode and holds the paddy bundles with one or both hands above the threshing cylinder. Two workers can also work together on the thresher. The device can also be operated by a worker while sitting on a seat/stool. The weight of such a thresher is about 40 kg, and the output is 45 kg/h with one person (OUAT 2014)
User	Male and female workers
Mode of operation	Treadling.
Design process	
Weight	The pedal operated paddy thresher is portable equipment, and two persons carry from place to place. It may be desirable to keep the weight of the equipment less than 30 kg (2 persons × 40% of the 10th percentile body weight of females). In the case of males, the weight can be preferably less than 36 kg (2 persons × 40% of the 10th percentile body weight of males). Since the thresher may be moved from place to place only once or twice a day, 50% of the body weight may be taken the limit. Thus, the weight of the thresher may be limited to 37.5 kg (2 persons × 50% of the 10th percentile body weight of females) for operation by both males and females
Height of the top of threshing cylinder	The height of the threshing cylinder should be more than 95th percentile metacarpal III height of males and less than 5th percentile elbow height of the females to avoid bending discomfort. These values are 770 and 890 mm, respectively. Lower the height less will be the physiological cost involved in the lifting of the crop for feeding. Therefore, a height of the top of pedal operated paddy thresher (top of the threshing cylinder) should be 770 mm
Pedal height	For rotation of the cylinder, the operator applies force by the leg at pedal during downwards movement when it is at the topmost position. Because of the inertia of the rotating cylinder, pedal again moves upwards, and the worker applies downwards force by the leg, and the movement continues. Agrawal (2008) reported that the strength in downwards leg force is maximum when the foot is at 0.16 times of the stature. The 5th percentile values of the body height of males and females are 1523 and 1420 mm. Hence, for maximum leg force, the foot height should be 244 and 227 mm for male and female workers. Taking 0.6 times of popliteal height as pedal height for maximum force application (Shrimali 2005), the foot height for male and female workers may be taken as 221 and 205 mm, respectively. Considering these data, the pedal height as 220 mm may be appropriate. It may give a stroke length about 210 mm, leaving 10 mm clearance between the ground and the pedal at the bottom-most position
Safety cover and guards	Thresher covers and guards help the operator from hitting of grains on the face/eye and also limiting the throw of grains from threshing (*refer to* Fig. 2.13). The gap between the cylinder top and cover should be about 300 mm to hold the paddy bundles comfortably. Safety guards should be provided on all rotating parts
Lifting and holding handles	Proper handles should be provided on both sides of the thresher for holding, lifting, and carrying purposes. The desirable shape of the handle is cylindrical or slightly oval. Since both male and female workers operate the equipment, the 95th percentile value of middle finger palm grip diameter of male workers becomes the lower limit, and the 5th percentile grip diameter (inside) of the female workers as the upper limit (i.e. ranging within 35 and 38 mm, respectively). For women workers, however, the 95th percentile value of their middle finger palm grip diameter is 32 mm, which may be taken as the recommended handle grip diameter

(continued)

(continued)

Threshing: Pedal-operated paddy thresher (Fig. 12.8)	
Force limit for pedalling	The 5th percentile value of leg strength in a standing posture of male workers was 414 N at 0.6 of popliteal height (Shrimali 2005). Taking 2/3rd of this value, the corresponding power for female workers would be 276 N. Taking 30% as the force limit for continuous operation, the force required at pedal should not exceed 83 N for females and 124 N for male workers
Dust	During the operation of pedal-operated threshers, the dispersion of crop-related dust in the working zone, and also in the vicinity are often higher than the recommended limits of total and respirable dust concentration (i.e. 10 and 5 mg/m^3). In such a situation, workers may be provided with suitable dust masks to avoid breating hazards

Fig. 12.9 Power-operated thresher, with feeding chute/hopper at different heights

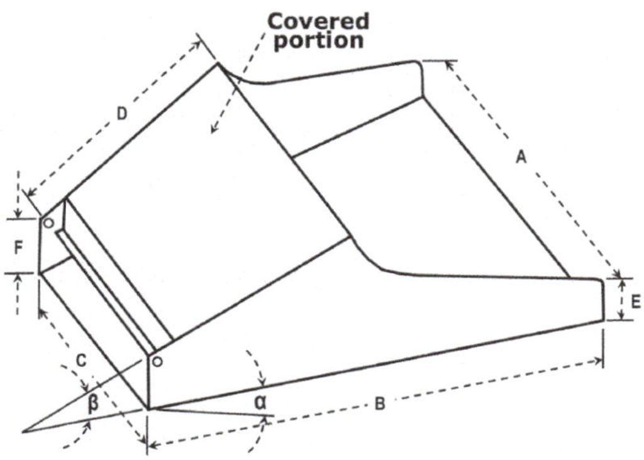

Fig. 12.10 Dimensions of the feeding chute. (A) Width of feeding chute at the front (500 to 650 mm); (B) Length of feeding chute (900 mm, minimum); (C) Width of feeding chute at rear (200 to 220 mm); (D) Length of the covered portion (450 mm, minimum, 550 mm for chaff cutter type thresher); (E) Height of side cover at the front (50 to 60 mm); (F) Height of side cover at rear (125 to 200 mm); (α) The angle of feeding chute with horizontal (0° to 15°); (β) The angle of lift of covered portion (10° to 30°)

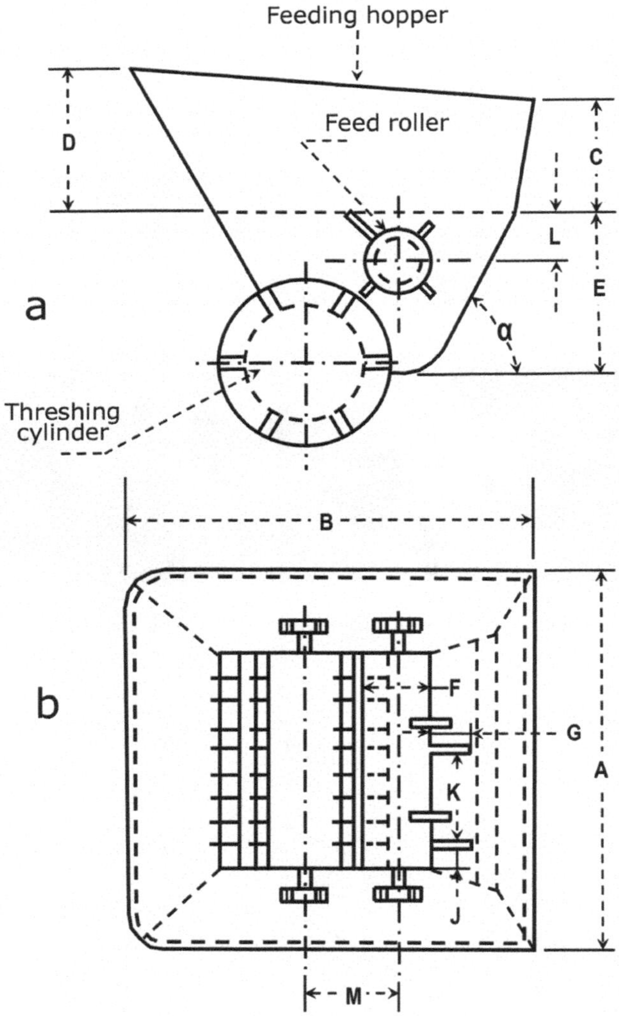

Fig. 12.11 **a**, **b** Design characteristics of the hopper of a power thresher. A (Width of feeding hopper at the top) = 400 mm more than the threshing cylinder; B (Length of feeding hopper at the top) = 740 mm, minimum; C (Height of top of hopper from feed roller on operator side) = 150 mm, minimum; D (Height of top of hopper from feed roller on another side) = 545 mm, minimum; E (Height of top of feed roller from the centre of the threshing cylinder) = 560 mm, minimum; F (Effective diameter of feed roller) = 300 mm; G (Height of peg on feed roller = 100 mm; L (Vertical distance between the centre of feed roller to the curved portion of hopper) = 125 mm; M (Horizontal distance of centres of the threshing cylinder and feed roller) = 450 mm; and α (Angle of feeding hopper with horizontal) = 55°. **c**, **d** Recommended dimensions of the feeding hopper. AH (acromial height), AR (arm reach), M (male), F (female), P (percentile)

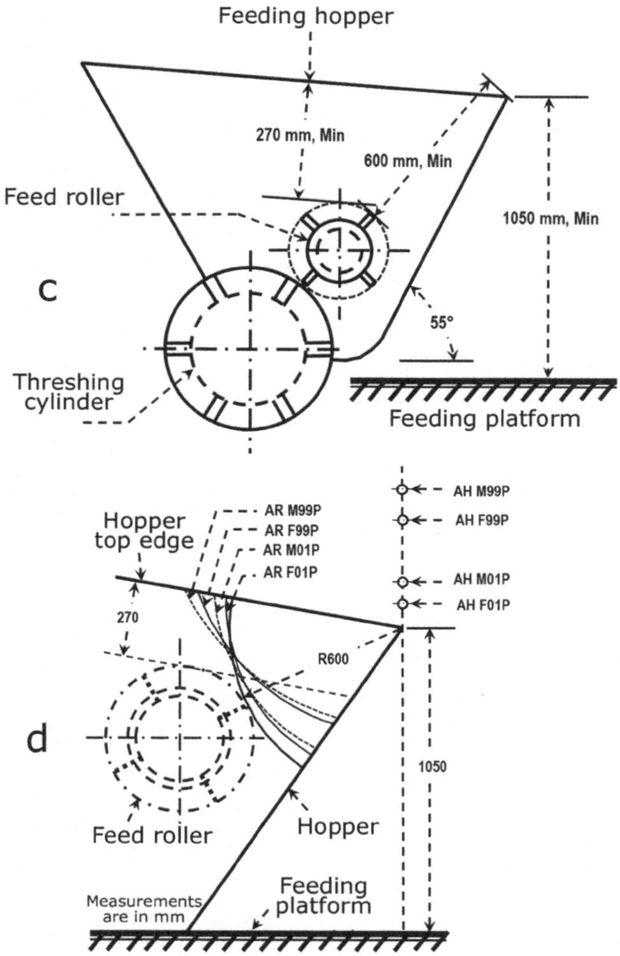

Fig. 12.11 (continued)

Threshing: Power-operated thresher (Fig. 12.9)	
Description	Thresher machine separates grains from earheads/plant stalks and gets cleaned grains and straw at the respective output spouts. A typical power thresher consists of a threshing cylinder, concave, flywheel, blower/aspirator assembly, straw rack/sieve assembly, and a feeding chute/hopper or a conveyor mounted on a robust iron structure. These machines are operated by an electric motor or diesel engine mounted on the frame or from a tractor. Power to various moving components is transmitted with the help of a belt and pulleys from the power source. The feeding chute/hopper height from the ground varies from 1 to 2.5 m depending on the construction and capacity of the machine. When the height of the feeding chute/hopper is more, a stable platform is provided for the operator to do the feeding. Also, all the rotating parts are required to be covered with proper safety guards. The power thresher categories (including maize dehusker sheller) have been recognized as the dangerous machines under the Dangerous Machine Regulation Act 1983 by Govt. of India (Gite et al. 2006). Therefore, all the threshers manufactured and used in the country must conform to the provisions given in IS 9020 (power threshers – safety requirements) (BIS 2002)
User	Male and female workers.
Mode of operation	Operated by an electric motor, diesel/petrol engine, or tractor in stationary mode
Design process	
Height of feeding chute/hopper	Humans adopt a natural posture for minimal strain on the musculoskeletal structure. It is essential, therefore, that the body orientation in the workplace should be as optimal as possible, and thereby reduce bodily strains in work operations. The human dimensional considerations are the critical requisites for man-machine compatibility. Both male and female workers are engaged in the feeding of the crop in the power-operated thresher or maize dehusker activity. The acromial height (shoulder height) and iliocristale height (waist height) may be considered in deciding the height of feeding chute/hopper. For minimum discomfort during feeding, the height of feeding chute should be lower than the 5th percentile acromial height for the female workers (i.e. 1050 mm). The height may be more than the 95th percentile iliocristale height of the male workers (i.e. 1173 mm) to avoid bent postures during feeding. Regarding the thresher operation shown in Fig. 12.9 a, b, c, the recommended feeding chute/hopper height should be 1050 mm, and not more than 1173 mm. A stable supplementary platform provided in some machines for operators to reach the chute/hopper height
The dimension of the feeding chute	A typical feeding chute, as recommended in IS 9020 (BIS 2002), is shown in Fig. 12.10. The essential dimensions are the length of the chute (B), length of the covered portion (D), and the angle of lift of the covered portion. As it is a matter of safety, the length of the chute (B) is decided considering the 99th percentile arm reach from the wall of male workers (i.e. 975 mm). The 99th percentile wall to acromial length is 144 mm. Therefore, the maximum arm length arrives at (975–144 = 831 mm). The minimum length of the feeding chute length is recommended as 900 mm, giving a margin for inadvertent human error during the operation. The length of the top cover (D) is decided based on the coronoid fossa to hand length, and its 99th percentile value is 470 mm. Considering the angle of lift of covered portion as 10°, the minimum length of the covered portion arrives at 478 mm ≈ 480 mm
Kumar et al. (2002) reported the minimum width of feeding chute at the rear as 220 mm, and the optimum angle of feeding chute with horizontal as 15° to 25°	
The dimension of feeding hopper	Nowadays, hopper units come with star wheel or feed roller type feeding mechanism, in the case of 7.5 kW or higher power rated threshers. IS 9020 specifies the dimensions of these types of feeding systems (Fig. 12.11a, b). Despite providing hopper with star wheel or feed roller type feeding mechanism, there have been severe accidents like the workers' hand get entangled with star wheel/feed roller. Therefore, the recommendations have been put forward (Gite 2017), as shown in Fig. 12.11c, d. (a) The height of the feeding hopper from the platform should be 1050 mm (based on 95th percentile iliocristale height). (b) The minimum distance from the edge of the hopper (where the operator stands) to feed roller tip should be 600 mm. The distance is arrived at from the acromial height and arm length (99th percentile acromion to the arm length) of the workers. (c) The minimum distance between the top of the hopper to the feed roller tip should be 270 mm
Safety	All rotating parts should be provided with safety covers. No parts of the body come in contact with moving machine parts. If the power transmission to thresher pulley is from tractor PTO, the belt from PTO to pulley should be well guarded. During the thresher operation, the noise level at the operator's ear level may exceed 90 dB (A). As a good practice, all workers should be provided with earmuffs to protect them from excessive noise exposure. Inevitably, the threshing operation generates enormous clouds of dust in the working zone and also in the vicinity. The situation demands that all workers wear suitable dust masks to protect their health

Winnowing—Paddy Winnower

Fig. 12.12 Hand-operated paddy winnower

Winnowing: Hand-operated paddy winnower (Fig. 12.12)	
Description	The hand-operated paddy winnower consists of mainframe, handle, gear mechanism, volute case, fan, hopper, and outlets for clean grain and chaff. One worker operates this machine in cranking mode in standing posture, and another worker puts the grains into the hopper for cleaning. The device can also be performed while sitting on a stool or a bench. The weight of the winnower is about 35 to 40 kg, and the output is about 242 kg/h of clean paddy grains (Singh and Gite 2007)
User	Male and female workers
Mode of operation	cranking
Design process	
Weight	The hand-operated paddy winnower can be moved from place to place by two persons. It is desirable to retain the weight of the equipment less than 30 kg (2 persons × 40% of the 10th percentile body weight of female workers). In the case of male workers, the weight may preferably be less than 36 kg (2 persons × 40% of the 10th percentile body weight of male workers). However, the lifting and carrying of the machine maybe once or twice in a day. Therefore, 50% of the bodyweight may be taken as the limit, and accordingly, the winnower weight limit arrives at 37.5 kg (2 persons × 50% of the 10th percentile body weight of female workers)

(continued)

(continued)

Winnowing: Hand-operated paddy winnower (Fig. 12.12)	
Height of feeding hopper	The height of the feeding hopper should be lower than the 5th percentile value of acromial height (shoulder height, i.e. 1173 mm) for minimal discomfort during feeding. Also, to avoid the bent posture during feeding, the height should be more than the 95th percentile iliocristale height (waist height, i.e. 1050 mm). Therefore, the suggested feeding hopper height ranges within 1050 to 1170 mm. For setting the hopper height in the recommended range, an adjustable, stable platform may be a part of the machine to meet the requirement
Height of crankshaft	Generally, the height of crankshaft for hand-operated rotary equipment can be taken as equal to iliocristale height (waist height), and the 5th percentile iliocristale height of female workers is 820 mm. Kharade et al. (2010) reported that the height of crankshaft for hand-operated rotary equipment should be 0.9 times of the elbow height. The 5th percentile elbow height for female workers is 890 mm. Accordingly, the crankshaft height arrives at 800 mm. Thus, a crank height with 800–820 mm is recommended for hand-operated paddy winnower.
Cranking speed and power transmission system	The optimum speed of crank rotation for winnower was found to be 65 rpm to get an airspeed of 3.2 m/s for cleaning of paddy grains (Singh and Gite 2007)
Shape and size of the handle (cylindrical)	Both male and female workers operate the equipment. The 95th percentile value of middle finger palm grip diameter of male workers (38 mm) is taken as the lower limit, and the 5th percentile grip diameter (inside) of the female workers (35 mm) is taken as the upper limit, respectively. Since women workers mostly use the winnower, the recommended handle grip diameter is about 35 to 38 mm.
Handle length	For deciding the handle length, the 95th percentile value of handbreadth across thumb is taken, which is 102 mm for female and 110 mm for male workers. Adding some clearance required, the length of the handle arrives at 120 mm. The cylindrical handle should have a through-hole in the centre along the length. It will rotate on its axis during cranking for smooth operation
Feed rate and capacity of the machine	In hand-operated cranking equipment, female and male workers can sustain a power output of about 45 to 50 W, for long duration with proper rest pauses. The feed rate should be adjusted accordingly.
Safety covers	All the rotating parts must have safety covers. No parts of the human body come in contact with moving machine parts
Force limit for cranking	The force limit for cranking operation, referring to the 5th percentile values of torque strength with the preferred hand of women in a standing posture, is 21 Nm. The limit for cranking torque may be 30% of torque strength, i.e. 6.3 Nm for continuous operation. With the minimum crank length as 260 mm, the cranking force would arrive at 25 N
Dust	The winnowing operation, by its very nature of the activity, generates large clouds of dust in the working zone. It would be appropriate that all workers wear suitable dust masks during the activity

Decortication—Groundnut Decorticator

Fig. 12.13 Groundnut decorticator operation in **a** sitting and **b** standing modes

Decortication: Groundnut decorticator (sitting and standing type) (Fig. 12.13)	
Description	Sitting or standing types of groundnut decorticators are similar in design. The device is used for shelling pods to get kernels. Main components of the decorticator include a frame, concave, handles, and shoe assembly. In sitting type decorticator, the device is fixed on a platform, and, additionally, a sitting stool is provided for the worker to sit and operate the unit. The weight of the machine is about 12 to 14 kg. A person puts about 1-3 kg pods in the decorticator and operates the handle by preferred hand or with both sides in push/pull mode. A worker can shell about 30 to 35 kg of pods per hour using such devices (Singh et al. 2007)
User	Female workers in sitting mode, and both male and female workers in standing mode
Mode of operation	Oscillation
Design process	
Weight of the equipment	The groundnut decorticator is portable equipment, and therefore, the weight of the machine should be less than 15 kg, referring to the bodyweight of female workers

(continued)

(continued)

Decortication: Groundnut decorticator (sitting and standing type) (Fig. 12.13)	
Height of the equipment	For the sitting type equipment, which is operated by female workers, its total height (up to the top of the handle in vertical position) should be less than 811 mm (i.e. 5th percentile acromial height sitting, 470 mm plus 5th percentile popliteal height sitting, 341 mm). For the standing mode groundnut decorticator operated by the male as well as female workers, its total height (up to the top of the handle in vertical position) may not be more than 1173 mm (5th percentile acromial height of female workers). It should not be less than 770 mm (95th percentile metacarpal height of male workers). Here, the desirable height of the handle is chosen that gives the least fatigue, and 0.7 to 0.8 times of the acromial height is the suitable handle height for work (Gite and Yadav 1989). Accordingly, the handle height may be adjustable between 0.8 times of the 5th percentile acromial height of female workers (i.e. 938 mm) and 0.7 times of the 95th percentile acromial height of male workers (i.e. 1034 mm). For the fixed type of handle, the height may be kept as 1000 mm
Width of the decorticator	As per ergonomics guidelines, the lateral distance between feet for comfortable sitting should not exceed 500 mm. Therefore, the machine width should not exceed 430 mm, considering that the 95th percentile value of heel breadth of female workers is 70 mm. The width of the decorticator used in standing mode may be decided based on the capacity of the equipment and the force limit of the workers
Length of decorticator	In the case of sitting type units, the distance between the seat to the centre of the handle (SRP to the centre of handle) should be equal to or less than the 5th percentile shoulder grip length of female workers (i.e. 570 mm). The range of the sitting type decorticator is generally less than the standing type. For standing type unit also, the length of the machine depends on the shoulder grip length. Ideally, the decorticator length is decided based on the total to-and-fro movement of the handle, and it should be less than 5th percentile shoulder grip length of female workers (i.e. 570 mm)
Size of the handle grip	For the length of the handle grip in sitting type unit, the 95th percentile value of handbreadth is taken (i.e. 105 mm). For the grip diameter, the 95th percentile value of middle finger to palm grip diameter and the 5th percentile value of grip diamter (inside) of female wor ker may be taken as the upper and lower limits, i.e., 32 and 35 mm, respectively. For the standing type equipment, the 95th percentile value of middle finger palm grip diameter of male workers is the lower limit, and the 5th percentile grip diameter (inside) of the female workers is the upper limit. Accordingly, the handle grip diameter recommended is 35 mm. The minimum length of the handlebar should be more than 250 mm (2 hands x handbreadth across thumb plus 30 mm clearance). Also, the handle crossbar length may not be more than 325 mm (5th percentile of the deltoid breadth of female workers)
Size of seat	For sitting type decorticator, the seat height should be less than the 5th percentile popliteal height of females in the seated position (i.e. 341 mm). Taking the stool/bench placed on a platform of about 20 mm thick, the height of the stool should be less than 321 mm. Considering the 95th percentile hip breadth sitting, the width of the seat should be 371 mm
Force application	Generally, a worker puts about 1 to 3 kg pods in the decorticator and operates the handle by preferred hand or both hands. Here, the worker applies push/pull force in sitting posture with arm fully extended/flexed. The 5th percentile value of push and pull strength with both hands are 84 and 101 N, respectively. Taking 50% as the limit, the power required to operate the decorticator should not exceed 42 N in pushing and 51 N in pulling mode. The 5th percentile values of push and pull force in sitting posture by right/left hand are 30 and 42 N, respectively. Therefore, the power required at the handle grip should be less than 30 N in push mode and 42 N in pull mode so that the equipment is suitable for 95% of the female workers

Tree Climbing—Coconut Tree Climber

Fig. 12.14 Coconut tree climbing; **a**, **b** traditional methods, **c** tree climber, and **d**, **e** climbing using the tree climber

Coconut tree climber (Fig. 12.14)	
Description	The innovative tree climber substantially obviates the tedious and risky method of coconut tree climbing (Fig. 12.14). A person can climb a coconut tree comfortably, using the device. The upper and the lower frames of the climber (Fig. 12.14c) are alternately raised by the user to reach an elevation on the tree. The upper structure is a tubular frame consisting of a rear rigid base section and a front adjustable tree gripping section. The rigid base section bears a seat, front support rail, concave rear rail, and side rails. The upper and lower frame members are connected by a safety strap to prevent accidental slippage of the lower frame down the tree or any inadvertent disengagement with the user's feet while ascending and descending. The extension of the adjustable 'U' shaped member serves as a handle to enable the user to ascend or descend the tree. Several rubber fasteners/rollers attached at strategic places secure the climber—the rollers provided with the climber grips the tree firmly to avoid dislodging and slipping. A canvas side bag fitted with a cross rail is a toolbox for safekeeping of hand tools and knives. To ascend the coconut tree, the user places one's feet on the lower frame, rests the bodyweight on the seat of the upper frame, and pulls the lower frame upwards using the feet. The user raises the body with the support of the feet on the lower frame and uses the hands to raise the upper frame to the waist-high position (Fig. 12.14d, e). The backward steps allow descending the tree. The weight of the climber is about 14 kg, and the heart rate during work is about 127 beats/min (TNAU 2009)
User	Male and female workers
Mode of operation	Hand and foot operation

(continued)

(continued)

Coconut tree climber (Fig. 12.14)	
Design process	
Weight	A coconut tree climber is a portable unit. The weight of the equipment should be less than 15 kg (40% of the 10th percentile of bodyweight of female workers)
Equipment components	*Width of the rigid base section of the upper frame* The width of the rigid base section is arrived at from the 95th percentile hip breadth (i.e. about 372 mm) for male and female workers. With a clearance of 25 mm on either side, the internal width of the rigid frame comes to (372 mm + 50 mm) = 422 ($\approx$ 430 mm). *Length of the canvas belt* Two canvas belts connect two sides of the upper and lower frames. The length of the belt is according to the 95th percentile metacarpal III height (i.e. 770 mm) of male workers as the lower limit. Also, the belt should be less than the 5th percentile olecranon height (i.e. 865 mm) of female workers. A length of 840 mm may be taken for these connecting belts. *Length of cushions on the cross rail of the lower frame* The length of the cushions is based on the 95th percentile of foot breadth (110 mm) of male workers and about 5 mm extra space on both sides of the feet, and thus the length may be taken as 120 mm. *The horizontal distance between front and rear cross rails of the lower frame* During climbing, the feet move the lower frame upwards when the maximum medial protuberances of the feet are kept below the front cross rail, and the heel portions are kept above and behind the second cross rail. The horizontal distance between the front and rear cross rails would be the midpoint of the 5th percentile foot length of female workers (i.e. 0.5 times of 199 mm = 100 mm). These cross rails have cushions at the point of foot touching. *Size of upper handrails* Essentially, the equipment has to be lightweight for smooth shifting between trees. During the climbing process, the worker keeps feet on the lower rail, holds the handle, and pulls the upper frame upwards, while in the backward steps for coming down, the climber lowers the upper frame downwards. The size of the rails should be as minimum as possible and robust enough to sustain the body load. Thus, pipe rails matching to the middle finger palm grip diameter of 20 mm are recommended

Pruning—Secateurs

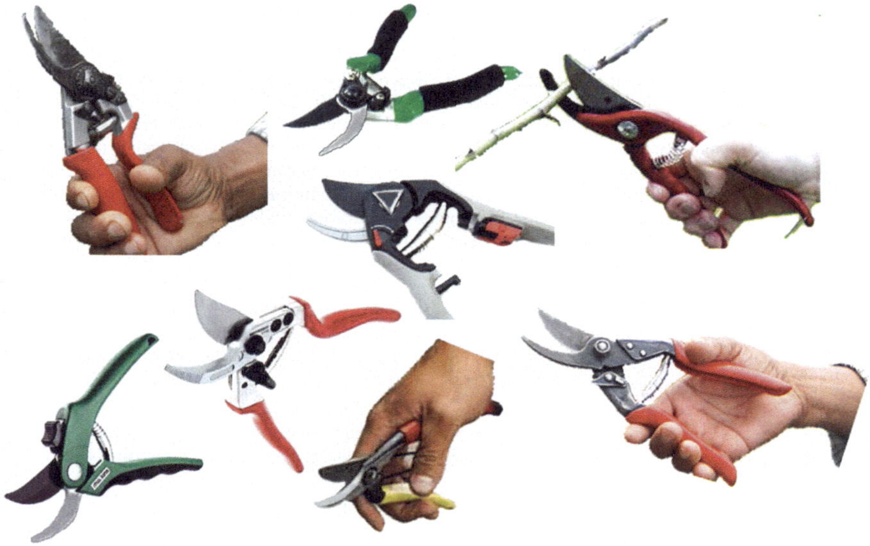

Fig. 12.15 Pruning secateurs of different designs

Pruning Secateurs (Fig. 12.15)	
Description	Pruning secateurs, garden/orchard tool with two short sharp blades, is used in pruning/cutting of plant/tree stems and small branches up to 15 mm diameter. There are two types of secateurs, namely, bypass or hooked type and anvil type. In pruning, bypass secateurs are generally used, which have the blade sharpened on one side, and it cuts past a curved metal surface. For hard/woody branches, anvil type secateurs are used. The components of the pruning secateurs are handles, blades, spring, and locking devices. A person holds the handles of the secateurs in palm grip and applies force to cut the plant/tree branch. The weight of typical pruning secateurs is about 300 g. A person may do about 40 to 50 repetitive actions per min during pruning of an apple tree (CSKHPKV 2016)
Mode of operation	Push (handgrip)
Design process	

(continued)

(continued)

Pruning Secateurs (Fig. 12.15)	
Size of secateurs	The length of the tool is taken between the tip of the blade to the far end of the handle in a closed position. Generally, the blade length is about 40 mm, and the end of the blade to the pivot point is about 60 mm distance. Thus, the pivot to the centre of the cutting blade distance is about 40 mm. By taking the handbreadth across thumb of the 95th percentile of male workers, the handle length may be set at more than 110 mm. Taking 2.5 times as the mechanical advantage, the distance between the pivot and centre of handle grip may be about 100 mm. By adding 50% of the handbreadth, the axis to the far end of the handle may be 100 plus 55, i.e. 155 mm, and thus, the total length of secateurs maybe 155 plus 60 = 215 mm. The size of secateurs generally varies from 150 to 225 mm, as per IS 3494 (BIS 1978/1999). The handle should not be too large, but the blade opening angle should be broad. When fully opened, the distance between secateurs' handles at grip point may not be more than the 5th percentile maximum grip length of users (i.e. 93 mm for male workers). Therefore, the distance between secateurs' handles at grip point when fully opened should be less than 93 mm. Plastic/rubber coating of the handle material should be durable, and it should not be too slippery or rubbery. The handles should not have sharp edges, and the tool can also be provided with a rolling lower handle to help reduce strain on the wrist
Springs and locking arrangement	The springs allow the blades to return to its initial position with every operation. The locking arrangement locks the secateurs when not in use. In the locked position, the edges should be fully closed. The locking mechanism should be provided in a way that the device can be operated by one hand
Force required to operate secateurs	Grip force is applied on the secateurs when a branch of the plant is placed in between the blades. The 5th percentile handgrip force for right and left hands are 206 and 195 N for male workers. Taking 30% of muscular strength for continuous work, the corresponding limits for grip force exertion maybe 62 N and 59 N. The weight and length of secateurs should preferably be as minimum as possible for minimal discomfort in the palm and wrist. The materials of the tool and dimensions of the blade and handle determine the weight of secateurs. It is desirable to have an optimal length for more mechanical advantage. For cutting branches of 10 to 13 mm diameter, the force required at the cutting point ranged from 60 to 115 N (Leppanen et al. 1998). Taking 115 N as the cutting force requirement and 2.5 times as the mechanical advantage the force needed at the handle grip maybe about 46 N, which is within the limiting value (i.e. less than 59 N)

Chaff Cutting—Chaff Cutter

Fig. 12.16 **a** Hand-operated and **b** power-operated chaff cutters

Chaff cutting: Hand-operated and power-operated chaff cutter (Fig. 12.16)	
Description	A chaff cutter is used to chop green/dry fodder into small pieces to feed animals. A typical hand-operated chaff cutter includes a flywheel mounted with curved blades, feed rollers, handle, stand, and safety features such as flywheel locking device, front safety guard, knife blade cover, gear cover, and a feeding chute. One worker operates the machine, and the other feeds the fodder to be chopped. The output of hand-operated chaff cutter is about 100–150 kg/h, and the mean working heart rate of the workers is about 125 beats/min (PAU 2012; Kumar et al. 2010, 2013). A power-operated chaff cutter is operated by an electric motor or diesel engine or by a prime mover such as tractor/power tiller through belt pulley transmission. The chaff cutter consists of a feed roller, flywheel/cylinder, feeding chute/conveyor, transmission system, blades, cover plate, shear plate, safety guards, and warning roller. A person feeds green/dry fodder and gets it chopped into pieces of 10–40 mm size. A commercially available power-operated chaff cutter, having two hp electric motor, fitted with safety features, can give an output of about 690 kg of chaff/h (PAU, 2012)
User	Male and female workers
Mode of operation	The hand-operated chaff cutter is operated in cranking mode, whereas the power-operated unit is operated by an electric motor or diesel/petrol engine
Design process	
Feeding chute height	In both types of chaff cutters, the height of feeding chute should be lower than the 5th percentile acromial height (shoulder height) of female workers. The height should be more than the 95th percentile iliocristale height (waist height) of male workers to avoid bent posture. These values are 1173 mm and 1050 mm, respectively. Therefore, the trough height is recommended as 1050 mm. Where the optimal trough height is constrained to achieve due to machine design considerations, a stable platform may be added to meet the requirements
Crankshaft height	In hand-operated chaff cutter, the height of crankshaft can be taken as the 5th percentile iiliocristale height of female workers (i.e. 820 mm). Also, it should not be less than 95th percentile metacarpal III height of male workers (i.e. 770 mm), and therefore, the recommended crankshaft height is 800–820 mm
Handle shape and size	Generally, the shape of the handle is cylindrical. As the hand-operated equipment is operated by both male as well as female workers, the 95th percentile middle finger palm grip diameter of male workers is the lower limit. The 5th percentile grip diameter (inside) of the female workers is the upper limit. These values are 38 mm and 35 mm, respectively. The recommended handle grip diameter is 35 mm. The handle generally consists of a mild steel rod with a hollow wooden grip placed on it. The diameter of the handle should be a minimum of 15 mm, and the wooden grip diameter should be about 35 mm. The 95th percentile value of handbreadth across thumb is taken for the length of the handle (i.e. 102 mm for female and 110 mm for male workers). Generally, the chaff cutter is operated by both hands. Therefore, the length of handle may be taken as $110 \times 2 + 100$ (for clearance) $=$ 320 mm
Cranking speed and force limit	As mentioned elsewhere, female and male farmworkers can sustain a power output of about 45 W and 60 W for a long duration work, for the equipment operated in cranking mode. Tiwari and Singh (2004) mentioned that the average operating speed of the hand-operated chaff cutter is 50 rpm when one knife is used, and 35 rpm when using two knives. A provision of a flywheel of a suitable size can smoothen the operation. The force limit for cranking operation is based on the 5th percentile torque strength with both hands in standing posture (i.e. 25 Nm for female workers). Taking 30% of the value, the limit for cranking is 7.5 Nm for prolonged operation. For the crank length as 250 mm, the cranking torque at the handle grip is 30 N

(continued)

(continued)

Chaff cutting: Hand-operated and power-operated chaff cutter (Fig. 12.16)	
Safety	As per IS 7898 (BIS 2001/2012) and IS 15542 (BIS 2005), the hand-operated, as well as power-operated chaff cutters, must be provided with flywheel locking device, front safety guard/knife blade cover, gear cover and a feeding chute of 900 mm length with 450 mm cover on it PAU (2012). Also, all the rotating parts should be provided with safety covers. During operation of a power-operated chaff cutter, many times, the noise level at the operator's ear level may exceed 90 dB (A). The earmuffs are essential for the workers to protect them from excessive noise exposure. The emission of a huge cloud of dust is apparent during the chaff cutting operation of dry fodder. Providing suitable respiratory masks to workers would help to protect them from respiratory hazards.

References

Agrawal, K. N. (2008). Design modification of pedal operated paddy thresher for north eastern hilly region of India. Unpublished Ph. D. Thesis. Indian Institute of Technology, Kharagpur.

BIS. (1978/1999). IS: 3494 specification for pruning secateur. Bureau of Indian Standards, New Delhi. India.

BIS. (2001/2012). IS: 7898. Manually-operated chaff cutter - Specification (Second Revision). Bureau of Indian Standards, New Delhi. India.

BIS. (2002). IS 9020, Power Threshers – Safety Requirements. Bureau of Indian Standards, New Delhi.

BIS. (2005). IS: 15542. Power-operated chaff cutter- Safety requirements. Bureau of Indian Standards, New Delhi.

CIAE (2014). Research highlights 2012–14. All India corordinated research project on Ergonomics and Safety in Agriculture. No. CIAE/2014/176. Central Institute of Agricultural Engineering, Bhopal.

CSKHPKV. (2016). *Annual report*. Chaudhary Sarwan Kumar Himachal Pradesh Krishi Vishvavidyalaya, Palampur, India: All India Coordinated Research Project on Ergonomics and Safety in Agriculture.

Dixit, A. K., Singh, S., & Singh, S. (2006). Ergonomic studies on power operated knapsack sprayer. *Journal of Research, 43*(4), 323–327.

FAO. (1995). *FAO basic guidelines for the selection of portable (operator carried) agricultural pesticide sprayers*. Rome: Food and Agricultural Organization of United Nations.

Gite, L. P. (1996). Some Investigations on a Side Lever Operated Knapsack Sprayer from Mechanical and Ergonomical Considerations. Unpublished Ph. D. Thesis, Indian Institute of Technology, Kharagpur, India.

Gite, L. P. (2017). Final report of the emeritus scientist project on development of ergonomical design guidelines for agricultural tools, equipment and work places, Central Institute of Agricultural Engineering, Bhopal, India.

Gite, L. P., & Yadav, B. G. (1989). Anthropometric survey for agricultural machinery design: an Indian case study. *Applied Ergonomics, 20*(3), 191–196.

Gite, L. P., Pharade, S. C., & Majumder, J. (2006). Revision of dangerous machine (Regulation) Act-1983. Agricultural Engineering Today, 30(3, 4), 44–55.

Hopfen, H. J., & Biesalski, E. (1953). Small Farm Implements (FAO Development Paper No. 32). Food and Agriculture Organization of the United Nations. Rome.

Kharade, H. S., Deshmukh, M., Tiwari, P. S., & Gite, L. P. (2010). Optimum crankshaft height and crank length for hand operated rotary maize seller. *PKV Research Journal, 34*(2), 177–180.

Kumar, A., Mohan, D., Patel, R., & Varghese, M. (2002). Development of grain threshers based on ergonomic design criteria. *Applied ergonomics, 33*(5), 503–508.

Kumar, A., Singh, J. K., & Singh, C. (2010). Ergonomic Assessment of a Manual Fodder Cutter. *Journal of Agricultural Engineering, 47*(4), 1–9.

Kumar, A., Singh, J. K., & Singh, C. (2013). Prevention of chaff cutter injuries in rural India. *International journal of injury control and safety promotion, 20*(1), 59–67.

Leppanen, G., Kallionpaaand, M., & Vilkki, M. (1998). *Ergonomic evaluation of garden secateurs* (p. 383). Global Ergonomics.

Mondal, T. (2019) Situational Analysis and Usability Assessment of Farm Mechanization in Small and Marginal Holdings, Ph.D thesis, Ramakrishna Mission Vivekananda Educational and Research Institute (RKMVERI), Belur, Howrah.

MPUAT. 2012. Biennial Report 2010–12. All India coordinated research project on ergonomics and safety in agriculture. Maharana Pratap University of Agriculture and Technology, Udaipur, India.

Nag, P. K., Goswami, A., Ashtekar, S. P., & Pradhan, C. K. (1988). Ergonomics in sickle operation. *Applied Ergonomics, 19*(3), 233–239.

OUAT. (2014). *Drudgery reducing tools and equipment for female agricultural workers.* Bhubaneswar, India: Orissa University of Agriculture and Technology.

Pandey, M. M., Majumdar, K. L., & Singh, G. (1997). Farm machinery research digest. All India coordinated research project on farm implements and machinery. No. CIAE/97/69. Central Institute of Agricultural Engineering, Bhopal.

PAU. (2012). *Progress report. All India coordinated research project on ergonomics and safety in agriculture.* Ludhiana, India: Punjab Agricultural University.

Shrimali, H. (2005). Ergonomical study on an agricultural foot sprayer. Unpublished Ph.D. Thesis, Maharana Pratap University of Agriculture and Technology, Udaipur, India.

Singh, S. P. (2012). Physiological workload of farm women while evaluating sickles for paddy harvesting. *Agricultural Engineering International: CIGR Journal, 14*(1), 82–88.

Singh, S. P., & Gite, L. P. (2007). Ergonomical evaluation of a hand operated paddy winnower by women workers. *Journal of Agricultural Engineering, 44*(4), 67–71.

Singh, S. P., Gite, L. P., Agrawal, N. & Majumder, J. (2007). Women friendly improved farm tools and equipment. No. CIAE/2007/128. Central Institute of Agricultural Engineering, Bhopal.

Sutjana, D. P. (2000). Use of serrated sickle to increase farmer's productivity. *Journal of Human Ergology, 29*(1–2), 1–6.

Tiwari, P. S., & Singh, K.P. (2004). Design of chaff cutter. In A. C. Varshney, P. S. Tiwari, S. Narang, & C. R. Mehta (Eds.), *Data book for agricultural machinery design.* Book No. CIAE/2004/1. Central Institute of Agricultural Engineering, Bhopal, India.

Tiwari, P. S., Gite, L. P., Pandey, M. M., & Shrivastava, A. K. (2011). Pedal power for occupational activities: Effect of power output and pedalling rate on physiological responses. *International Journal of Industrial Ergonomics, 41*(3), 261–267.

TNAU. (2009). Progress report 2007–09. All India coordinated research project on ergonomics and safety in agriculture. AEC&RI. Pub; 59/2009. Tamil Nadu Agricultural University, Coimbatore, India.

TNAU. (2012). Progress report 2010–12. All India coordinated research project on ergonomics and safety in agriculture. No. AEC&RI/Pub.14/2012. Tamil Nadu Agricultural University, Coimbatore, India.

Zakiuddin, K. S., Sondawale, H. V., Modak, J. P., & Ceccarelli, M. (2012). History of human powered threshing machines: A literature review. In *Explorations in the history of machines and mechanisms* (pp. 431–445). Dordrecht: Springer.

Chapter 13
Ergonomics Application in Design of Workplace of Tractors and Power Tillers

Introduction

In developed as well as developing countries, a tractor is one of the most crucial farm machines for field operations, transport, and other stationary operations. Generally, the tractor, a four-wheel unit, is operated by male workers, though, at many places, women also drive tractors and operate various equipment. The nature of tasks on a tractor necessitates many actions to be performed by the operator, which puts varying physiological demands on the body. These tasks include steering of the tractor; looking backwards to observe and control the machine/implement; and applying force to operate the clutch, brake, and hydraulic control levers. Therefore, the design consideration of workplace of tractor operators is vital for operational effectiveness. The critical design parameters of the operator's workplace include the operator's seat, controls, and attenuation of vibration and noise.

Several studies have been carried out on various aspects of ergonomics and human factors of tractor design and operation. The factors of importance encompass issues related to riding comfort, visibility, location of controls, ease of operating controls, and environmental design of the tractor cab. Emphasis has been placed on the tractor seat to meet primary objectives, namely, the comfort, health, and safety of the operator. The comfort refers to the considerations such as seat dimensions and their adjustments, cushioning materials, and the perception of the convenience of the operators. Health refers to the long-term spinal support, seat ergonomics, and terrain-induced vibration attenuation. Safety refers to the ability of the seat to keep an operator *in position* during an accident. Proper tractor seat design helps in modifying the distribution of load of the body structure to reduce the operator's discomfort. The tractor seat has to withstand the forces coming on it during field operations and still maintain their dimensional stability, firmness, and essential characteristics that enable the seat to retain its contoured shape (Mehta and Tewari 2000). Two-wheel tractors (also called as power tillers) are widely used in land preparation in some regions. Here, a worker walks behind the unit in carrying out various operations. Sometimes a seat is

P. K. Nag and L. P. Gite, *Human-Centered Agriculture*, Design Science and Innovation,
https://doi.org/10.1007/978-981-15-7269-2_13

also provided for the operator to sit and carry out the operation. Ergonomics considerations in the design of the operator's workplace on the tractor are necessary to reduce drudgery and occupational health problems, and enhance productivity and safety of workers (*refer to* Chap. 9, Health hazards in farming). The engineers and designers use anthropometric and muscle strength data of user population, biomechanics principles as well as other ergonomics guidelines to arrive at design specifications. The illustrations given here are to show the design process from an ergonomics perspective and are primarily based on the data of the Indian population (Gite 2017), as given in Chap. 6. The readers may use ones' country-specific anthropometric data for the purpose.

- **Tractor—**Operators' workplace
  - Dimensions of the operator's seat.
  - Location, dimensions, and actuating force limits of controls.
  - Physical work environment and safety.
- **Power tiller (Two-wheel tractor)**
  - Location, dimensions, and actuating force limits of controls.
  - Physical work environment and safety.

Tractor: Operator's Work Place—Dimensions of Operator Seat

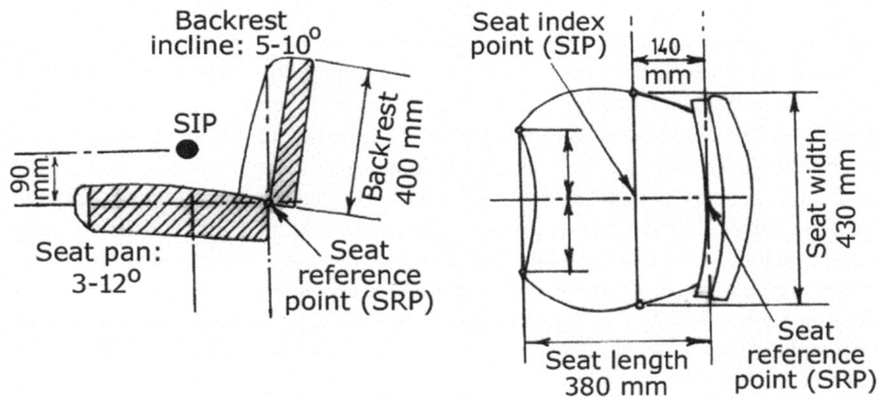

Fig. 13.1 Tractor seat dimensions

Tractor: Dimensions of operator's seat (Fig. 13.1)	
Description	In driving a tractor, the seat acts as an interface between the tractor and operator. Therefore, the tractor seat is a recognizable component that influences the comfort, performance, health, and safety of the operator. Each tractor operator develops unique postural dynamics with the circumstances of the workplace, e.g. constraints of space, tractor controls, and also vibratory impacts. The concentration of forces exerted on the intervertebral structures, joints, and ligaments is critical to the risk of spinal and paraspinal discomfort, including inflammation of tendons and tendon sheaths, chronic muscle pain, and joint degeneration. The tractor operator-seat dynamics is greatly influenced by the features of seat, such as seat height and width, backrest, armrest, slope of the seat pan, concavity, cushion details, and seat adjustment range. As mentioned earlier, in seat design, there are two essential considerations, namely, seat reference point (SRP) and seat index point (SIP). As per IS:11806 (BIS 1994), the SRP is a point in the central longitudinal plane of the seat where the tangential plane of the lower backrest and a horizontal plane intersects. For workplace design purposes, the SIP is the intersection on the central vertical plane through the seat centerline of the theoretical pivot axis between the human torso and thighs, as per IS 11113 (BIS 1999/2004) and ISO 5353 (ISO 1995/2016). The SIP usually is 90 mm above and 140 mm in front of SRP refer to IS12343 (BIS 1998) (*refer to* Chap. 11, Fig. 11.2)
User	Male workers
Design process	
Seat height	Proper seat height allows the user to place the feet firmly on the supportive surface or foot controls. It is measured from footrest to front of the seat surface with 65 kg (IS 11113) weight on the tractor seat. The seat height should be lower than the distance from footrest to the underside of the thigh (i.e. popliteal height sitting). The 5th percentile value of popliteal height sitting of male farmworkers is 369 mm. Mehta et al. (2008) suggested that the seat height on Indian tractors may be kept as 380 mm, including allowances for shoes
Seat length	The seat length has to be such that the operator can get lumbar support with the backrest during tractor driving. As per IS 12343 (BIS 1998), seat length in front of SIP is the horizontal distance parallel to the longitudinal plane of the tractor measured from the front edge of the seat cushion (offset 150 mm on either side of the longitudinal centerline) to the vertical transverse plane containing the SIP. However, the total seat length is taken as the horizontal distance parallel to the longitudinal plane of the tractor measured from the front edge of the seat cushion (offset 150 mm on either side of the longitudinal centerline) to the vertical transverse plane containing the SRP. The total seat length should not be more than the 5th percentile buttock-popliteal length keeping in view that there should be enough clearance between the back of the lower leg and front edge of the seat of the operator. For a seat length higher than the 5th percentile buttock-popliteal length, the operator cannot suitably use the backrest. The 5th percentile buttock-popliteal length of Indian male workers is 382 mm. Therefore, the total seat length recommended will be 380 mm, and the seat length in front of SIP as 240 mm

(continued)

(continued)

Tractor: Dimensions of operator's seat (Fig. 13.1)	
Seat pan	*Seat pan width* It is the horizontal distance between the outside edges of the seat surface measured in a plane perpendicular to the median plane of the seat or the width measured along a horizontal transverse line passing through the SIP. For the driver's comfort and convenient posture change, the seat width should be wider than the hip breadth sitting of the 95th percentile operator (i.e. 372 mm), and should not be less than 450 mm. However, based on anthropometric considerations of Indian farmworkers, this value arrives at 430 mm, making allowances for clothes and space for free movement *Seat pan tilt* It refers to the angle of the seat pan to the horizontal. A seat pan that is tilted backwards can produce effects. For example, by the force of gravity, the operator's back is moved towards the backrest and supports back muscles. Any marginal inclination of the seat pan at the front helps to prevent the slippage out of the seat. The IS 12343 (BIS 1998) recommends a tilt of 3° to 12° backward for the seat pan. The concavity of the tractor seat pan should have a radius of curvature of 750 mm for Indian operators for better contact and more uniform pressure distribution at the seat–operator interface (Tewari and Prasad 2000)
Seat backrest	*Backrest width* Backrest stabilizes posture by freeing shoulder space for retraction of the shoulder blades, reduction in muscle activity, and low back flexion. The 5th and 95th percentile interscye breadths of male farmworkers are 251 mm and 360 mm, respectively. Therefore, trapezoidal shape backrest may be preferred for the tractor seat with a smaller length on the upper part to have free hand movement. The seat backrest width at the bottom and top should be 380 mm and 275 mm, respectively, including the allowances for operators' clothes *Backrest height* The backrest height is measured above the compressed seat, if padding is present. A high backrest prevents full mobility of the arms and shoulder in rear viewing and operation of the hydraulic control lever in a tractor. The 5th percentile sitting acromial height of male farmworkers is 500 mm. Thus, with a clearance of 100 mm to have free shoulder movement, the height of backrest can be taken as 400 mm *Backrest inclination* The inclination of the backrest to the seat pan (a) prevents the operator from slipping forward and (b) helps him to lean against the backrest with the lower (lumbar) part of his back and sacrum supported. In order that the trunk-thigh angle is in comfort range when sitting, the slope angle as per IS:12343 should be in the range of 95–105°. Researchers have specified the trunk-thigh angle as 90 to 115° (Hansson and Pettersson 1980; Courtney and Evans 1987; and Göbel and Luczak 2000). The backrest inclination, accordingly, recommended is 95° to 100° from horizontal *Backrest concavity* The tractor seat backrest with a radius of curvature of 300 mm is recommended by Tewari and Prasad (2000) for Indian operators for better contact between the back of the body and backrest

(continued)

(continued)

Tractor: Dimensions of operator's seat (Fig. 13.1)	
Seat armrest	The armrest should not interfere with basic body movements. The armrest must be wider apart than the seat, with a minimum of 25 mm on each side of the seat. The armrest should be at the 5th percentile elbow rest height above the seat surface, which is 167 mm and should project 250–300 mm forward from the back of the seat
Seat cushion	The seat surface should be cushioned (25–50 mm of compression). The cushion compressibility is expressed as indentation load deflection. The tractor seat cushion with synthetic rubber foam materials (thickness = 100 mm and $\rho = 70$ kg/m^3) and the seat backrest cushion with composite materials (layers of coir and medium density foam, thickness = 79 mm and $\rho = 47$ kg/m^3) were comfortable for Indian operators (Mehta and Tewari 2000)
Seat adjustments	The vertical and longitudinal adjustment of the tractor seat position allows accommodating most of the tractor operators. According to IS 12343, the minimum longitudinal and vertical adjustments to be provided from the mid-position of the seat are ±25 and ±30 mm, respectively. However, considering the popliteal height of the 5th and the 95th percentile users, the total seat adjustments in vertical direction work out to be 101 mm (±50.5 mm from mid-position). Analysis of different dimensions, such as the popliteal height, buttock-knee length, knee height, popliteal length, and foot length of the 5th and the 95th percentile users and the knee angle during work, also suggests the seat adjustability in the horizontal direction as 197 mm (± 98.5 mm from mid-position). Thus, a vertical adjustment of ± 50 mm and horizontal adjustment of ± 100 mm in the tractor seat from mid-position may accommodate 90% of Indian male operators

Tractor Controls

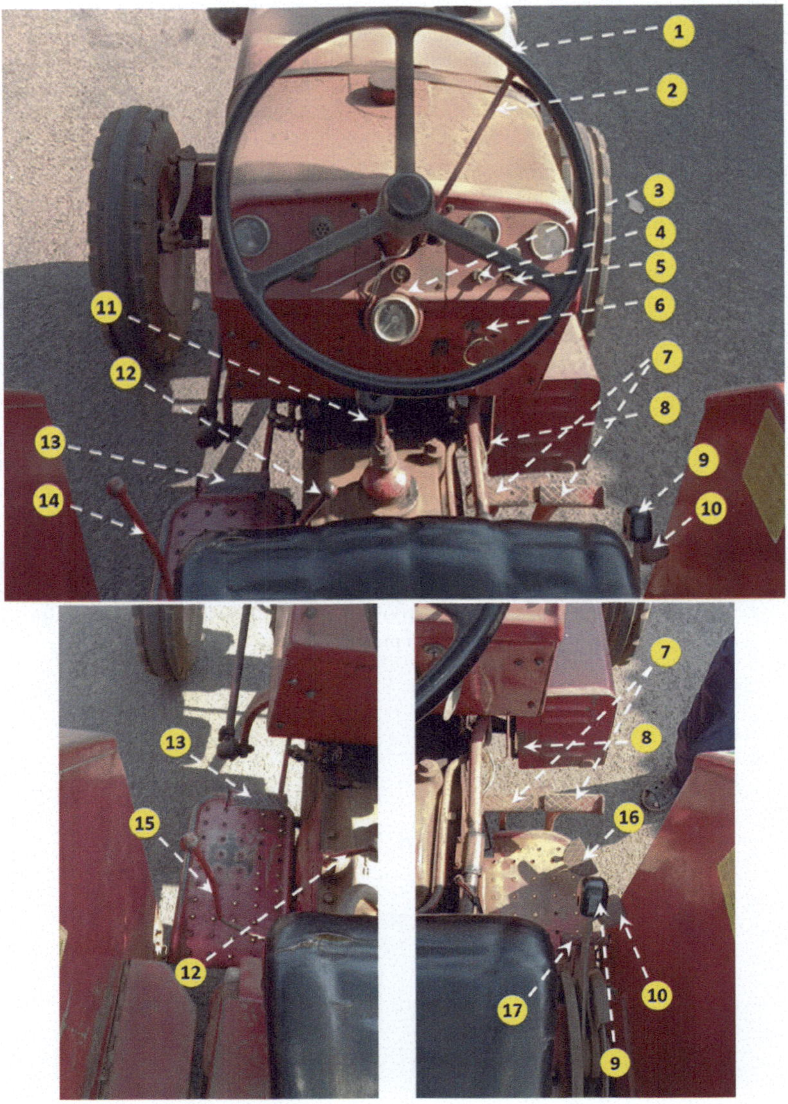

Fig. 13.2 Location of tractor controls. (1) steering wheel, (2) hand accelerator, (3) dashboard, (4) light control, (5) stop lever knob, (6) start button/key, (7) brake pedals, (8) parking lever, (9) position control lever, (10) draft control lever, (11) gear shift lever, (12) speed selection lever: High/low, (13) clutch pedal, (14 and 15) PTO gear control, (16) foot accelerator, (17) differential lock

Tractor: Controls (Fig. 13.2)			
Description	In a tractor, proper design of controls is essential for comfortable and safe operation. The design involves location and dimensions of controls as well as force limits for operating the controls. The size of the operators has apparent influence in the design of controls, such as pedals, levers, wheels, or knobs. The major tractor controls covered here are brake pedals, clutch pedal, accelerator pedal, differential lock pedal, steering wheel, hydraulic control levers, gear shift lever, speed selection lever, PTO lever, hand throttle lever, parking brake, fuel shut off control knob, seat adjustment lever, hazard switch, plough lamp switch, combination switch (side indicator, horn, and dipper operation lever), multi-position light switch, and starter switch		
Mode of operation	Leg-operated controls (push type), hand-operated levers (push or pull type), and hand-operated wheels (torque)		
Design process			
	Location of controls	Dimensions of controls	Actuating force limits
Brake pedals	In a typical tractor, two brake pedals are provided on the right-hand side, for the left- and right-side rear wheels. A lever joins these pedals whenever needed, i.e. travelling on the road. The location of brake and clutch pedals requires to be decided, with reference to the operator's seat. It is necessary to define a leg reach envelopes for the 5th and 95th percentile operators using buttock-popliteal and buttock-knee length, popliteal, and knee height (sitting), and foot length. Based on the leg reach envelope (*refer to* Chap. 11, Fig. 11.4), it is recommended that longitudinally, the brake pedals should not be nearer than 410 mm and farther than 493 mm from SIP. The vertical downward distance should not be more than 396 mm from SIP, including the downward movement of pedals. Taking a 90 mm downward movement of brake pedals, the vertical distance of the pedals should be 306 mm. Laterally, the brake pedals should be placed close to the centerline, but at least 90 mm away and within 215 mm from the centerline	Generally, brake pedals are operated by the right foot. The foot placed on the brake pedal is positioned in such a way that the point of maximum medial protuberance of one's foot aligns with the farthest edge of the pedal. The width of brake pedals should not be less than the 95th percentile foot breadth of Indian male farmworkers (i.e. 110 mm) Clark and Corlett (1984) suggested that the length of the brake pedal (dimension in the line of travel) maybe 75 mm. More the length less is the pressure coming on the lower limbs. With footwears, the pressures on leg and foot muscles get partly dissipated. At many times, there is a space constraint for the designers to locate the foot pedals of desirable length. The minimum length of brake pedals recommended is 40 mm. Taking the maximum force applied on the brake as 400 N on 44 cm^2 pedal area, the force exertion here will be 9 N/cm^2, which seems an acceptable limit (Gkikas et al. 2009)	The 5th percentile right leg strength of male farmworkers is taken as the maximum limit for actuating force to operate the brake, i.e. 251 N

(continued)

(continued)

Tractor: Controls (Fig. 13.2)			
Clutch pedal	The tractor clutch pedal is on the left side and operated with the left foot. Like in the case of the brake pedal, the precise location of the clutch pedal is based on the leg reach envelope of the 5th and 95th percentile operators (*refer to* Chap. 11, Fig. 11.4). The locational details of the clutch pedal are the same as that of the brake pedal	The clutch pedal is usually operated more frequently, and therefore its placement location is crucial. It should be positioned symmetrically with the brake pedal, with the centre of the axis of driver serving as the symmetrical axis for both. The dimensions of a clutch pedal of a tractor will be the same as that for the brake pedals	The 5th percentile left leg strength in sitting of male workers is 231 N. Since the clutch pedal is required to be operated more frequently, 50% of the 5th percentile left leg strength may be taken as the actuating force limit (115 N) for the clutch pedal
Accelerator pedal	The accelerator pedal regulates engine speed. Based on the leg reach envelope given in Chap. 11 (Fig. 11.4), the longitudinal distance of accelerator pedal from SIP should be less than 493 mm. The vertical downward distance should not be more than 396 mm below SIP, including about 30 mm downward movement of the accelerator pedal. The placement of the accelerator pedal to the right side of the brake pedal should not be farther than 426 mm from the centerline. This limit is taken from the sitting hip breadth of the 5th percentile operator and 30° as the limit for leg abduction	The accelerator pedal moves in such a direction that the vibrations and jolts induced by driving over obstacles do not cause the foot and pedal to shake and vibrate, and the engine speed remains steady. The vibration of the accelerator pedal can be prevented when the heel of the foot gets steady support. The accelerator pedal is a toe-operated (foot-operated) control. The length of the toe is the difference between foot length and instep length, and its 95th percentile value is 270 mm minus 210 mm = 60 mm, which is the minimum length of the accelerator pedal. The width of the accelerator pedal should preferably be more than 95th percentile foot breadth (110 mm). The force required to operate the accelerator pedal is generally less than 100 N. For the pedal size of 30 cm^2, the pressure on toe muscles of about 3.3 N/cm^2 is within the acceptable range (Gkikas et al. 2009)	The accelerator pedal regulates the engine speed. If the pedal pressure is too high, the energy required to operate the pedal continuously will be excessive. The 5th percentile right foot strength of male workers is 155 N, and the 30% of it is 47 N, which can be taken as the force limit. The weight of the lower leg is about 9% of body weight, and during operation, it is partly supported at the heel (i.e. 50%). Under such presumption, the force required in operating accelerator pedal should not be less than 32 N (i.e. 71.5 x 9.81 x 0.09 x 0.5), where 71.5 kg is the 95th percentile bodyweight of farmworkers. The accelerator pedal should be so designed that the force required for its operation is more than 32 N and less than 47 N

(continued)

(continued)

Tractor: Controls (Fig. 13.2)			
Steering wheel	The steering wheel is a frequently operated hand control to manoeuver the tractor movements. Pajnoo et al. (2012) recommended that the horizontal distance of the steering wheel from SRP should be 39 to 42% of body stature. Also, the vertical distance of the steering wheel from SRP should be 30% of stature to enable operators to apply maximum force in manoeuvering the tractor. Considering the limits arrived from the reach envelope (Chap. 11, Fig. 11.3), the centre of the steering wheel should be at least 318 mm far but not farther than 485 mm from SIP horizontally. In the case of vertical location, the centre point of the steering wheel should not be less than 186 mm and not more than 410 mm above SIP. In the case of lateral location, the centre of the steering wheel is generally along the centerline of the tractor. The angle of the steering column is 65°, ranging between 50° and 70° from horizontal	*Diameter* One of the functions of the steering wheel is to transmit the steering forces to the wheels. The small diameter may necessitate more force for steering operation, whereas too large diameter may create difficulty in operation. It would be appropriate that the steering wheel diameter is equal to 95th percentile biacromial breadth (i.e. 420 mm). *Rim width* The rim diameter of the steering wheel may be designed based on the 95th percentile value of middle finger palm grip diameter (38 mm) and the 5th percentile inside grip diameter (39 mm) of male workers	The 5th percentile torque strength with both hands (sitting) is 36 Nm (force 173 N with a lever arm length of 0.21 m). Taking 30% as the limit, the force required in steering wheel operation should not be more than 52 N
Hydraulic control levers	Every tractor is provided with two hydraulic control levers, as position control and draft control levers. These levers are mounted on the right-hand side of the operator by the side of the seat. The levers are operated using a palm grip or finger grip. The operating knobs of these levers should not be lower than 143 mm from SIP (Chap. 11, Fig. 11.3)	The hydraulic control levers are operated frequently during field movements. Due to space limitation for placing the levers, the knobs of the levers are rectangular type, with width and length being between 20 and 40 mm	The force required in operating the levers should not exceed the 5th percentile right-hand push and pull force in sitting posture (47 and 54 N, respectively). Preferably the actuating forces should be lower than 50% of these values, i.e. 24 N for the push and 27 N for pull operation

(continued)

(continued)

Tractor: Controls (Fig. 13.2)			
Parking brake lever	The parking brake lever control is operated only infrequently. Therefore, it may be located either in the optimum area or in a limiting area	The handle grip is cylindrical. The 95th percentile handbreadth across thumb may be taken for the design of length of the parking brake lever (i.e. 110 mm). The diameter of the lever may be about 38 mm	The 5th percentile right-/left-hand push and pull strength in sitting may be taken as the limiting force for operating the parking brake on a tractor. The force required for the operation should not exceed 46 N, considering the left- and right-hand strength limits during pushing and pulling
Gear shift lever/speed selection lever/PTO lever/hand throttle lever	The gear shift lever and hand throttle lever (hand accelerator) are used most frequently. These controls should be located close to the steering wheel to minimize the time for hand movements and to prevent excessive movements of the body	A round knob or knob with flathead is preferred for the gear shift lever, speed selection lever, and PTO lever. The 5th percentile grip span of males is taken as the diameter of the gear shift knob and speed selection lever knob (i.e. 45 mm). For hand throttle lever, a cylindrical knob of 38 mm diameter and 110 mm length is recommended	Like in the case of the parking brake lever, the 5th percentile right-/left-hand push and pull strength in sitting may be taken as the limiting force to operate gear shift lever, speed selection lever, and PTO lever. On average, the force required to operate of these controls should be less than 46 N
Other hand-operated controls	Other hand-operated controls include fuel shut off control knob, seat adjustment lever, hazard switch, plough lamp switch, combination switch (side indicator, horn, and dipper operation lever), multi-position light switch, and starter switch. These controls should be located within the operators' workspace, as defined by the hand reach envelope given in (Chap. 11, Fig. 11.3)	Dimensions of these controls may be decided based on space availability for their placement and standard components available in the market	*Seat adjustment lever* This lever is operated occasionally. The 5th percentile right-hand pull strength (sitting) may be taken as the limiting force for operation of seat adjustment lever on a tractor, which is 54 N

(continued)

(continued)

Tractor: Controls (Fig. 13.2)			
Differential lock pedal	Differential lock pedal is a control generally placed at the right side of the seat and is operated by the heel of the right foot. The differential lock pedal is kept just above the floor surface. As mentioned earlier, the recommended seat height from floor (footrest platform) is 380 mm maximum. Therefore, the location of differential lock pedal may be kept at 450 mm downwards from SIP (keeping 20 mm for pedal movement) and between 90 mm and 426 mm from the centerline on the right-hand side (considering 30° limits for leg abduction)	The width of the differential lock pedal width should be preferably equal to the 95th percentile heel breadth, which is 73 mm. The length of the pedal may be taken as 30 mm for comfortable operation	The 5th percentile right leg strength sitting of male workers is 251 N, which should not exceed in operating differential lock pedal
Footrest/platform and footstep	The operator's feet must be correctly supported on a platform for the comfortable and smooth operation of a tractor. Based on the 5th percentile popliteal height (sitting) of the operator, the footrest/platform should not be lower than 470 mm from SIP. For easy entry and egress, it is necessary to provide a footstep on the left-hand side. If the height of the tractor platform is too high, the height of footsteps from the ground should preferably be not more than 450 mm. The size of the footstep will depend on 95th percentile foot length (270 mm) and foot breadth (110 mm), respectively		

Physical Environment and Safety

Fig. 13.3 Tractor safety and gadgets **a** crossbar seat suspension, **b** rollover protective structure, **c** tractor cabin with wide visibility

Tractor: Physical environment and safety (Fig. 13.3)	
Description	The physical environment in a workplace includes vibration, noise, dust, and thermal conditions. A tractor operator is subjected to these stress agents while carrying out various field operations. As most of the tractors in developing countries do not have cabins, the operators are exposed to severe health hazards. It is necessary to protect the operators' health by keeping the levels of the stress agents within permissible limits by following proper design and operational options.
Design process	
Vibration	The tractor operator gets the whole-body vibration through the seat during travelling over different terrains. The hand-arm vibrations are primarily transmitted through the steering wheel. The vertical components of the vibration (a_z) are more severe, as compared to the horizontal (a_x and a_y) as well as rotational modes. The safe exposure limits (8 h daily exposure) for whole-body vibration, as per ISO 2631/1 (ISO 1997/2017), are 0.63 and 0.45 m/s^2 in vertical and horizontal directions, respectively. Improved seat suspension having crossbar linkage mechanism with proper spring and damper has been attempted to keep the seat vibration within acceptable limits. Putting vibration isolators (rubberized isomers and piezoelectric materials embedded in rubber isomers) below the seat is another useful option to reduce transmission of vibration to the body (Manian et al. 2004; CIAE 2012). Regarding hand-arm vibration, the permissible limit as per ISO 5349 (ISO 2001a, b) is 2.1 m/s^2. If the vibration is greater, isolators can be incorporated in the steering wheel system to reduce the vibration. In some countries, suspended axles or suspended cab are provided on the tractor to protect the operator from vibration
Noise	The tractor operators are exposed to high noise levels. ISO 1999 (ISO 2013/2018) suggests a maximum permissible noise level as 90 dB(A) at operators' ear level for 8-h daily exposure. The tractor noise can be brought down by providing correctly designed muffler (Kumar et al. 2018) Despite the intervention, if high noise level persists, the operators may be provided with personal protective equipment (e.g. earmuffs, earplugs) to reduce ones' noise exposure. Providing a cabin is another option to reduce the operator's noise exposure

(continued)

(continued)

Tractor: Physical environment and safety (Fig. 13.3)	
Dust	In many farm operations such as tillage, threshing, and combining, emission of the enormous cloud of dust is very common. As per OSHA (2019), the total dust and respirable dust in the workplace should not exceed 10 and 5 mg/m^3 of air, respectively. In the tractors with air-conditioned cabin, the dust exposure to operators is less. Preferably, the tractor drivers during field operation should wear a suitable dust mask
Thermal conditions	A tractor cab protects the operator from thermal stress as well as noise and dust. The acceptable ambient temperature of comfort in tractor cabin would be somewhat higher in the summer than in the winter, being 23 to 27 °C and 20 to 25 °C, respectively (humidity ranged between 30 and 60%). In dry tropics, the upper limit of ambient temperature for the comfort zone can be 30 °C. An air velocity of 0.2 to 0.5 m/s is considered optimum for a comfortable environment (ASHRAE 2013/2017)
Safety	Most tractor fatalities and injuries result from overturning accidents while working in the field or during transport on the road. The tractor having safety cabins or rollover protective structures (ROPS) can minimize the risks of accidents, refer to standards ISO 3411 (ISO 2007/2016) and BS EN 15695-1 (BSI 2009/2017). Hsiao et al. (2005) applied anthropometric criteria for the design of tractor cabs and protection frames. The tractor ROPS should comply with ISO 5700 (Static ROPS Test Standard) (ISO 2013). Other safety features to be provided on a tractor include slow-moving vehicle (SMV) emblem, proper lights with turning indicators, neutral safety switch, power take-off shaft guard, and sufficient ballasting on the front side to arrest the overturning during travel on uphill. Also, the entry and egress pathway of the tractor should be well designed for the operator to ride or come down from the tractor smoothly

Power Tiller: Controls

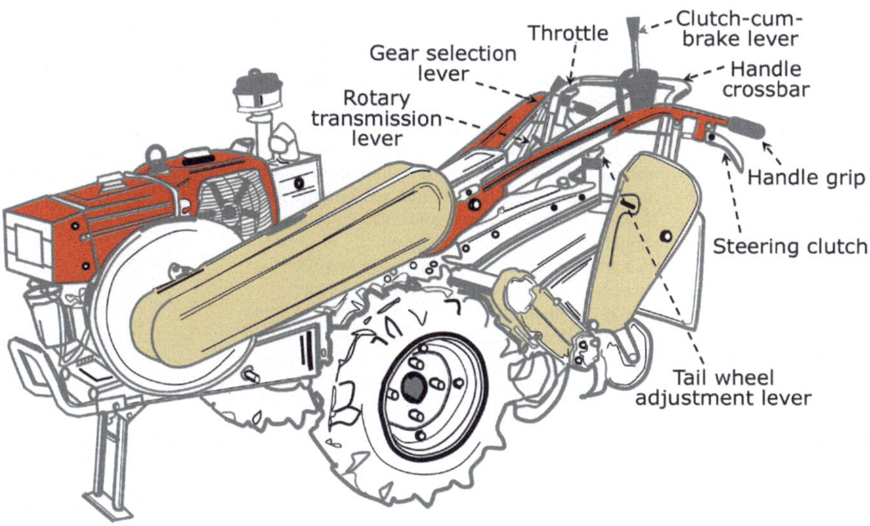

Fig. 13.4 Location of power tiller controls

Power tiller: Controls (Fig. 13.4)			
Description	Power tiller, also called the two-wheel tractor or hand tractor, is self-propelled equipment having a rotary cultivator as its integral part. The equipment can be used as a prime mover after removing its rotary cultivator part and attaching a trailer for transport. Controls of a typical power tiller include handle, steering clutch, clutch-cum-brake lever, gear shift lever, rotary transmission lever, throttle (accelerator) lever, and tail wheel adjustment lever		
User	Male workers		
Mode of operation	Walking type unit with hand-operated controls—push, pull, and rotary (torque) mode		
Design process			
	Location of controls	Dimensions of controls	Actuating force limits
Handles	*Height of handle* The worker puts the machine in gear, operates the clutch, walks behind it, and manoeuvre the unit during field operation. For comfortable operation, the height of handle should be between 95th percentile metacarpal III height (770 mm) and the 5th percentile elbow height (941 mm) of male workers. The design of the power tiller should have the above height range. *Distance between two handles* The worker operates the power tiller by standing between two handles provided on the machine. Sometimes the worker walks within these handles. The distance between the two handles should be such that the clearance available exceeds the 95th percentile bideltoid breadth (475 mm) of a male worker. With a clearance of 100 mm on both sides, the minimum distance between two handles should be 475 plus 200 = 675 mm	*Handle grip and crossbar* The desirable shape of handle grip is cylindrical. As per biomechanics principles, hand placement on handle grip in mid (neutral) position is desirable. The 95th percentile handbreadth across thumb is taken to decide the length of handle grip (i.e. 110 mm). As mentioned earlier, the handle grip diameter is arrived at from the 95th percentile of the middle finger to palm grip diameter (18 and 38 mm), and the 5th percentile of inside grip diameter (39 mm). For the tapered handle grip, the diameter of the smallest and largest section may be within 30 to 38 mm. In a uniform cross-section grip, its value should be less than 38 mm. In most power tillers, a cylindrical handle crossbar (about 38 mm diameter) is provided at about 250 to 300 mm from the edge of the handle grip	*Force limit on handles* The 5th percentile push and pull strength with both hands in standing posture are 133 and 140 N, respectively. The force required to manoeuvre the power tiller should not exceed the above-stated strength levels. The operator may not require to exert force on the handle continuously. Therefore, 50% of these values, i.e. 67 N push and 70 N pull can be taken as the limits for comfortable operation

(continued)

(continued)

Power tiller: Controls (Fig. 13.4)			
Clutch-cum-brake lever/ gear selection lever/ rotary transmission lever	The location of the clutch-cum-brake lever should be such that its farthest position (when the clutch is engaged) should be within the range of the 5th percentile operator. Similarly, the gear selection and rotary transmission lever should have to be within the range of the 5th percentile operator. The distance of the farthest position of these levers from the handle crossbar should not be more than 455 mm. From the edge of the handle grip, these controls should not be more than 750 mm, since the handle crossbar position is about 250 to 300 mm from the edge of the handle grip	The clutch-cum-brake lever is attached at a fulcrum. As the length increases, the force required to operate the lever decreases. Too long a lever may create operational difficulty. The length of the lever is generally within 200 to 300 mm. The dimensional details of different types of levers are the same as that of the tapered or cylindrical handle grip of power tiller described above. Sometimes a spherical knob is provided, where the diameter should not exceed the 5th percentile grip span (45 mm)	Different functional categories of levers are operated by one hand in push/pull mode. The force exerted in the standing posture by one hand is about 4/5th of the force that can be exerted by two hands (Davis and Stubbs 1980). This would suggest that the force required to operate these levers should not exceed 107 N push or 112 N pull (80% of the 5th percentile push or pull strength). Since the operator exerts a force on the levers only intermittently during the operation, 50% of the force exertion, i.e. 54 to 56 N push or pull can be taken as the limits for comfortable operation
Throttle lever *(Accelerator)*	The throttle lever in power tiller is mounted on the handle near the handle grip. This is a finger-operated control, and therefore its diameter generally ranges between 5 and 10 mm. The control operation of the lever is similar to other levers mentioned above, and the limit of force exertion may be at about 54 to 56 N		
Steering clutch	The steering clutch levers are fixed on the handle near the handle grip. The 5th percentile maximum grip length is 93 mm. That is, the steering clutch lever should be mounted in a way that the total distance from the top of the handle grip to the bottom of the steering clutch lever at grip point is less than 93 mm	The clutch levers are curved type, and also tapered for convenience of operation. The diameter ranges between 5 and 20 mm at two ends. The length of the steering clutch is about 150 mm (30 mm to 40 mm more than the handle grip length for fixing purposes)	The steering clutch levers are on the right as well as left hands. The 5th percentile grip strengths of right and left hands are 206 and 195 N, respectively. Taking 30% of the muscle strength for prolonged operation, the limit of force exertion can be taken as 60 N

(continued)

(continued)

Power tiller: Controls (Fig. 13.4)			
Tailwheel adjustment lever	The shoulder grip length at elbow height for the 5th percentile operator is 599 mm. Therefore, the horizontal distance of the farthest point of tail wheel adjustment lever from the end of the handle should not be more than 600 mm	The tail wheel adjustment lever is operated in rotary mode, for moving the wheel up and down. This is executed through the operation of a crank. More the length of the crank less is the force required in operation. Due to space limitation, the crank length is kept at 100 to 150 mm. The spherical knob is provided on the top of the crank. The knob diameter should be less than the 5th percentile grip span, i.e. 45 mm	The operation of the tail wheel adjustment lever is infrequent. The torque required at the knob should be less than 13 Nm (50% of the 5th percentile torque strength of the preferred hand in standing posture)

Power Tiller: Physical Environment and Safety

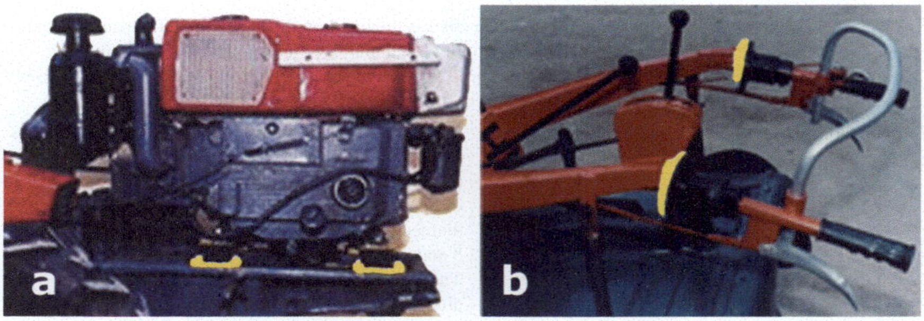

Fig. 13.5 Yellow marked vibration isolators **a** beneath the engine and **b** at handlebar of a power tiller

Power tiller: Physical environment and safety (Fig. 13.5)	
Description	The operator of a power tiller walks behind and manoeuvres the machine for field operation. Some power tillers are also have sitting provision. Here, the operator is exposed to severe hand-arm vibration and high noise, including dust exposure, climatic stress, and physical drudgery of operation. The operators' health must be protected by keeping the levels of stress agents within permissible limits by following proper design and operational options

(continued)

(continued)

Power tiller: Physical environment and safety (Fig. 13.5)	
Design process	
Vibration and noise	In a power tiller operation, one mainly gets hand-arm vibrations through the handle of the unit. These vibrations are generated by the engine and also due to terrain and operational factors. The safe exposure limits (8 h daily exposure) for hand-arm vibration as per ISO 5349 (ISO 2001a, b) is 2.1 m/s^2. The vibration isolators, made of steryl butadiene rubber or other suitable material, mounted between (a) engine and chassis, (b) power tiller frame and handlebar, and (c) handle frame and the handle grip can substantially reduce vibration transmission to the operator (Kathirvel et al. 2004). Power tillers produce high noise during tillage and other operations. The noise level can be brought down by providing correctly designed muffler and fitting the various components rigidly with the frame. In high noise situations, it would be appropriate to provide the operators with personal protective equipment (e.g. earmuffs, earplugs) to reduce ones' noise exposure. The selection of earmuffs and earplugs should be based on the intensity and frequency cut-off
Safety	Many power tiller accidents happen due to contact of the rotary tiller with leg or other body parts. Proper safety cover of the rotary tiller protects the workers from the risk of accidents. Soil throw and dust emission during farm operation are recognized health hazards to workers and others in the vicinity. Wearing dust mask by all concerned workers during field operation, e.g. tillage in dryland partially protects ones' health. In the tropical climate, the thermal stress to the workers can be significantly reduced by changing the work schedules, as described in Chap. 5. Sometimes, a seat is provided in the power tiller with the intent to reduce ones' drudgery of work, due to walking during field operation or while doing road transport . Care is necessary to see that the design of the seat is compatible with the user population, and no harmful transmission of vibration to the whole body takes place

References

ASHRAE. (2013/2017). Standard 55. Thermal environmental conditions for human occupancy. American Society of Heating, Refrigerating and Airconditioning Engineers, Atlanta.

BIS. (1994). IS 11806. Earth-moving machinery, and tractors and machinery for agriculture and forestry – seat reference point. (Now Withdrawn). Bureau of Indian Standards, New Delhi.

BIS. (1998). IS 12343. Agricultural tractors -Operators seat - Technical requirements. Bureau of Indian Standards, New Delhi.

BIS. (1999/2004). IS 11113. Earth-moving machinery, and tractors and machinery for agriculture and forestry – seat index point. Bureau of Indian Standards, New Delhi.

BSI. (2009/2017). BS EN 15695-1, Agricultural tractors and self-propelled sprayers. Protection of the operator (driver) against hazardous substances. Cab classification, requirements and test procedures. British Standards Institution, London.

CIAE. (2012). Research Highlights 2010-12. All India Coordinated Research Project on Ergonomics and Safety in Agriculture. Central Institute of Agricultural Engineering, Bhopal, India.

Clark, T. S., & Corlett, E. N. (1984). *The ergonomics of workspaces and machines: a design manual.* Taylor & Francis.

Courtney, A. J., & Evans, W. A. (1987). A preliminary investigation of bus cab design for Cantonese drivers. *Journal of Human Ergology, 16*(2), 163–171.

Davis, P. R., & Stubbs, D. A. (1980). Force limits in manual work. IPC Science and Technology Press, Guilford.

Gite, L. P. (2017). Final Report. Emeritus scientist project on Development of Ergonomical Design Guidelines for Agricultural Tools, Equipment and Work Places, Central Institute of Agricultural Engineering, Bhopal, India.

Gkikas, N., Richardson, J. H., & Hill, J. R. (2009). A 50-driver naturalistic braking study: Overview and first results. In *Contemporary Ergonomics 2009: Proceedings of the International Conference on Contemporary Ergonomics 2009* (p. 423). CRC Press.

Göbel, M., & Luczak, H. (2000). Ergonomic design of urban bus driver workplaces. *Advances in ergonomics, occupational health and safety* (pp. 67–73). New Delhi: New Age International (P) Limited.

Hansson, J. E., & Pettersson, B. (1980). An ergonomic checklist for transport and materials handling machinery. *An ergonomic checklist for transport and materials handling machinery.* The National Institute of Occupational Health, The Building Industry Labour Research Foundation, The Logging Research Foundation and The Royal College of Forestry), Sweden.

Hsiao, H., Whitestone, J., Bradtmiller, B., Whisler, R., Zwiener, J., Lafferty, C., et al. (2005). Anthropometric criteria for the design of tractor cabs and protection frames. *Ergonomics, 48*(4), 323–353.

ISO. (1995/2016). ISO 5353 - Earth-moving machinery, and tractors and machinery for agriculture and forestry—Seat index point. International Organization for Standardization, Geneva.

ISO. (1997/2017). ISO 2631/1. Evaluation of human exposure to whole-body vibration—Part 1: General requirements. International Organization for Standardization, Geneva.

ISO. (2001a). ISO 5349 (Part 1). Mechanical vibration—Measurement and evaluation of human exposure to hand-transmitted vibration—Part 1: General requirements. International Organization for Standardization, Geneva.

ISO. (2001b). ISO 5349 (Part 2). Mechanical vibration—Measurement and evaluation of human exposure to hand-transmitted vibration — Part 2: Practical guidance for measurement at the workplace. International Organization for Standardization, Geneva.

ISO. (2007/2016). ISO 3411. Earth-moving machinery—Physical dimensions of operators and minimum operator space envelope. International Organization for Standardization, Geneva.

ISO. (2013). ISO 5700, Tractors for agriculture and forestry—Roll-over protective structures—Static test method and acceptance conditions. International Organization for Standardization, Geneva.

ISO. (2013/2018). ISO 1999. Acoustics – Determination of occupational noise exposure and estimation of noise-induced impairment. International Organization for Standardization, Geneva.

Kathirvel, K., Manian, R., sam, B., & Gite, L. P. (2004). Anti-vibration devices for increased comfort of power tiller operators. Technical bulletin. AICRP on Human Engineering and Safety in Agriculture, TNAU, Coimbatore.

Kumar, N., Tewari, V. K., & Singh, G. (2018). Performance evaluation of tractor mufflers for human comfort. *Journal of Agricultural Engineering, 55*(2), 1–12.

Manian, R., Kathirvel, K., Balasankari, P. K., & Mehta, C. R. (2004). *Technical bulletin. Anti-vibration device for increased comfort of tractor operators.* All India coordinated project on ergonomics and safety in agriculture. TNAU, Coimbatore.

Mehta, C. R., & Tewari, V. K. (2000). Seating discomfort for tractor operators–a critical review. *International Journal of Industrial Ergonomics, 25*(6), 661–674.

Mehta, C. R., Gite, L. P., Pharade, S. C., Majumder, J., & Pandey, M. M. (2008). Review of anthropometric considerations for tractor seat design. *International Journal of Industrial Ergonomics, 38*(5–6), 546–554.

OSHA.(2019). Permissible Exposure Limits – Annotated Tables. https://www.osha.gov/dsg/annotated-pels/index.html. Occupational Safety and Health Adminastration. United States Deptt. Of Labour, Washington DC., U.S.A.

Pajnoo, R. K., Saha, K. P., & Mehta, C. R. (2012). Optimum location of steering wheel for Indian tractor operators. *Agricultural Engineering Today, 36*(2), 10–14.

Tewari, V. K., & Prasad, N. (2000). Optimum seat pan and back-rest parameters for a comfortable tractor seat. *Ergonomics, 43*(2), 167–186.

Part VI
Health Services in Agriculture

Chapter 14
OHS Services and Management in Agriculture

Introduction

The need for systems approach to occupational health and safety (OHS) services and management has increasingly been recognized worldwide in all types and sizes of occupational pursuits. The standards bodies, commercial organizations, and industry associations adopted national as well as sectoral OHS management frameworks. Besides, the UN agencies, like ILO and WHO, proposed global strategies for access to OHS services by small enterprises, rural farming, and informal sectors of employment (Goldstein et al. 2001; ILO 2004a; WHO 2007). The ILO Convention 161 (ILO 1985) on OHS services, and the WHO global strategy on occupational health for all, envisages extending services to all sectors of working people. Despite these endeavours, the vast global workforce in farming, amounting to nearly one billion, is grossly deprived of having access to OHS services. The ILO convention 184 (ILO 2001a) addresses OHS issues in agriculture. The WHO initiative on the global plan of action on workers' health, the ILO global strategy on OHS, and the ILO Convention 187 (ILO 2006) on the promotional frame-work address the contemporary requirements.

The concept of occupational health and safety management system (OHSMS), however, has become increasingly popular in the organized sectors of industries to assure compliance with requirements of OHS controls. The emphatic endeavour is visible in applying management framework to implement cost-effective OHS and assisting governments, employers, and employees towards the prevention of workplace injuries and diseases (Robson et al. 2007; Fernández-Muñiz et al. 2009; Stolk et al. 2012). Several standards organizations, accreditation bodies, and the national regulatory agencies have deployed OHSMS aiming at formalizing and implementing the OHS management framework in the enterprises and bringing visibility to its business objectives (Nag 2002; Nag and Nag 2005; Saksvik and Quinlan 2003; Cambon 2007). This contribution elucidates the core characteristics of OHSMS of selected national and sectoral models. It analyses the effectiveness and challenges of the management systems in the diverse sectors of employment, including farming.

P. K. Nag and L. P. Gite, *Human-Centered Agriculture*, Design Science and Innovation,
https://doi.org/10.1007/978-981-15-7269-2_14

OHS Services and Management—A Systems Approach

Since long, there has been an endeavour in defining and characterizing the concept of OHS services and management (Frick and Wren 2000; Alli 2001; Stolk et al. 2012). On the one hand, the OHS services have been viewed as an essential service of sound occupational health methods for the protection of people's health at work, for promotion of health, well-being, and workability, as well as for prevention of ill health and accidents. The OHS management, on the other, takes the systems approach to deliberate linking and logical sequencing of processes and elements of the system that govern the way for a repeatable and identifiable way of managing OHS. Ideally, the system of management of OHS distinguishes from traditional OHS services by being more proactive, process-based, and better internally integrated with the core functions of the organization. That is, to align OHS objectives with business objectives of an organization, integrate OHS programmes and systems into the business systems. It establishes a logical framework upon which to build an OHS programme, devises a set of policies, targets, plans, and procedures, provides an auditable reference for performance benchmarking, and thereby establishes a continual improvement framework towards organizational effectiveness. In other words, an OHSMS is an integrated set of organizational elements involved in the continuous cycle of planning, implementation, evaluation, and continual improvement, directed towards the abatement of occupational hazards in the workplace.

Needless to emphasize that enterprises of all kinds are under constant pressure to comply with OHS regulations. Accordingly, various OHSMS models and schemes have been in use now, with sector-specific applicability. This contribution brings a brief account of different OHSMS models that are applied in large enterprises and also in SME, and in informal sectors. The agricultural sector in a worldwide view is grossly deprived of OHS services. The contribution draws an account of the structure and requirement to developing an unifying approach of OHS management framework in agriculture.

OHSMS Schemes and Models

Various OHSMS schemes and models have been in use now, with relative applicability to diverse sectors. The outlines of different OHS management systems are given in Table 14.1. The management elements of many schemes and models, e.g. BS 8800 (BSI 1996); OHSAS 18001 (OHSAS project group 1999/2007); ILO-OSH 2001 (ILO 2001b); ISO 45001 (ISO 2018), have been explicitly ascribed to Edwards Deming's plan-do-check-act (PDCA) management cycle (Deming and Edwards 1982) as depicted in Fig. 14.1. In the *Plan* phase, an enterprise establishes the objectives and processes necessary to deliver results under its OHS policy. In the *Do* phase, the enterprise shows the implementation of the stated processes. The *Check* phase refers to monitoring, measuring, and reporting of results, and verifying

Table 14.1 Brief outlines of the OHS management systems

BS 8800 (1996)/ ISO 14001 (2015)	OHSAS 18001 (2007)	ISO 45001:2018	ILO-OSH (2001)	SA 8000 (2008)	ANSI Z10 (2012)
HS(G)65 approach 4. OHSMS initial status review **4.1 OHS policy** **4.2 Organizing** (responsibilities, organizational arrangements, documentation) **4.3 Planning and implementing** (risk assessment, legal and other requirements, management arrangements) 4.4 Measuring performance 4.5 Audit 4.6 Periodic status review **ISO 14001 approach** 1-3 Scope, normative references, and terms and definitions 4 Context of the organization 4.1-4.3 Understanding the organization, the needs and expectations of interested parties, and determining the scope of EMS 4.4 Environmental management system 5 Leadership 5.1 Leadership and commitment 5.2 Environmental policy 5.3 Organizational roles and responsibilities 6 Planning	1-3. Scope, references, terms & definitions 4 OH&S management system requirements 4.1 General requirements 4.2 **OH& S policy** 4.3 **Planning** (Hazard identification, risk assessment and determining controls, legal and other requirements, objectives and programme(s) 4.4 **Implementation and operation** (resource, roles, responsibility and authority, competence, training and awareness, communication, participation and consultation, documentation, control of documents, operational control, emergency preparedness, and response) 4.5 **Checking** (performance measurement and monitoring, evaluation of compliance, incident investigation, nonconformity, corrective and preventive action, control of records, internal audit)	1-3. Scope, references, terms & definitions 4 Context of the organization 4.1-4.2 Understanding the organization and its context; the needs and expectations of workers and other interested parties 4.3 Determining the scope of the OH&S Management system 4.4 OH&S management system 5 Leadership and worker participation 5.1 Leadership and commitment 5.2 OH&S policy 5.3 Organizational roles, responsibilities and authorities 5.4 Consultation and participation of workers 6 Planning 6.1 Actions to address risks and opportunities 6.2 OH&S objectives and planning to achieve them 7 Support 7.1 Resources 7.2 Competence 7.3 Awareness 7.4 Communication 7.5 Documented information 8 Operation 8.1 Operational planning and control 8.2 Emergency preparedness and response 9 Performance evaluation	2. National framework for OSH-MS (national policy and tailored guidelines) **3. OSHMS policy** (OSH policy, employee participation) **Organizing** (responsibility, accountability, competence and training, documentation, communication) **Planning and implementation** (initial review, system planning, development and implementation, OSH objectives, hazard prevention, control measures, management of change, emergency prevention, preparedness and response, procurement, contracting) **Evaluation** (performance monitoring and measurement, *investigation of work-related injuries*, OSH performance, audit, management review) **Action for improvement** (preventive and corrective action, continual improvement)	**Social accountability requirements** 1. Child labour 2. Forced and compulsory labour 3. Health and safety 4. Freedom of association & right to collective bargaining 5. Discrimination 6. Disciplinary practices 7. Working hours 8. Remuneration **9. Management systems** 9.1 Policy 9.2 Management representative 9.3 SA8000 Worker representative 9.4 Management review 9.5-9.6 Planning and implementation 9.7-9.10 Control of suppliers/ subcontractors and subsuppliers 9.11-9.12 Addressing concerns and taking corrective action 9.13-9.14 Outside communication and stakeholder engagement 9.15 Access for verification 9.16 Records	1-2 Scope, purpose & application, and definitions **3 Management leadership & employee participation** 3.1 Management Leadership (OHSMS, policy, responsibility, and authority) 3.2 Employee participation **4 Planning** 4.1 Initial and ongoing reviews 4.2 Assessment and prioritization 4.3 Objective 4.4 Implementation plans and allocation of resources **5 Implementation and operation** 5.1 OHSMS operational elements (risk assessment, the hierarchy of controls, design review, management of change, procurement, contractors, emergency preparedness) 5.2 Education, training and awareness 5.3 Communication 5.4 Document and record control **6 Evaluation and corrective action** 6.1 Monitoring and Measurement 6.2 Incident Investigation 6.3 Audits 6.4 Corrective and Preventive Actions 6.5 Feedback to the Planning Process

(continued)

Table 14.1 (continued)

BS 8800 (1996)/ ISO 14001 (2015)	OHSAS 18001 (2007)	ISO 45001:2018	ILO-OSH (2001)	SA 8000 (2008)	ANSI Z10 (2012)
6.1 Actions to address risks and opportunities 6.2 Environmental objectives and planning 7 Support 7.1-7.5 Resources, competence, awareness, communication, and documented information 8 Operation 8.1 Operational planning and control 8.2 Emergency preparedness and response 9 Performance evaluation 9.1 Monitoring, measurement, analysis and evaluation 9.2 Internal audit 9.3 Management review 10 Improvement (Nonconformity and corrective action, Continual improvement)	4.6 **Management review**	9.1 Monitoring, measurement, analysis and performance evaluation 9.2 Internal audit 9.3 Management review 10 Improvement 10.1 General 10.2 Incident, nonconformity and corrective action 10.3 Continual improvement		**Note:** SA 8000 management system is included herewith, considering its proximity and relevance to OHSMS.	**7 Management review** (review process and review outcomes) **Appendices (A-O):** A — Policy statements B — Roles and responsibilities C — Employee participation D - Planning-identification, assessment, and prioritization E — Objectives/ implementation plans F — Risk assessment G — Hierarchy of control H — Management of change I - Procurement J - Contractor safety and health K - Incident investigation L — Audit M - Management review process

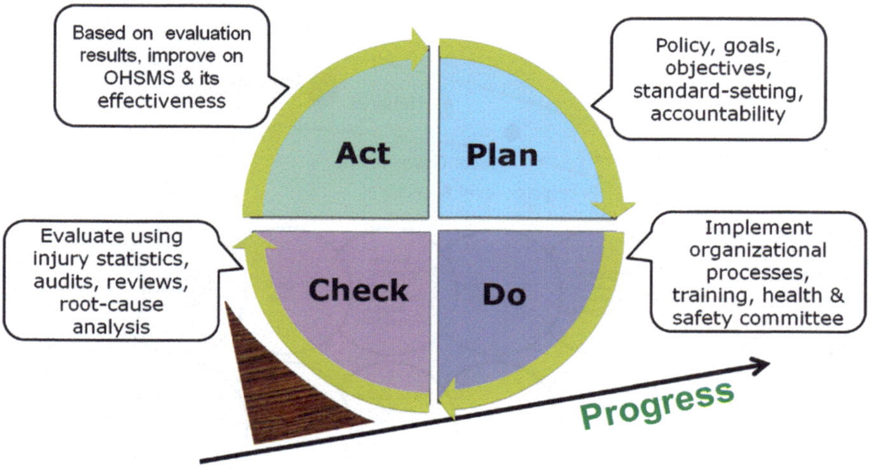

Fig. 14.1 Edwards Deming's plan-do-check-act (PDCA) management cycle

whether the planned objectives have been met. On evaluation, in the *Act* phase, suitable corrective measures are implemented to improve results further.

An organization implements OHSMS for the systematic identification, evaluation, and prevention or control of workplace hazards. In harmony with the ISO 9001 (ISO 2008) and ISO 14001 (ISO 2015) systems the guidance and documentation on OHSMS might follow a pyramidal structure; that is,

(a) *The top-level is the OHSMS manual, referring to the scope and policy of the management system. The orderly arrangement of interdependent activities depicts the processes of an organization.*
(b) *The second level is made up of the standard operating procedures (SOPs)—task procedures, health and safety procedures, risk assessment, and specific instructions to the processes and products.*
(c) *The third level refers to the records management that essentially means the documented evidence of OHS performance.*

The structural content of the OHSMS documentation may be organized, depending on the model or scheme advocated (Nag 2002). The OHSMS variables and dimensions that are central to a systems approach are thematically shown in Fig. 14.2. A brief description of the variables and documentation needs, for example, are given in Table 14.2.

Framework of the OHSMS Models

Brief outlines of different national and international OHSMS country-specific models and standards are mentioned in Table 14.3. The management elements of

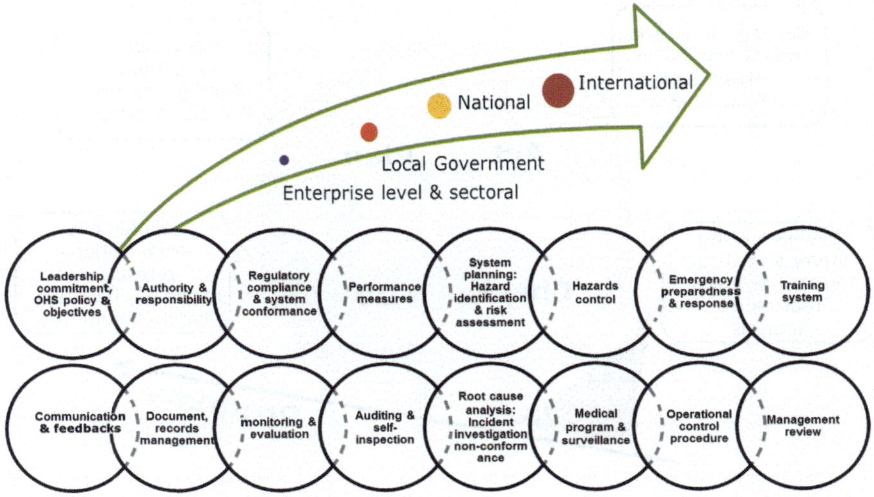

Fig. 14.2 OHSMS variables and dimensions

several models correspond with the Edwards Deming's plan-do-check-act (PDCA) management cycle, as depicted in Fig. 14.1

British HSE Guidance and BS 8800—Under the regulatory provisions of the British Health and Safety Act, the HS (G) 65 (HSE 1993/2013) approach provided the basis to employers on managing health and safety (Fig. 14.3). The BS 8800 (BSI 1996) on OHSMS was based on two approaches, i.e. the HSE guidance—HS (G) 65 and ISO 14001 standards, to improve OHS performance in organizations and build responsible image within the marketplace. Structurally, the BS 8800 contains annexes on practical guidance, including planning and implementing, risk assessment, and measuring performance.

OHSAS 18001:1999/***2007***—A specification standard, occupational health and safety assessment series (OHSAS 18001:1999) was evolved from different national standards and proprietary certification schemes, and further revised in 2007 (BSI 1999/2007; Kausek 2007), making it significantly closer to the ILO-OSH (ILO 2001a, b). The OHSAS 18001 specification standard sets out the minimum requirements (Table 14.1) for a best practice OHSMS and applies to organizations of all types and sizes and accommodates diverse geographical, cultural, and social conditions. The schematic of the system elements of OHSAS 18001 requirements is illustrated in Fig. 14.4. The objective outlined in OHSAS 18001 was to e*stablish an OHSMS to eliminate or minimize risk to employees and other parties who may be exposed to OHS risks associated with its activities; implement, maintain, and continually improve an OHSMS; assure itself of its conformance with its stated OHS policy; demonstrate such conformance to others; seek certification of its OHSMS by an external organization, and make self-determination and declaration of conformance with the OHSAS specification* (BSI 1999/2007). The OSHAS 18002-2008 (BSI 2000/2008) guidance

Table 14.2 The OHSMS dimensions for documentation

Dimensions	Description	Documentation (indicative)
Leadership commitment, OHS policy, and objectives	Commitments to OHS development are operationally defined. The formal OHS policy reflects the leadership commitment in setting measurable goals and objectives towards meeting OHS performance expectations of the organization	Identify the core elements of OHSMS on a time horizon (e.g. to improve existing OHS features; reduce the risk levels; ascertain the frequency of undesired incident(s); recognize disorders and diseases
Accountability, responsibility, and authority	The organization defines the roles of personnel who are involved in OHSMS functions. Executive ownership and accountability are critical (e.g. budget, health, and safety committee, medical services, emergency preparedness, training)	A management representative is designated, making him accountable for the establishment of prevention and health promotion programmes, through anticipation, recognition, evaluation, and control of OHS hazards. Roles and responsibilities shall be documented and communicated to employees and other relevant parties
Regulatory compliance and system conformance	Government and non-government regulations and standards to which an organization subscribes impose requirements on OHSMS. Broadly, both regulatory compliance and conformance to other system requirements lay the groundwork for a systematic management approach	Emphasis is on the framework of action on OHS, to recognize the regulatory requirements, disclosure of information, backup enforcement, and prosecution. The documented procedure elucidates how well the organization complies with legal OHS requirements. – recording the results of legal OHS compliance evaluations; – identifying activities that might be affected by applicable legal and other requirements; – monitoring the control interventions, consequent to changes in OHS legislation

(continued)

quotes the specific requirements from OSHAS 18001 and gives relevant guidance. The broader objectives of OHSAS 18001 are further embodied in the ISO 45001 (2018), for wider acceptance.

ILO-OSHMS: (ILO 2001a, b)—Recognizing the positive impact of introducing OHSMS at the enterprise level, the International Labour Office brought out a consensus model—ILO-OSH, that reflects the ILO values of advancing the objectives of labour conventions, ratified by member nations. Some related conventions are 155—Occupational safety and health (ILO 1981), 161—occupational health services (ILO 1985), 162—asbestos (ILO 1986), 167—safety and health in construction (ILO 1988), 170—Chemicals (ILO 1990), 174—prevention of major industrial accidents (ILO 1993), 176—safety and health in mines (ILO 1995), 184—safety and health in agriculture (ILO 2001a), and others. As mentioned, the ILO convention 187 defines the objective of strengthening occupational health services, expanding and improving its coverage, content, and activities (ILO 2006).

Table 14.2 (continued)

Dimensions	Description	Documentation (indicative)
Performance measures	For valid and reliable OHS performance measurements, the indicators, variables, measurement units, and their logical relationships are established. The distinction has been made between the leading and trailing indicators	A documented procedure is required to monitor and measure OHS performance. It outlines the parameters that determine, for example, whether the OHS policy and objectives are achieved, risk controls have been implemented, lessons learned from OHSMS failures (accidents and illnesses), the effectiveness of training, communication and consultation programmes, and information for management review
System planning for hazard identification and risk assessment	The organization requires a documented OHS management programme (strategies and plan of actions), encompassing all significant OHS hazards in its domain. Primarily, the hazard is defined as a source or situation with the potential to cause harm, such as injury or ill health, damage to property, and workplace environment	– the emphasis is on identifying hazards and determining the level of risks associated with the hazards; – describing the measures to monitor and control the risks, mainly those not tolerable; – OHS objectives and actions to reduce identified risks, and follow-up monitoring activities; – ascertaining core competency and training requirements to implement the control measures
Hazard Control	The methods of the hazard control system cover administrative controls, engineering controls, or by using personal protective equipment (PPE). The hazard control covers the management of hazardous agents, e.g. radioactive materials, noise, heat, cold, lasers, hazardous wastes	Emphasize on hazard identification and risk assessment. Records of hazard control should reflect the principles of eliminating hazards and reducing risks (reducing the likelihood of occurrence or potential severity of injury or damage)
Emergency preparedness and response	This refers to how the organization prepares for and responds to emergencies and accidents. It develops plans and procedures to cope with them, test its planned responses and seeks to improve the effectiveness of emergency responses	– identification of potential accidents and emergencies; – personnel for emergency response (e.g. firefighters, first aid staff, toxic spillage specialists); – documented emergency plans, including evacuation procedures; – interface and communication with external emergency responders; – availability of vital information, e.g. plant layout drawings, hazardous material data; – emergency equipment list (alarm systems, means of escape, fire fighting equipment, first aid equipment, test records); – records and review of the practice drills, and recommended actions arising from the reviews

(continued)

Table 14.2 (continued)

Dimensions	Description	Documentation (indicative)
Operational control procedure	The organization establishes operational control procedures to identify risks (including those introduced by contractors or visitors) and records the instances where any failure might lead to accidents or deviation from OHS policy and objectives. The procedures are to be reviewed periodically	– *Purchase or transfer of goods and services and use of external resources (e.g.* purchase or transfer of hazardous materials; handling of machinery; provisions for new plant or equipment, and OHS competency of contractors); – *Hazardous tasks* (e.g. identification of hazardous tasks; predetermination of working methods; pre-qualification of personnel; permit–to–work systems to control exposures to hazardous tasks); – *Hazardous materials (e.g.* inventory identification, Materials Safety Data Sheet (MSDS), storage provisions, and control of access;); – *Maintenance of safe plant and equipment* (e.g. segregation and control of access; testing of OHS related equipment; local exhaust; ventilation systems; medical facilities; PPE; machine guarding; shutdown, fire suppression; radiological sources; safeguards, and monitoring devices)
Training system	The training is an integral component of OHS management, having qualified personnel to provide OHS services to the organization	– an effective procedure ensures competency requirements for individual roles; – analysis of training needs and programmes for employees; and – training and evaluation records. The emphasis is on providing training to all participants and should preferably take place during the working hours
Communication System/ feedback channels	An effective communication system with defined feedback channels facilitates participation in good OHS practices. A viable communication system should identify how and to whom information on the functioning of the OHSMS will be transmitted	A defined procedure is required for worker participation and consultation with contractors and other parties for appropriate communication and feedback. – formal management for employee consultations through OHS committees; – employee involvement (and employee OHS representatives) in hazard identification, site inspection, risk assessment, incident investigation, and control; – OHS briefings for employees and other interested parties; – notice boards containing OHS performance data and newsletter
Document and records management	A well-functioning document and record management system is a crucial indicator of whether the OHSMS is currently in conformance; that is, documentary evidence that the OHSMS operates effectively, and the processes have been carried out under safe conditions	All documents and data critical to the operation of the OHSMS should be identified and controlled. The document control procedure should define the controls for the identification, approval, issue, and removal of OHS documentation, together with the control of OHS data

(continued)

Table 14.2 (continued)

Dimensions	Description	Documentation (indicative)
Monitoring and evaluation	The monitoring and evaluation of an OHSMS encompass performance monitoring, auditing, and self-inspection, incident investigation, root cause analysis, the availability of medical surveillance and health programmes, and management review	– procedure(s) for monitoring and measuring, and evidence of the results of implementing OHSMS; – measuring equipment lists, measurement procedures; critical equipment lists, including equipment inspection checklists; – calibration records, maintenance activities and results; – OHSMS checklists, medical surveillance programmes
Auditing and self-inspection	The planned auditing (internal and external) allows reviewing the organization's OHSMS effectiveness and conformity to the specification it subscribes. Generally, the OHSMS requires consultation on the selection of auditors, and the person undertakes the audits should be impartial and objective	– OHSMS audit plan and procedures; – OHSMS audit reports, including nonconformance reports, corrective action requests; – closed-out nonconformance reports; – evidence of the reporting of the audit results to top management. The OHSMS models are a mix of auditable and non-auditable standards. One may opt to have its OHSMS audited against a certification standard. The process of certification offers independent verification and auditing that an organization becomes more self-regulatory in promoting OHS
Incident investigation, nonconformance—root cause analysis	Incident investigation refers to the activities in determining the origin and cause(s) of accidents and incidents. The organization shall have documented procedures for reporting and evaluating/ investigating accidents, and nonconformances. Root cause analysis allows in moving up the causal chain in detecting, analyzing, and eliminating potential causes of nonconformities	– incidents investigation reports, including hazard identification, risk assessment, and risk control reports; – nonconformance register and reports; – evidence of evaluation of corrective and preventive actions taken; – management review inputs
Medical surveillance and health programmes	This refers to the activities associated with providing OHS services and developing the operation of a health surveillance and health promotion programme. The OHSMS models evolved on the ISO 14001 (ISO 2015) templates do not explicitly mention the elements of the medical programme; however, these aspects provide feedback to the hazard control system	For medical surveillance, the professional may consider the ILO (2002) list of occupational diseases, concerning the exposure characteristics of workers. WHO (2007) proposed a national policy framework to deal with all aspects of workers' health, including primary prevention of occupational hazards, protection and promotion of health at work, employment conditions, and a better response of the health systems to workers

(continued)

Table 14.2 (continued)

Dimensions	Description	Documentation (indicative)
Management review	A periodic management review involves evaluating the ability of OHSMS to meet the needs of the organization, its stakeholders, employees, and regulating agencies	– minutes of the review meeting; – revisions to the OHS policy and objectives in light of audit results, changing circumstances and commitment to continual improvement; – corrective/improvement actions, with assigned responsibilities and targets for completion; review of corrective action; – emphasis on the planning of future internal OHSMS audits
Continual Improvement	The OHSMS builds on the concept of continual improvement to eliminate or minimize the potential risks of workplace injury and illness	The enterprise demonstrates, through appropriate documentation, the processes of OHSMS to achieving continual improvement
Integration	Integration can be viewed from both macro- as well as micro-level perspectives. The macro perspective emphasizes a systems approach to OHSMS as a cost-effective operational framework within the scope of the business. The micro-level perspective considers integration based on the model elements. Besides, the relative effectiveness of OHSMS models depends on its compatibility with other management systems in practice	An enterprise certified to process-based models of ISO 9001:2008 or ISO 14001:2015 has already in place the generic framework to integrate health and safety, and other management systems. Compliance with ISO 45001 (2018) standard on health and safety would require the integration of OHSMS elements into the overall policy and management arrangements of the enterprise

Structurally, the ILO-OSH model is distinctively different from the OHSAS 18001, by the unique focus of the ILO model in developing national OHS policy and guidelines, including tailored enterprise-level management framework (Fig. 14.5). For enterprise-level applications, the ILO-OSH model is constituted by sixteen elements that are grouped under five sections, namely, policy, organizing, planning and implementation, evaluation, and action for improvement (Fig. 14.6). This model has an emphasis on the dimensions, such as employee participation, training at all levels, injury investigation, and health programmes (Table 14.1).

SA 8000 (Social Accountability Standard)—SA 8000 (SAI 2008/2014) seeks to guarantee the fundamental rights of workers and workplace conditions (Fig. 14.7). The requirements of workers, such as child labour, forced labour, health and safety, right to collective bargaining, discrimination, disciplinary practices, working hours and remuneration, have been defined concerning national laws, international human rights norms and the principles of the tripartite ILO conventions. The SA 8000 is an auditable framework of the management system. It enables enterprises to develop and enforce policies and procedures to manage those issues which it can control and influence. Besides recognizable well-structured system framework, the SA 8000 requires compliance of provisions of related ILO conventions and recommendations on OHS.

Table 14.3 Brief outlines of the country adopted OHS management systems

US OSHA programme (1970)	Australia/New Zealand		Spain (UNE 81900)	Norway (96/402803)	The Netherlands (NPR 5001)	Ireland (NSAI SR 320) (1997)	Japan (JISHA 1997)
	AS/NZS 4801 (Australia/New Zealand) (1997)	SafetyMAP (1994/2002)					
VPP (1988) a. Management commitment and planning b. Hazard assessment c. Hazard correction and control d. Safety and health training e. Employee participation **Safety and Health Program (1989)** a. Scope and application b. Basic obligation c. Management leadership and employee participation d. Hazard assessment	4.1 Commitment and policy (leadership and commitment; initial OHS review, OHS policy) 4.2 Planning (hazards (identification, assessment, and control of risk, legal and other requirements, objectives and targets, performance indicators, management plans) 4.3 Implementation (ensuring capability, support action, risk assessment and control, contingency preparedness and response) 4.4 Measurement and evaluation	1 Building and sustaining a commitment 2 Documenting strategy 3 Design and contract review 4 Document control 5 Purchasing 6 Working safety by the system 7 Monitoring standards 8 Reporting and correcting deficiencies 9 Managing movement and materials 10 Collecting and using data 11 Auditing of management systems 12 Developing skills and competencies.	4.1 OHS policy 4.2 OHS management system 4.3 Responsibilities (management and staff responsibility and resources, management review) 4.4 Risk evaluation (record of legal, regulatory and other requirements, hazard identification, Risk evaluation and control, maintenance of risk control measures) 4.5 Prevention planning (OHS management program) 4.6 OHS handbook and documents	4.1 External requirements (stakeholder, legal requirements) 4.2 Management responsibility (leadership commitment, policy, and objectives, system documentation) 4.3 Management system structure (organization, documentation) 4.4 Communication (internal and external) 4.5 Planning (human resource management, establishing targets, management program, management of change, loss prevention, contingency, and emergency	4.3 Maintenance and development of OHSMS 5 First and periodic review of OHS status (initial and periodic management review) 5.3 Risk assessment (initial and periodic risk assessment) 6 Decision-making and planning (OHS policy, plan of action) 7 Organization and implementation 7.1 Organization (responsibilities, authority, resources, OHS targets, consultation with staff,	4. OHSMS principles 5. Initial status review 6.1 OHS policy 6.2 Planning (legal and other requirements, objectives and targets, planning for OHS) 6.3 Implementation and operation (structure and responsibility, consultation, training, awareness and competence, communication, documentation, operational control, management and control of contractors, emergency preparedness and response) 6.4 Checking	0. Corporate policy and the CEO's policy 1. Establishment's policy, director's Policy 2. OHS management programs 3. OHS management manuals 4. OHS management organization system 5. OHS committees 6. Guidance to subcontractors 7. OHS education 8. OHS daily activities 9. Machinery and equipment safety 10. Chemical safety 11. Working environment

(continued)

Table 14.3 (continued)

and control e. Information and training f. Program evaluation g. Dates h. Definitions.	(inspection, testing, and monitoring; audits, corrective and preventive action) 4.5 Review and continual improvement.		4.7 Performance control (active control, verification, ensuing control, non-conformity and corrective actions) 4.8 Safety and health records 4.9 Evaluation of OHSMS (audits, review of OHSMS).	preparedness) 4.6 Operational and functional management (marketing, development, procurement, processing) 4.7 Measuring and evaluation (process performance, evaluation, assessment, audit, review, records) 4.8 Improvement (nonconformance handling, preventive and corrective action, improvement).	representatives, qualification, training and information, expert assistance, information management, internal/ external information, and communication) 7.2 Implementation of the plan of action 8 Performance measurement (review OHSMS, evaluation of the plan of action).	and corrective action (monitoring and measurement, accidents and incidents of non-compliance with OHSMS - corrective and preventive action, audit, records) 6.5 Management review.	12. Work management 13. Medical surveillance 14. Promotion of physical and mental health 15. Comfortable workplaces 16. Measures for workers in need of special care 17. Traffic safety of commuters 18. Analysis of accident causes 19. Emergencies.

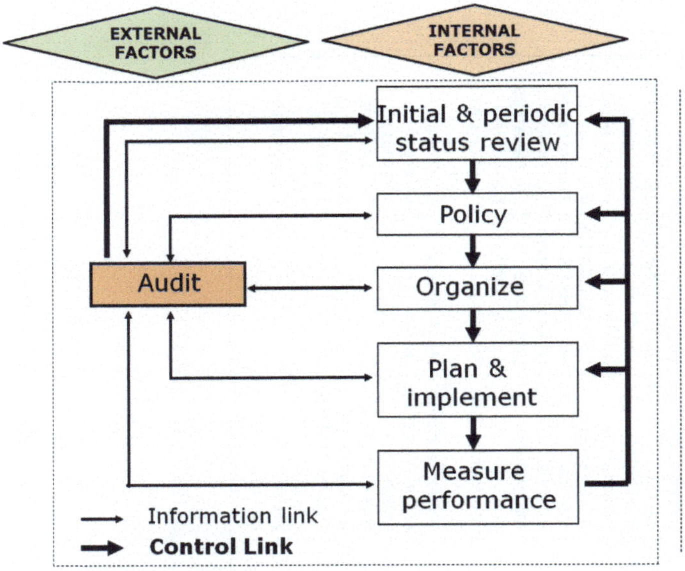

Fig. 14.3 British Health and Safety HS (G) 65 approach to managing health and safety

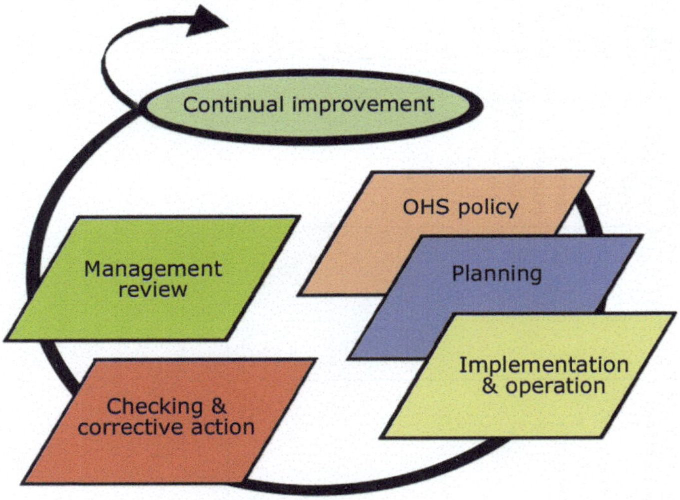

Fig. 14.4 System elements of OHSAS 18001 requirements

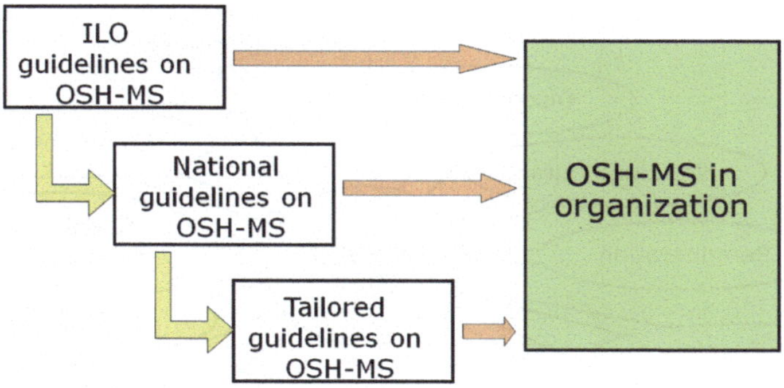

Fig. 14.5 ILO model in developing national OHS management guidance

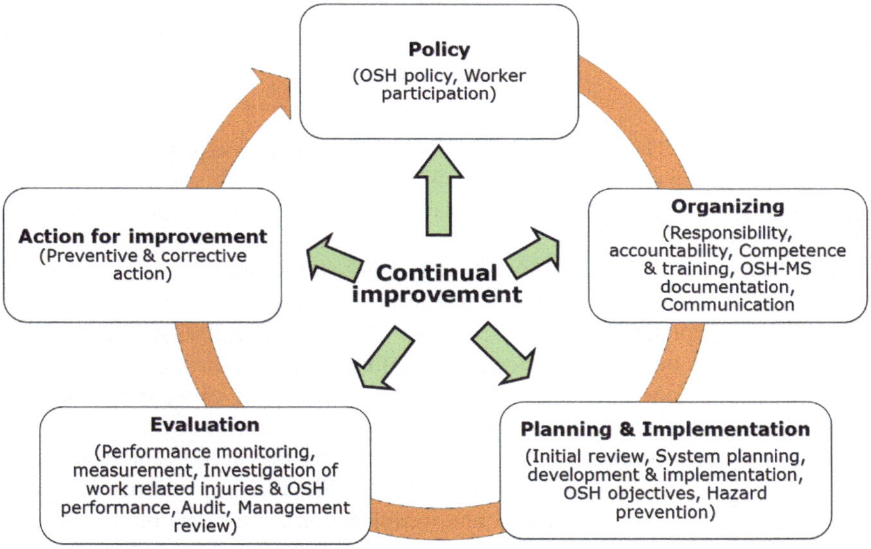

Fig. 14.6 System elements of ILO-OSH model

ANZI Z10 (ANSI 2005*)*—The American standard provides a new blueprint of OHSMS for widespread benefits in health and safety, productivity, financial, performance, quality, and other business objectives. On the primary template of Deming's P-D-C-A cycle of continual improvement, the Z10 framework includes the conventional system elements of BS 8800, as described in Table 14.1. The fifteen appendices elucidate the dimensions of the system elements, such as policy statements, roles and responsibilities, assessment and audit information. Under the implementation and operation of the system, the hierarchy of controls sets enterprises to adopt risk reduction steps through elimination, the substitution of less hazardous materials/operation/equipment, engineering controls, warnings, administrative controls,

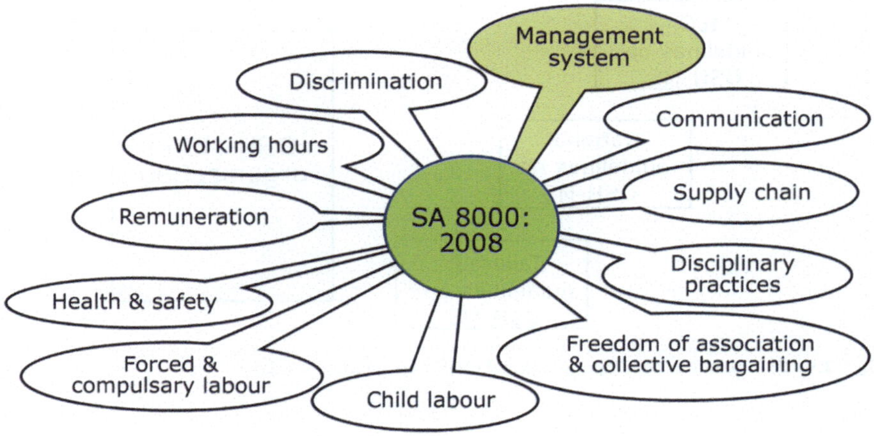

Fig. 14.7 SA 8000 (Social Accountability Standard) on the fundamental rights of workers and workplace conditions

and the use of PPE. The management review covers the review process, outcomes, and follow-up that may be conducted at least once every year. This would ensure that proper actions are taken following evaluation and improvements are continuously done.

ISO 45001: 2018—The OHSAS 18001: 1999 (revision 2007) as a model of OHSMS has been widely practised in as many as 127 countries, and hundreds and thousands of enterprises have been certified to OHSAS 18001:2007. Concerns have often been voiced on its international credibility since the scheme has not been passed through the formal standards development process. The voluntary forms of the specification documents create considerable confusion in the OHSMS certification across the sectors of business. The vivid need for ratifying OHSMS internationally led to establishing a new standard as ISO 45001 (ISO 2018) to improve OHS for everyone on local, national, regional and to implement in both the developed and developing countries effectively.

The ISO 45001: 2018 (Occupational health and safety management systems—Requirements with guidance for use) is a single document with specifications and an interpretation guidance annexure. The specification clauses follow the same high-level framework as in ISO 9001 and ISO 14001, such as (1) scope, (2) normative references, (3) terms and definitions, (4) context of the organization, (5) leadership and worker participation, (6) planning, (7) support, (8) operation, (9) performance evaluation, and (10) improvement. Annex A provides the general guidance on the interpretation of ISO 45001. Clause 3 of the standard describes the definitions such as *worker/employee* and *workplace,* and about changing the term *contractor* to an *external provider;* Clause 4 elucidates the context of the organization, needs, and expectations (requirements) of workers and interested parties. The standard emphasizes leadership and management roles, including worker participation (Clause 5). The top management requires to demonstrate its commitment and actions to OHS performance and outcomes. No OHSMS is valid unless the organization's OHS aspects receive visibility and is embodied in the overall management system. Clause 6 (planning) indicates the necessary actions to address OHS risks, identify hazards and incidents, and defines the organizational OH&S objectives and planning. Clause 7 highlights the requirements on resources, competence, and awareness, including documented information, communication, participation, consultation, and representation. Clause 8 is centred on the hierarchy of control, management of change, outsourcing, procurement, contractors, emergency preparedness, and response. Clause 9 refers to monitoring, measurement, analysis, and the evaluation of compliance, with a concomitant challenge on the aspects of internal audit and management review. Clause 10 elaborates on the incident investigation, nonconformity, corrective action, and continual improvement.

The intent behind the framework is to maintain alignment with other ISO management systems for organizations to establish integrated management systems covering multiple disciplines. Those familiar with the specifications of different OHSMS models may recognize that the core of the ISO 45001: 2018 primarily draws material from OHSAS 18001:2007, ANSI Z10:2012 standard, and other specifications. The challenge of the new standard is to better define the contexts and functions of an organization to look beyond its immediate OHS issues (e.g. internal employees, contractors, and suppliers), and how their work might affect the neighbours in the surrounding area. In other words, the new standard should allow organizations to determine the breadth of application pertaining to them. Emphasis is lacking on the environmental and biological monitoring and diagnostic criteria of occupational diseases and disorders.

Sectoral OHSMS Models

Several sector-specific auditable, as well as non-auditable OHSMS frameworks, have been evolved for application, namely, in chemicals, petrochemicals, construction, oil and gas installations, and some of those are given in Table 14.4. While the models have a vast difference in the structure and content, they have a distinct resemblance of using P-D-C-A framework in managing the workplace environment, products as well as health and safety.

The MASE/UIC (2007), Manuel d'Amélioration Sécurité Enterprise, was originated in the 90s, led by Exxon, to better structure the security approaches and industrial hygiene in petrochemical sites. The MASE has also been used in other industrial sectors, such as steel, nuclear. MASE defines the minimum requirements of a prevention system revolving around (a) the commitment of the management of the company, (b) competence and professional qualifications of the staff, (c) the preparation and organization of work, (d) controls, and (e) continuous improvement. Whereas the DT 78 (MASE/UIC 2007) allows a security agreement between the French Union of

Table 14.4 Brief outlines of the sectoral OHSMS schemes

MASE/ DT78 (2007)	GEHSE (2004)	Responsible Care Management System (UK) (1998)	E&P Forum (oil and gas companies) (1995)
1. Policy of SHE (responsibility for the direction, Policy of safety, health and environment (SHE), Indicators and expected dissemination, Planning of SHE, Organization of resources, Animation, Communication, analysis of causes of accidents and incident) 2. Human resources (training and professional qualification, new entrants, the appointment of personnel, competency, skills empowerment, system appropriation) 3. Operation (preparation, organization intervention, material and facilities, plan of prevention, matters of subcontractors, management documentation) 4. Controls (audits, inspection, and verification) 5. Continual improvement, management review	**2. Policy and personnel** (HSE policy, training specifications, Empowerment of stakeholders) **3. Planning and operation** (Planning details, Plan of prevention, identification, and analysis of risk, Treatment of Modifications, Work Permit, Emergency preparedness) **4. Subcontractors/ outsourcing** **5. Measures of performance** (safety, medical surveillance, revisit of experience) **6. Audits and controls** (Inspection and audits, Corrective Actions and anomalies, Management review)	**2-3. Leadership commitment, Policy** **4. Identifying requirements** (regulatory and other requirements, risk assessment, defining programme) **5. Planning** (HS&E objectives, setting targets for improvement, planning for control, performance criteria, emergency preparedness) **6. Organization** (structure and responsibility, MR, resources, competence, documentation, communications) **7. Implementation and control** (people, purchasing, contractors, manufacturing, storage, transportation, distribution, management of change) **8. Monitoring** (measurement, checking, and inspection, internal audit, improvements, records) **9. Management review**	**1-2. Leadership commitment, Policy, and strategic objectives** 3. **The organization, resources, and documentation** (structure and responsibilities, resources, competence, contractors, communication, documentation control) **4. Evaluation and risk management** (identification, evaluation, recording of hazards and effects, objectives and performance criteria, risk reduction measures) **5. Planning** (asset integrity, procedures, work instruction, management of change, contingency, emergency planning) **6. Implementation and monitoring** (activities, monitoring, records, non-compliance, corrective action, incident reporting, follow-up) **7. Auditing and reviewing**

Chemical Industries and employees to set up the OHS system framework. Currently, MASE and DT 78 standards are one, to put in place a common approach for safety improvement among European enterprises, especially in the chemical sector. The external auditor assesses the compliance documentation and conditions and accordingly grants authorization for 1 to 3 years, according to the level and scale of implementation. Corollary to MASE certification, the GEHSE (2000/2004) is the specification guide on hygiene, safety, and environment for outside contractors involved in hydrocarbon deposits or small oil facilities, including service stations. It is generally present in parts of the French territory, not subject to MASE.

Responsible Care—The US Chemical Manufacturers Association designed the programme of Responsible Care (ACC 1988), including the health and safety code in management practice. This is a multidisciplinary means to enable member enterprises to promote OHS among employees, contractors, and the public, and to protect the environment. The British Responsible Care management system includes a guidance and self-assessment document and describes the critical requirements of OHS and environment management system for incorporation into the management systems of an enterprise. The responsible care does not explicitly address medical programme and surveillance; however, the guidance includes the requirements of British HSE—HS (G) 65 (HSE 1993/2013), Eco-Management and Audit Scheme, EMAS (EC 1993), ISO 14001 (ISO 2015) and BS 8800 (BSI 1996).

E&P Forum and other business charters (oil and gas companies)—The Exploration and Production Forum (E&P Forum) guidelines (Downey 1995) of the international association of oil and gas companies refer to OHSMS and environmental management systems in exploration and production operations worldwide. The management system addresses core structural issues for programmes and procedure development, including controlled documentation. The structure and content of the E&P guidelines are sufficiently generic to make adaptable to different types of enterprises, systems, and cultures.

Global Acceptance of OHSMS Models and Programmes

The revelation of alarming statistics of work-related accidents or diseases (about 2.3 million people die every year as per ILO 2020) reaffirms the necessity of a strong OHS management aspects, systems, and services across the sectors of employment. This would allure culture of positivity at workplaces, and in turn, mitigate health risks and accidents, avoid legal entanglements, and reduce insurance costs. Concerns are vivid on the international credibility of the practised OHSMS schemes and models, for synergized application in the developed as well as in the developing countries. For the benefit of the readers, this contribution schematically categorized the existing OHSMS models and schemes like the following:

- Mandatory OHSMS with regulatory and advisory measures;
- Voluntary OHSMS standards with and without certification support;

– Good OHS practice guides, including self-regulatory OHSMS schemes; and
– Sector-specific OHSMS applications.

Mandatory OHSMS with Regulatory Measures

Based on a set of OHS management principles, the OSHMS arisen from government legislation and advisory, are primarily the mandatory approach to make the regulations uniformly applicable to all types of enterprises, including small workplaces. The European Union Directive 89/391/EEC (EU-OSHA 1989; Walters et al. 2002), for example, defines employers' responsibilities in OHS management, including evaluation of risks to health and safety of workers, implementation of preventive measures, and integration of measures into the activities of the enterprises at all hierarchical levels.

The Scandinavian countries make obligatory to integrate preventive actions of OHS into the management system of all enterprises and public institutions. The Norwegian internal control legislation provides a mandatory OHSMS for regulating the Norwegian offshore industries. The Labour Inspectorate adopts inspection methods and strategies, from the traditional on-site inspections to a more system-oriented approach. With the involvement and active participation of top management. Norges Standardisengsforbund (1996) No. 96/402803 facilitates OHS operation within the framework of statutory requirements, and evaluation of an organization's operations against the established international standards. The standard covers multiple issues, such as process quality influencing the quality of products and services, protection of the environment, including protection against pollution and waste, OHS, as well as safety of products and services.

The OHS Act 2000, New South Wales, Australia (NSW 2000), requires that the employers must identify, assess, and eliminate or control risks and consult with employees on OHS matters to provide a safe workplace. The OHS Act is implemented through WorkCover as the industry codes of practice, as referred to in the Workplace Injury Management and Workers Compensation Act 1998 (NSW 1998). Indonesia Ministry Regulation (1996/2008) directs to implement OHSMS in enterprises with more than 100 workers and having hazardous potential in the workplace risk besides issuing certificates to industries. The provision includes an independent audit of OHS once every three years, as per the Regulation PER.18/MEN/XI/2008 on the Implementation of an Audit for OHSMS. In Singapore, the mandatory implementation of OHSMS applies to shipyards, specified construction worksites, and three classes of factories in the manufacturing sector. In Hong Kong, under the Safety Management Regulation (1999), the *general duty provisions* have imposed that place the maintenance of OHS as a continuous legal responsibility on the proprietors or contractors of industrial undertakings to develop, implement and maintain a safety management system. The OHS Council, as a statutory body, is responsible for safety audit schemes, including auditor accreditation (Govt. of Hong Kong (2000).

The Workwell audit in Ontario (Canada) is a particular mandatory OHSMS intervention that is administered through the Workplace Safety and Insurance Board (Bennett 2002). This is required for workplaces with poor performance in terms of workers' compensation claims or legislative compliance. The Brazilian Ministry of Labor guidelines (NR-9 1997) promulgates the obligatoriness for implementation of an environmental risk prevention programme (PPRA) by employers (Govt. of Brazil 1997). However, the regulation lacks core OHSMS elements such as management commitment, OHS goals, and objectives, performance measures, programmes and procedures, documentation, records management, and auditing. Brazilian standard NBR 10004 (Brazilian Association of Technical Standards 2004) covers issues on safety and occupational medicine.

Voluntary OHSMS Standards With and Without Certification Support

The voluntary OHSMS standards or guidelines, with formalized specifications (e.g. BS 8800, OSHAS 18001), were first promoted by commercial organizations and industry associations, targeting large enterprises to implement good corporate citizenship and employee welfare. The OHSAS 18001:2007, as an auditable standard, was designed to align and integrate with the widely adopted ISO 9000/ISO 14000 standards. National voluntary OHSMS(s) exist in many countries, with or without certification purposes (Table 14.3). No official estimates are available as regards to certification of different voluntary OHSMS models; however, the certifying bodies that represented OHSAS 18001:1999/2007 had about greater 3/4th of the market place certification in ISO 9001/ISO 14001. The new ISO 45001: 2018 standard is in the making to replace the OHSAS certification scheme. Anticipation is alive that the ISO 45001 would receive more comprehensive visibility and acceptance among the stakeholders for synergized and unified applications across organizations in the developed as well as in the developing countries.

The Australian government has been active in advancing national OHS strategies by creating a mixture of regulatory mandate and voluntary adoption of OHSMS by the enterprises (Saksvik and Quinlan 2003). Audit systems have been introduced, incorporating the essential elements of OHSMS into their criteria (e.g. WorkSafe Plan, Western Australia; TriSafe Management Systems Audit, Queensland; Safety Achiever Business System, South Australia; SafetyMAP (1994/2002), Victoria). The SafetyMAP is used in health and safety organization (HSO, Victoria) certification programme, at three levels—initiation, transition, and advanced level of certification. The joint accreditation system of Australia and New Zealand (Standards Australia 1997/2001) controls certification of AS/NZS 4801 and 4804 standards on OHSMS.

The United States OSHA encourages states and territories to develop their safety and health plans under the Occupational Safety and Health Act, 1970 (OSHA 1970). The voluntary protection programme (VPP) was promulgated to recognize and approve worksites with exemplary OHS management programmes, that designated by the VPP levels, like *Star, Merit, and Demonstration* (OSHA 1988). There is an emphasis that an average VPP worksite has a lost workday incidence rate of 52% below the average of its industry. The Safety and Health Programme Management guidelines (OSHA 1989) identify four general elements—management commitment, employee involvement, worksite analysis, hazard prevention and control, safety, and health training. The guidelines aimed at providing systematic organizational policies, procedures, and practices to protect employees from, and allowing them to recognize and control workplace hazards, specific job hazards, and potential hazards. The ANSI Z10 (2005/2012) is a voluntary consensus standard on OHSMS, and this performance standard applies to organizations of all sizes and types.

The Netherlands Normalisatie-Instituut (1997) standard NPR 5001) includes functions to (a) guarantee OHS management through risks prevention approach, (b) encourage people to control risks, build relationships with stakeholders and meet the statutory OHS obligations, and (c) promote improving OHS, with integration to the business management continually. Whereas the Dutch Labour Inspectorate enforces actions, every employer is obliged to have a contract with a certified OHS Advisory Service (Arbodienst) to plan and execute risk assessment and carry out a periodic evaluation of the OHS situation. Like the Dutch standard, the Irish code of practice on OHSMS (National Standards Authority of Ireland 1997) under the provision of the Irish Safety, Health and Welfare Act 1989 (Govt. of Ireland 1989), is not intended for certification purposes.

Asociacion Espanola de Normalizacion y Certificacion, UNE 81900 (Spanish Association of Standardization 1996) documented general rules on OHSMS (i.e. the auditable specification, audit process, and a vocabulary document). Spanish certification standard is sufficiently generic to be adaptable to any organization regardless of its size or the nature of its activities. The Polish national programme, PL 9407-1996, included a collection of best practices and models to deal with OHS management in small and medium-sized enterprises (Abad et al. 2002). The British Safety Council (2013) has a unique voluntary Five-Star audit model to objectively evaluate the OHS system(s) and arrangements against current best practice techniques. The specification covers sixty-six system elements under five sections of the audit (policy and organization—750 points; strategy and planning—1125 points; implementation and operation—1500 points; performance measurement—1125 points; evaluation and review—500 points), totalling a maximum numerical value of 5000 points. Two management indicators (leadership and continual improvement) are evaluated as the elements and also as scoring areas within other elements. The cumulative scoring of an enterprise-level management system is expressed in a numerical safety grading system on 100 percentage points.

Good OHS Practice Guides and Self-regulatory OHSMS Schemes

Specific OHS practice or service guides are necessarily management systems to assist enterprises to establish responsible image within the marketplace. For example, BS 8800 guidance on OHSMS stands apart to show how OHS performance can be improved, and the system of management can be integrated with other aspects of the business. Though the specifications of BS 8800 are not used for certification purposes, the influence of its framework was visible in OHSAS 18001 and ILO-OSH, and also ISO 45001. The ILO-OSH model is not intended for certification purposes; however, the critical dimensions, such as employee participation, medical surveillance, and health programmes, are embedded in the ILO-OSH model.

On the other hand, self-regulatory OHSMS schemes with hybridization of voluntary motives and legislative requirements have been adopted in many countries and promoted through statutory bodies. Based on ILO-OSH model, the State Administration of Work Safety, People's Republic of China issued OHSMS guidelines, and the guidance committee of the State Commission of Economy and Trade Trial Standards developed guidelines and accreditation of organizations. The Occupational Disease Control Act and the State Administration of Work Safety law came into force in 2002 and 2003, respectively (Wu and Chi 2015).

The Malaysian OHS Act (Govt. of Malaysia 1994) extends the legal basis in protecting workers to all sectors of economic activity. Under the Act, the National Council for Occupational Safety and Health is a tripartite forum to develop and review national OHS strategies and programmes. In Japan, under the Industrial Safety and Health Act (JISHA 1997; Kogi and Kawakami 2002), the Ministry of Health, Labour, and Welfare are responsible for establishing industrial accident prevention programmes. The Japan Industrial Safety and Health Association (JISHA) guides to establish OHSMS to enterprises of all scales and types of industry . The Korean Occupational Safety and Health Agency (KOSHA), under the Industrial Safety and Health Act (1990, amendment 2000), promotes OHSMS and assists enterprises. The Government emphasizes ensuring OHS regulations and promoting self-management to reflect effective OHS practices (ILO 2004b).

The Indian Factories Act 1948 with amendments (India code 2020) and other legislations provide enabling steps to include OHS management in the registered enterprises. The rule provisions cover hazardous industries and processes, OHS surveys, notifiable diseases, accidents, and employee participation in safety management. The Directorate General of Factory Advice Service and Labour Institutes (DGFASLI), in coordination with the central and state governments, introduce OHSMS elements in the enterprise. Besides, the introduction of the IS 18001 and ISO 45001 by the Bureau of Indian Standards (BIS 2007) on OHSMS are indicative of the simultaneous voluntary movement towards OHSMS certification, through different accredited bodies. On the other hand, however, the informal sectors, including agriculture and construction, are vastly deprived of essential delivery of OHS practices. The informal enterprises employ a workforce several times larger than that in the registered enterprises. Therefore, an implied challenge remains for effective OHS transfer in this vulnerable sector.

Sector-Specific OHSMS Application

A quick scan of the Fortune 1000 listed corporations reveals a variety of unique OHS programmes that have been evolved and successfully integrated with the business functions of the enterprises. The auditable OHSMS schemes, such as Responsible Care, E&P Forum, MASE/DT 78, GEHSE systems have been widely practised in chemicals, petrochemicals, oil and gas installations. The E&P forum is comprehensive in addressing the traditional OHS issues, and core management system approaches for integration with other organizational systems. With the flexibility to comply with regulatory obligations in France, thousands of French enterprises have voluntarily adopted sector-specific OHSMS schemes, such as MASE/DT 78, to better deal with occupational risks linked to their activities (INRS 2004; Cambon 2007).

Challenges of OHSMS Programmes

In general, the OHSMS models are the enterprise-level frameworks. The analysis of impacts of OHSMS(s) on OHS management and intervention demands a comparison of objective and subjective outcome measures on a reasonable time horizon (Verbeek et al. 2009; Tompa et al. 2010). Robson et al. (2007) compared the voluntary and mandatory OHSMS(s). The criteria included are the following:

(a) implementation (change in workplace OSHMS);
(b) intermediate outcomes (awareness and better safety climate, increased hazards reporting by employees and their increased involvement on OHS issues);
(c) final OHS outcomes (decline in work-related injury/illness, reduction in loss of working hours); and
(d) socio-economic outcomes (decrease in costs related to disability, workers' compensation costs, short and long-term disability costs, firm insurance premiums or workplace productivity).

Fernández-Muñiz et al. (2009) undertook a literature review to formulate hypotheses aiming at identifying good practices in safety management, and further tested the hypotheses in 455 Spanish enterprises. The enterprises with long experiences in OSHMS(s) revealed more positive feedbacks in terms of OHS and economic outcomes, irrespective of voluntary and mandatory OSHMS interventions. Further study in 131 OHSAS 18001-certified enterprises (Fernández-Muñiz et al. 2012) noted that the top management's commitment and communication influence on the safety behaviour, safety performance, employee satisfaction, and competitiveness of the organization. Undoubtedly, the top management's commitment is critical for success in OHSMS implementation. The level of commitment, the frequency of unsafe conduct and rule violations by employees and subcontractors, and the rate of execution of corrective and preventive measures were reported as significant performance indicators in OHSMS performance evaluation in PCB manufacturing industries in Taiwan (Chena et al. 2009). Vinodkumara and Bhasib (2011) revealed from a survey in eight chemical industries in Kerala (India) that the employees had different perceptions about the safety variables, and the safety rules and procedures emerged as common predictors of safety behaviour in all different categories of industries.

There is an increasing trend among the companies and the certifying bodies in integrating and implementing more than one management systems that are mutually compatible and seamlessly managed and executed without duplication. Since no standard on integrated management systems (IMS) exists to date, there are structured endeavours in integrating audit programmes of multiple management systems, e.g. combining quality, environment, and health and safety (Nag 2002). Different categories of management systems have some structural resemblance to the logical flow of management elements. Therefore, the integration of the systems has perceptible limitations from the operational point of view. Salomone (2008) investigated sample Italian companies as regards the potential of an IMS. The analysis included some common aspects, such as real motivations (company image, costs saving), obstacles (unclear regulations, lack of financial support), driving forces, and external pressures that companies meet when implementing the management systems. Pheng and Kwang (2005) observed from a survey in 30 certified to IMS construction companies in Singapore, that the larger enterprises reaped more benefits and incurred lesser costs. A small number of companies strived for IMS because of regulatory requirements.

Research is scanty on the measures of OHSMS performance and effectiveness of the OHSMS models and schemes in securing healthy and safe workplaces. The evidence from large enterprises favours that an OSHMS brings better awareness on OHS goals and risk identification, and yields increased corrective measures, and positive socio-economic and health outcomes. Reduced incidences of accidents and work-related diseases, decreased absenteeism, and higher productivity are indicators of the system compatibility between worker protection and corporate competitiveness.

On the other hand, the situations in SME, informal sectors, and agriculture are excruciating (Nag and Nag 2003; Hasle and Limborg 2006; Hasle et al. 2009), particularly in the populous nations like China and India. Ncube and Kanda (2018) reviewed OSH legislation in 10 countries, namely, Botswana, Ghana, India, Kenya, Malawi, South Africa, Tanzania, Uganda, Zambia, and Zimbabwe. They observed that though the agricultural sector employed more than 50% of the workforce, most of these countries lacked OSH legislation on the sector, and the existing OSH laws, if any, were gender insensitive. They also mentioned that only 5 to 10% of workers in developing countries have access to adequate OSH services compared to 20 to 50% of workers in developed countries.

Undoubtedly, the OSH services and management are poor in industrially developing countries (Micheli and Cagno 2010; Simukonda et al. 2020) due to compelling circumstances of large employment share, and lack of economic and technological resources. Further, there is a blatant absence of links between the constituent players (regulatory agencies, employers/contractors, workers) in structuring contextual programmes and systems. The SA 8000 (SAI 2008/2014) is an auditable social accountability management standard and that may be potentially useful through the national OHS strategies. The system explicitly addresses the fundamental rights of workers and workplace conditions and requires compliance of provisions of international human rights norms and related ILO conventions and recommendations.

The case in the reference is the vast farming sector with small and marginal landholdings that are grossly deprived of essential needs and expectations of OHS transfer. The implied challenge of permeating of OHS to this vulnerable sector may be obviated through local infrastructure extensions, such as block development, primary health services, public network building on good practices, and training at all levels. The system designed as the basic occupational health services (BOHS) has received wide publicity (Rantanen 2005a, b) for its applicability in linking and sequencing of processes in managing workplace hazards in small and micro enterprises, and agriculture.

Basic Occupational Health Services (BOHS)

The BOHS systems framework was launched jointly by the WHO/ILO/ICOH (Rantanen 2005a, b; ILO 2013), as a step towards the occupational health services at the individual, community levels. That is, to make OHS accessible to people in the villages, and in the traditionally underserved areas of occupational pursuits, like in farming, cottage, and other small enterprises. The BOHS targets the essential services for the protection of people's health at work, promotion of health, well-being, and workability, as well as the prevention of ill health and accidents. The services are provided through the primary healthcare approach and adjusted to the national and local circumstances.

The logical flow of the BOHS cycle (Fig. 14.8) depicts the conventional action model of different other schemes described above, such as orientation and planning, identification of needs, assessment of problems, management of actions, evaluation of effectiveness, and revision of the programme for continual improvement. Thus, the structure and content of BOHS evolve as a functional process to target the worker-oriented as well as the work environment-oriented activities simultaneously. Accordingly, the design of a structured checklist may elaborate on the requirements of the activities and content of BOHS.

- *Orientation and planning* refer to orienting OHS personnel to plan on OHS services, according to the needs and circumstances of the enterprise and workplace;
- *Surveillance of the work environment* refers to regular walk-through surveys to identify the critical activities of BOHS, that is, the identification of hazardous

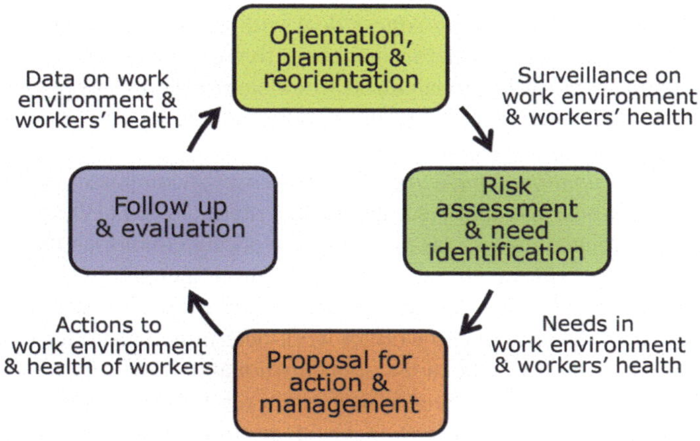

Fig. 14.8 The components of the BOHS cycle

exposures and conditions of work, identification and assessment of exposed workers;

- *Surveillance of workers' health* covers health examinations (pre-employment as well as periodic) to assess general health, and health impairment of a worker, relating to the work process, exposure to harmful agents, and cases of occupational diseases. This information is useful for suitable preventive actions, and the assessment of ones' working ability;
- *Assessment of health and safety risks* covers gathering information from the surveillance of the work environment, health surveillance, and other relevant sources for risk assessment;
- *Awareness about risks, preventive, and control actions, that is,* the employers are responsible for understanding workplace health hazards and risks better, and for educating workers on their risks avoidance, prevention, and protection, as well as on safe work practices. Legislation requires full information to the workers on risks and methods for protection and avoidance of risks;
- *Preventive actions for the management* includes measures for the employers and workers to execute risk management actions, and preventive and control measures to eliminate hazardous exposures and protect workers' health. Control measures should prevent unnecessary exposure during normal operating conditions, as well as during emergencies;
- *Prevention of accidents* is taken as the primary premise that accidental injuries are adverse health outcomes at work. Accident prevention demands surveillance of the work environment to identify and assess the magnitude of accident risk and factors that aggravate accident risks, such as exposure to chemicals, work stress, and shift work, and the health aspects that may affect accident risk;

- *First aid and emergency preparedness* deal with the readiness to first aid and training of workplace personnel in first aid activities. For emergency preparedness to significant hazards, preparedness would need expert OHS guidance about planning, training, equipment, first aid, and other emergency resources and access to emergency polyclinics and hospitals;
- *Diagnosis of occupational and work-related diseases* encompasses health surveillance, examinations, and approved diagnostic criteria. WHO (1985) conceptualized the work-related disease, as (a) work or working conditions that constitute the principal causal factor, (b) one or several causal agents (physical, chemical, biological and ergonomic) that trigger and aggravate the disease, and (c) the enhanced risk of diseases due to work or work-determined lifestyles. For details, one may refer to the ILO official list of occupational diseases (ILO 2002);
- *General health care, curative, and rehabilitation services* provide general preventive health services and guidance on health promotion activities to introduce healthy lifestyles. Authorized occupational health professionals associated with primary healthcare units are responsible for the diagnosis and treatment of occupational and work-related injuries and diseases and introducing rehabilitation measures;
- *Record keeping is* a general obligation of the employers to maintain documented records on health services extended to the workers, environmental and biological monitoring, and health and safety aspects of the workplace; and
- *Evaluation* requires a systematic analysis of the documented records about the provision of OHS services, and its effectiveness in the prevention of health and safety hazards. The periodic evaluation follows the cycle of continual improvement of activities and the re-orientation of OHS services.

Development of the OHS System in Agriculture

The sustainable essential services on occupational health in a sector of employment within the framework of the primary health care would demand a reasonably robust BOHS infrastructure. Generally, the occupational health services require a multidisciplinary team, comprising of a physician, nurse, hygienist, ergonomist, and psychologist. However, providing such a resource is beyond the scope of availability in rural agriculture, the self-employed, and the allied informal sectors. The apparent option lies in reinforcing the primary and secondary healthcare facilities that exist in the rural as well as in the remote settlements. This available facility forms the foundation and avenue for the OHS development for the farming sector, through suitable orientation and training of non-specialists in providing the BOHS. The essentiality lies in planning to lower the threshold for initiation and preserving the OHS competence and content at the starting point. The locally available human resources, such as nurses, grassroots level agents, may undergo a short training in OHS, associated with a primary/secondary healthcare unit. A well-designed short course of about eight weeks supplemented with periodic orientation on case studies is considered as a reasonable minimum.

The public authorities and local governments may be responsible for ensuring the availability of required infrastructure, resources, requirements, and competence of personnel concerning the size of workplaces (rural farming community), as well as on their geographical distribution. Receiving help from expert services constitutes an integral part of the service. The minimum service provision specific to a country, as stipulated by the ILO Convention 161 and 155, aims to target to varying local circumstances, expectations, and requirements. Vast farming sectors uniquely vary across the sectors of development, such as small and marginal farm holding, traditional farms, and mechanized farmhouses. The content of the BOHS may be relevant to the local farming issues such as accident risks, the severity of physical work, work safety, basic sanitation and hygiene, and the hazardous chemical, physical and biological risk factors. Besides, the services may take several optional forms, covering primary prevention, curative services, health promotion, and healthy work culture and climate. The competence of the frontline BOHS is unlikely to be sufficient to ensure a practical solution to the gamut of problems of farmworkers. The responsibility is inevitably on the government support system to impart the secondary level OHS support services about environmental monitoring, and analytical, clinical, psychological aspects, and to provide training and consultation services to BOHS personnel, large farmers, and other stakeholders.

Structure and Requirement in Agriculture

In developed countries (e.g. North American and European countries), agriculture has been notified as an industry, and therefore, provisions related to OHS services have become applicable to agriculture also. The OHSMS applied to the organized sector may be suitably incorporated in sectors of farming and agriculture that have been notified as an industry. The agricultural workforce is less than 10% of the total workforce in those countries. In the countries in Asia and Africa, however, agriculture grossly falls in the informal or unorganized sector, and workforce constitutes about 50 to 70% of the total workforce. Unlike organized sectors of industries, the regulatory provisions on the health and safety of farmworkers are not well defined, and in general, the sector is deprived of necessities. The majority of the agricultural workers (including farmers as well as landless labourers) are resource-poor, for whom the health and safety aspects get minimum attention. The OHSMS models and systems in the vastly populated agriculture in Asia and Africa would demand a more proactive role of the respective governments, matching local and regional requirements. A schematic of the agricultural system management in India is given in Fig. 14.9. The system can suitably incorporate the framework of BOHS to address the OHS issues of farmworkers. The existing system may be amended by the inclusion of a provision of a designated Directorate on Agricultural Health and Safety in the Ministry of Agriculture, with defined functions, such as:

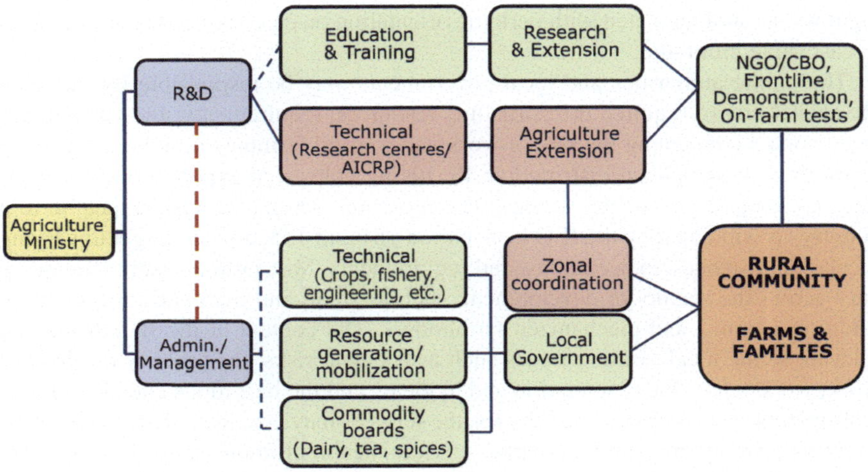

Fig. 14.9 Agricultural system management in India

- Serving as a technical arm to assist the Ministry of Agriculture in formulating national policies on OHS in agriculture;
- Rendering advice and carrying out support research activities for the administration of various acts/government policies;
- Coordinating technical and legal activities to facilitate uniform standards of enforcement of OHS in agriculture and allied sectors; emphasizing on building the competence of enforcement agencies;
- Coordinating social security benefits to farmers and farmworkers, and also compensation payment to agricultural accident victims;
- Educating and training farmers, farmworkers, farm machinery manufacturers, agricultural departmental officials, and other stakeholders on matters relating to OHS;
- Cooperating with international agencies, like ILO, WHO, FAO and advising the central government in the adoption of international standards concerning safety and health in agriculture;
- Training of foreign nationals and rendering expert advice to developing countries, and
- Gathering and disseminating information and material relating to safety and health in agriculture.

The system of management allows flexibility for a country or a specific sector of agriculture to customize the core OHS elements and adapts to their unique site-specific conditions. The degree of flexibility and viability of OHS in the farming sector would demand a unifying systems approach (Fig. 14.10). That is,

(a) the structured management commitment of the state regulatory agencies, local government, and rural development agencies;

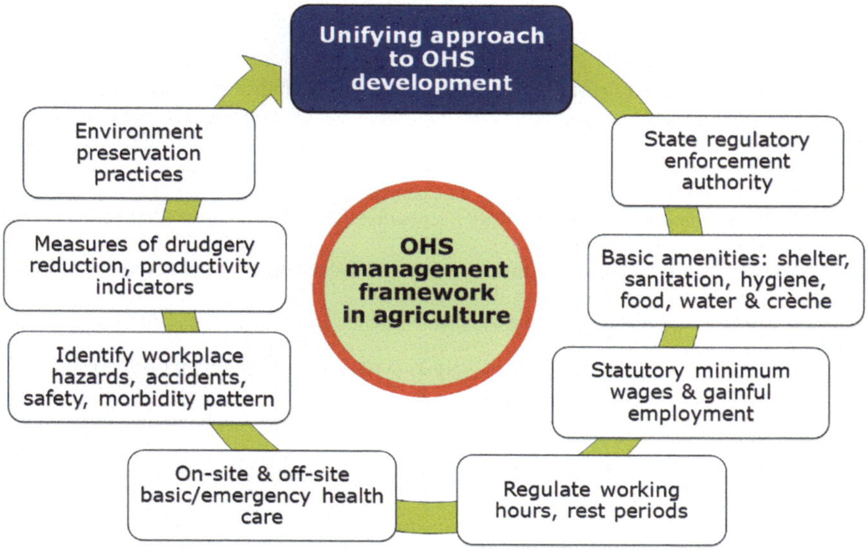

Fig. 14.10 A framework for agricultural workers' welfare and OHS management

(b) participation of farmers in system development and improvement, and suitably analyse the work and working environment;
(c) adoption of collective measures among the rural communities for hazard prevention and control of risk factors;
(d) education to all stakeholders on OHS issues, and bringing in medical management at the doorstep of the farmers; and
(e) introduction of appropriate performance indicators for system evaluation in terms of gains in productivity, quality, and morale among farm workers.

Understandably, the rural farming sector is devoid of established OHSMS framework. The apparent challenge remains in aligning the prevailing industrial practices in OHS delivery to the vulnerable agricultural sector. A suitable implementation of BOHS tailored to the country's agricultural system is a viable avenue for development and continual improvement OHS facility. The strategy may be directed towards evolving a programme development that includes the participation of multi-dimensional stakeholders to the range of OHS functions and activities in agriculture. With the participatory objective, the BOHS has the potential and promise to bring in innovative, cost-effective, simple, and flexible approaches for tangible development on health and safety in the rural sector.

References

Abad, J., Mondelo, P. R., & Llimona, J. (2002). Towards an international standard on occupational health and safety management. *International Journal of Occupational Safety and Ergonomics, 8*(3), 309–319.

ACC. (1988). *Responsible Care - Chemical manufacturing industry's environmental, health, safety and security performance initiative.* American Chemistry Council.

Alli, B. O. (2008). *Fundamental principles of occupational health and safety* (2nd ed.). Geneva: International Labour Organization.

ANSI. (2005/2012). Occupational health and safety management systems, ANSI Z10-2012. American National Standards Institute, Washington, DC.

Bennett, D. (2002). Health and safety management systems: Liability or asset? *Journal of Public Health Policy, 23*(2), 153–171.

Brazilian Association of Technical Standards. (2004). ABNT NBR 10004- Solid Waste Classification. Brazilian standards documentation. Rio de Janeiro.

British Safety Council. (2013). Five Star Occupational Health and Safety Audit. *Specification document, 2013,* 12.

BSI. (1996). *British Standard BS 8800: 1996: Guide to occupational health and safety management systems.* London.

BSI. (1999/2007). BS-OHSAS 18001. *Occupational health and safety management systems–Requirements,* Occupational health and safety assessment series. British standards Insdtitution, London.

BSI. (2000/2008). BS-OHSAS 18002:2000. Occupational health and safety management systems—guidelines for the implementation of OHSAS 18001 British Standards Institution, London.

BIS. (2007). *IS 18001- Occupational Health and Safety Management Systems - requirements with guidance for use.* New Delhi: Bureau of Indian Standards.

Cambon, J. (2007). *Vers une nouvelle méthodologie de mesure de la performance des systèmes de management de la santé-sécurité au travail* (Doctoral dissertation).

Chen, C. Y., Wu, G. S., Chuang, K. J., & Ma, C. M. (2009). A comparative analysis of the factors affecting the implementation of occupational health and safety management systems in the printed circuit board industry in Taiwan. *Journal of Loss Prevention in the Process Industries, 22*(2), 210–215.

Deming, W. E., & Edwards, D. W. (1982). *Quality, productivity, and competitive position* (Vol. 183). Cambridge, MA: Massachusetts Institute of Technology, Center for advanced engineering study.

Downey, I. L. (1995). E & P Forum- Health, Safety and Environmental Management System Guidelines. Society of Petyroleum Engineers, Aberdeen, United Kingdom.

EC. (1993). The Eco-Management and Audit Scheme, EMAS. The Council of the European Communities Regulation, No. L 168—(EMAS), 1836/93), European Commission Environment DG, Brussels Belgium.

EU-OSHA. (1989). Directive 89/391/EEC - OSH "Framework Directive". European Agency for Safety and Health at Work. Bilbao, Spain.

Fernández-Muñiz, B., Montes-Peón, J. M., & Vázquez-Ordás, C. J. (2009). Relation between occupational safety management and firm performance. *Safety Science, 47*(7), 980–991.

Fernández-Muñiz, B., Montes-Peón, J. M., & Vázquez-Ordás, C. J. (2012). Safety climate in OHSAS 18001-certified organisations: Antecedents and consequences of safety behaviour. *Accident Analysis and Prevention, 45,* 745–758.

Frick, K., & Wren, J. (2000). Reviewing occupational health and safety management: multiple roots, diverse perspectives and ambiguous outcomes. In K. Frick, P. L. Jensen, M. Quinlan, & T. Wilthagen (Eds.), *Systematic occupational health and safety management: Perspectives on an international development* (pp. 17–42). Bingley: Emerald Group Publ Ltd.

GEHSE. (2000/2004). Guide d'Engagement Hygiène, sécurité, environnement pour les entreprises extérieures intervenant dans les dépôts d'hydrocarbures ou les petits établissements pétroliers ou les stations-service, Association GEHSE, France.
Goldstein, G., Helmer, R., & Fingerhut, M. (2001). The WHO global strategy on occupational health and safety. *African Newsletter on Occupational Health and Safety, 11*(3), 56–60.
Govt. of Brazil. (1997). NR-9 (PPRA)—Environmental Risk Prevention Programme, 1997., Ministry of Labor guidelines, Brazil.
Govt. of Hong Kong. (2000). A Guide to the Factories and Industrial Undertakings (Safety Management) Regulation. Occupational Safety and Health Branch Labour Department. (www.hkosha.org.hk/Temp Doc/A Guide to Safety Management Regulation).
Govt. of Ireland. (1989). Safety, Health and Welfare at Work Act, 1989. http://www.irishstatutebook.ie/eli/1989/act/7/enacted/en/print.html.
Govt. of Malaysia. (1994). Laws of Malaysia, Act 514 -Occupational Safety And Health Act 1994.
Hasle, P., & Limborg, H. J. (2006). A review of the literature on preventive occupational health and safety activities in small enterprises. *Industrial Health, 44*(1), 6–12.
Hasle, P., Kines, P., & Andersen, L. P. (2009). Small enterprise owners' accident causation attribution and prevention. *Safety Science, 47*(1), 9–19.
HSE. (1993/2113). Managing for health and safety, HSG 65. Health and Safety Executive, London, U K.
India code. (2020). The Factories Act. 1948 (with amendment). Ministry of Labour, Govt. of India.
Indonesian Ministry Regulation. (1996/2008). Implementation of Occupational Health and Safety Management System (PER. 05/MEN/1996), Implementation of an Audit for OHSMS (PER.18/MEN/XI/2008). (https://betterwork.org/portfolio/7-1-osh-management-system-and-policy/#).
INRS. (2004). *De l'évaluation des risques au management de la santé et de la sécurité au travail.* Paris: Institut National de Recherche et de Santé.
ILO. (1981). ILO Convention No. 155, Occupational safety and health. International Labour Office, Geneva.
ILO. (1985). ILO Convention No. 161, Occupational health services. International Labour Office, Geneva.
ILO. (1986). ILO Convention No. 162, asbestos. International Labour Office, Geneva.
ILO. (1988). ILO Convention No. 167, safety and health in construction. International Labour Office, Geneva.
ILO. (1990). ILO Convention No. 170, Chemicals. International Labour Office, Geneva.
ILO. (1993). ILO Convention No. 174, Prevention of major industrial accidents. International Labour Office, Geneva.
ILO. (1995). ILO Convention No. 176, safety and health in mines. International Labour Office, Geneva.
ILO. (2001a). ILO Convention No.184, Safety and Health in Agriculture. International Labour Office, Geneva.
ILO. (2001b). Guidelines on Occupational Safety and Health Management Systems (ILO-OSH 2001), International Labor Office, Geneva (ISBN: 92-2-111634-4).
ILO. (2002). *ILO Recommendation R194 -List of Occupational Diseases.* Geneva: International Labour Office.
ILO. (2003). Basic Occupational Health Services (BOHS). Working Paper for the Joint ILO/WHO Com-mittee on Occupational Health, Dec., 9–12.International Labour Office, Geneva.
ILO. (2004a). *ILO Global Strategy on Occupational Safety and Health.* Geneva: International Labour Office.
ILO. (2004b). Promotional framework for occupational safety and health, Report IV (1). International Labor Office, Geneva.
ILO. (2006). ILO Convention No.187, Promotional framework for Occupational Safety and Health. International Labour Office, Geneva.

ILO. (2020). World statstics. The enormous burden of poor working conditions. https://www.ilo.org/moscow/areas-of-work/occupational-safety-and-health/WCMS_249278/lang–en/index.htm.
ISO. (2008). ISO 9001. Quality management systems—requirements. International Standards Organization, Geneva.
ISO. (2015). ISO 14001. Environmental management systems—requirements with guidance for use. International Standards Organization, Geneva.
ISO. (2018). ISO 45001: 2018. Occupational health and safety management systems–requirements with guidance for use. International Organization for Standardization, Geneva.
JICOSH. (1972). Industrial Safety and Health Law No. 57 of June 8, 1972. With latest amendment Law No. 25 of May, 31, 2006. Japan International Centre or Occupational Safety and Health. Tokyo.
JISHA. (1997). *Occupational Health and Safety Management System (OHS-MS): JISHA Guidelines*. Tokyo: Japan Industrial Safety and Health Association.
Kausek, J. (2007). *OHSAS 18001: Designing and implementing an effective health and safety management system*. Government Institutes.
Kogi, K., & Kawakami, T. (2002). Trends in occupational safety and health management systems in Asia and the Pacific. *Asian-Pacific newsletter on occupational health and safety, 9*(2), 42–47.
MASE/UIC. (2007). MASE/ DT78 UIC Common standard that promotes a common approach on industrial safety improvement. Manual for Improvement of Safety in Companies (MASE), and Union of Chemical Industries, France.
Micheli, G. J., & Cagno, E. (2010). Dealing with SMEs as a whole in OHS issues: warnings from empirical evidence. *Safety Science, 48*(6), 729–733.
Nag P. K. (2002). The Management Systems - Quality, Environment, Health and Safety: ISO 9000:2000, ISO 14001, OHSAS 18001. Quest Publications, Mumbai.
Nag, P. K., & Nag, A. (2003). A national priority on occupational health and safety management system. *ICMR Bulletin, 33*(11–12).
Nag, A., & Nag, A. (2005). World trends in occupational health and safety management systems. In *Handbook of Standards and Guidelines in Ergonomics and Human Factors, Ed.W. Karwowski* (pp. 441–469). CRC Press.
National Standards Authority of Ireland. (1997). Recommendation for an occupational health and safety (OH and S) management system. NSAI SR 320.
Ncube, F., & Kanda, A. (2018). Current status and the future of occupational safety and health legislation in low-and middle-income countries. *Safety and Health at Work, 9*(4), 365–371.
Nederlands Normalisatie-Instituut. (1997). Dutch Technical Report—Guide to an occupational health and safety management system, NPR 5001.
Norges Standardisengsforbund. (1996). Management Principles for Enhancing Quality of Products and Services, Occupational Health & Safety, and Environment, Norwegian standard, 96/402803.
NSW. (1998). *Workplace Injury Management and Workers Compensation Act 1998 No 86*. Australia: New South Wales Govt.
NSW. (2000). Occupational Health and Safety Act 2000 No. 40. New South Wales Govt, Australia.
OSHA. (1970). OSHA Act of 1970. Occupational Safety and Health Administration, Washington DC, USA
OSHA. (1988). Voluntary Protection Programs- Federal Register No. 53:26339-48.12/4/1988 Occupational Safety and Health Administration (OSHA), USA.
OSHA. (1989). *Safety and Health Program Management Guidelines, Standard—Federal Register, 29 (CFR), 1926 Subpart C, 54(18), 3094–3916*. Washigton, D.C.: Occupational Safety and Health Administration.
OHSAS Project Group. (1999/2007). OSHAS 18001. Occupational health and safety management systems– Requirements. Occupational Health and Safety Assessment Series. OHSAS Project Group Secretariat, London.
OHSAS Project Group. (2008). OSHAS 18002. Occupational health and safety management systems– Guidelines for the implementation of OSHAS-18001:2007. Occupational Health and Safety Assessment Series. OHSAS Project Group Secretariat, London.

Pheng, L. S., & Kwang, G. K. (2005). ISO 9001, ISO 14001 and OHSAS 18001 management systems: integration, costs and benefits for construction companies. *Architectural Science Review, 48*(2), 145–151.

Rantanen, J. (2005a). Basic occupational health services: A WHO/ILO/ICOH guideline. 2nd ed. Helsinki (Finland): Finnish Institute of Occupational Health, p. 19.

Rantanen, J. (2005b). Basic occupational health services-their structure, content and objectives. *SJWEH Supplements, 1,* 5–15.

Robson, L. S., Clarke, J. A., Cullen, K., Bielecky, A., Severin, C., Bigelow, P. L., et al. (2007). The effectiveness of occupational health and safety management system interventions: a systematic review. *Safety Science, 45*(3), 329–353.

SafetyMap. (1994/2002). Auditing Health and Safety Management Systems. WorkSafe Victoria, Melbourne, Australia.

SAI. (2008/2014). SA 8000, Social Accountability International, New York, USA.

Saksvik, P. Ø., & Quinlan, M. (2003). Regulating systematic occupational health and safety management: comparing the Norwegian and Australian experience. *Relations Industrielles/Industrial Relations, 58*(1), 33–59.

Salomone, R. (2008). Integrated management systems: experiences in Italian organizations. *Journal of Cleaner Production, 16*(16), 1786–1806.

Simukonda, W., Manu, P., Mahamadu, A. M., & Dziekonski, K. (2020). Occupational safety and health management in developing countries: a study of construction companies in Malawi. *International Journal of Occupational Safety and Ergonomics, 26*(2), 303–318.

Spanish Association of Standardization. (1996). *UNE 81900, Prevention of occupational risks: General rules for implementation of an occupational safety and health management system.* Madrid, Spain: UNE- Spanish Association for Standardisation.

Standards Australia. (1997/2001). AS/NZS 4804/4801. Occupational health and safety management systems – specification with guidance for use. Standards Australia, Sydney/ Standards NewZealand, Wellington.

Stolk, C. V., Staetsky, L., Hassan, E., & Kim, C. W. (2012). Management of occupational safety and health: an analysis of the findings of the European Survey of Enterprises on New and Emerging Risks (ESENER). European Risk Observatory Report. *Management of occupational safety and health: an analysis of the findings of the European Survey of Enterprises on New and Emerging Risks (ESENER). European Risk Observatory Report.*

Tompa, E., Verbeek, J., Van Tulder, M., & de Boer, A. (2010). Developing guidelines for good practice in the economic evaluation of occupational safety and health interventions. *Scandinavian Journal of Work, Environment & Health,* 313–318.

Verbeek, J., Pulliainen, M., & Kankaanpää, E. (2009). A systematic review of occupational safety and health business cases. *Scandinavian Journal of Work, Environment & Health,* 403–412.

Vinodkumar, M. N., & Bhasi, M. (2011). A study on the impact of management system certification on safety management. *Safety Science, 49*(3), 498–507.

Walters, D., Wilthagen, T., & Jensen, P. L. (2002). Introduction. In D. Walters (Ed.), *Regulating health and safety management in the European Union: A study of the dynamics of change.* Brussels: Presses Interuniversitaires Européennes.

World Health Organization. (2007). Workers' Health: Global Plan of Action. Sixtieth World Health Assembly Agenda item 12.13. http://www.who.int/occupational_health/WHO_health_assembly_en_web.pdf.

WHO. (1985). Identification and control of work-related diseases: report of a WHO expert committee. Geneva, World Health Organization.

Wu, F., & Chi, Y. (2015). Regulatory system reform of occupational health and safety in China. *Industrial Health, 53*(3), 300–306.

Author Index

P. K. Nag and L. P. Gite, *Human-Centered Agriculture*, Design Science and Innovation,
https://doi.org/10.1007/978-981-15-7269-2

M

N

S

T

U

Subject Index

P. K. Nag and L. P. Gite, *Human-Centered Agriculture*, Design Science and Innovation,
https://doi.org/10.1007/978-981-15-7269-2

T

The manufacturer's authorised representative in the EU is Springer Nature Customer Service Centre GmbH, Europaplatz 3, 69115 Heidelberg, Germany. If you have any concerns regarding our products, please contact ProductSafety@springernature.com

Printed and bound by CPI Group (UK) Ltd, Croydon, CR0 4YY
30/09/2026
02225504-0017